Interactive Student Edition

Reveal
ALGEBRA 1®
Volume 1

Mc
Graw
Hill

mheducation.com/prek-12

Cover: (t to b, l to r) pearleye/E+/Getty Images, Merfin/iStock/Getty Images,
fototrav/E+/Getty Images, rickyd/Shutterstock

Send all inquiries to:
McGraw-Hill Education
8787 Orion Place
Columbus, OH 43240

ISBN: 978-0-07-662599-4 (*Interactive Student Edition*, Volume 1)
MHID: 0-07-662599-0 (*Interactive Student Edition*, Volume 1)
ISBN: 978-0-07-899743-3 (*Interactive Student Edition*, Volume 2)
MHID: 0-07-899743-7 (*Interactive Student Edition*, Volume 2)

Printed in the United States of America.

5 6 7 8 9 10 LMN 27 26 25 24 23 22 21 20

Contents in Brief

Reveal AGA® Makes Math Meaningful...

Interactive Student Edition

Learning on the Go!

The flexible approach of *Reveal Algebra 1, Reveal Geometry,* and *Reveal Algebra 2 (Reveal AGA)* can work for you using digital only or digital and your *Interactive Student Edition* together.

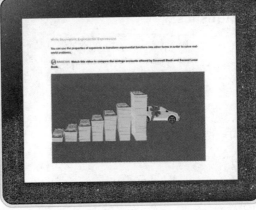

Student Digital Center

...to Reveal YOUR Full Potential!

Reveal AGA® Brings Math to Life in Every Lesson

Reveal AGA is a blended print and digital program that supports access on the go. You'll find the *Interactive Student Edition* aligns to the Student Digital Center, so you can record your digital observations in class and reference your notes later, or access just the digital center, or a combination of both! The Student Digital Center provides access to the interactive lessons, interactive content, animations, videos, and technology-enhanced practice questions.

Write down your username and password here

Username: _____

Password: _____

Go Online!
my.mheducation.com

Web Sketchpad® Powered by The Geometer's Sketchpad®- Dynamic, exploratory, visual activities embedded at point of use within the lesson.

Animations and Videos – Learn by seeing mathematics in action.

Interactive Tools – Get involved in the content by dragging and dropping, selecting, highlighting, and completing tables.

Personal Tutors – See and hear a teacher explain how to solve problems.

eTools – Math tools are available to help you solve problems and develop concepts.

Module 1

Expressions

Module 2
Equations in One Variable

Module 3
Relations and Functions

Module 4
Linear and Nonlinear Functions

TABLE OF CONTENTS

Module 5
Creating Linear Equations

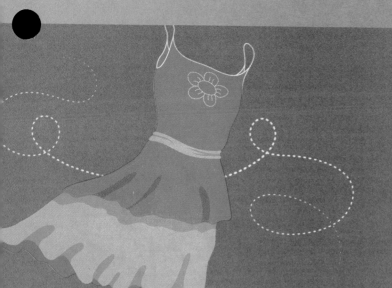

Copyright © McGraw-Hill Education

Module 7
Systems of Linear Equations and Inequalities

Module 8
Exponents and Roots

Module 9
Exponential Functions

TABLE OF CONTENTS

Module 11
Quadratic Functions

Module 12
Statistics

Expressions

What Will You Learn?

Place a check mark (✓) in each row that corresponds with how much you already know about each topic **before** starting this module.

KEY

👎 — I don't know.   👍 — I've heard of it.   👍 — I know it!

	Before			After		
	👎	👍	👍	👎	👍	👍
write numerical expressions						
evaluate numerical expressions						
use the order of operations						
write algebraic expressions						
evaluate algebraic expressions						
identify properties of equality						
apply the Identity and Inverse Properties to evaluate expressions						
apply the Commutative, Associative, and Distributive Properties to evaluate expressions						
write and evaluate absolute value expressions						
use descriptive modeling to describe real-world situations						
choose a level of accuracy appropriate to limitations on measurements						

📘 Foldables Make this Foldable to help you organize your notes about expressions. Begin with three sheets of notebook paper.

1. **Fold** three sheets of paper in half along the width. Then cut along the crease.

2. **Staple** the six half-sheets together to form a booklet.

3. **Cut** five centimeters from the bottom of the top sheet, four centimeters from the second sheet, and so on.

4. **Label** each tab with a lesson number.

1   2   3   4

What Vocabulary Will You Learn?

Check the box next to each vocabulary term that you may already know.

- □ absolute value
- □ accuracy
- □ additive identity
- □ additive inverses
- □ algebraic expression
- □ base
- □ closed
- □ coefficient

- □ constant term
- □ define a variable
- □ descriptive modeling
- □ equivalent expressions
- □ evaluate
- □ exponent
- □ like terms
- □ metric

- □ multiplicative identity
- □ multiplicative inverses
- □ numerical expression
- □ reciprocals
- □ simplest form
- □ term
- □ variable
- □ variable term

Are You Ready?

Complete the Quick Review to see if you are ready to start this module.
Then complete the Quick Check.

Quick Review

Example 1

Write $\frac{24}{40}$ in simplest form.

Find the greatest common factor (GCF) of 24 and 40.

factors of 24: 1, 2, 3, 4, 6, 8, 12, 24

factors of 40: 1, 2, 4, 5, 8, 10, 20, 40

The GCF of 24 and 40 is 8.

$\frac{24 \div 8}{40 \div 8} = \frac{3}{5}$ Divide the numerator and
denominator by their GCF, 8.

Example 2

Find $2\frac{1}{4} \div 1\frac{1}{2}$.

$2\frac{1}{4} \div 1\frac{1}{2} = \frac{9}{4} \div \frac{3}{2}$ Write mixed numbers as improper fractions.

$= \frac{9}{4} \times \frac{2}{3}$ Multiply by the reciprocal.

$= \frac{18}{12} = \frac{3}{2}$ or $1\frac{1}{2}$ Simplify.

Quick Check

Write each fraction in simplest form.

1. $\frac{24}{36}$

2. $\frac{34}{85}$

3. $\frac{5}{65}$

4. $\frac{64}{88}$

Evaluate.

5. $6 \times \frac{2}{3}$

6. $\frac{7}{8} - \frac{1}{6}$

7. $\frac{3}{8} \div \frac{1}{4}$

8. $\frac{1}{3} + \frac{3}{4}$

How did you do?

Which exercises did you answer correctly in the Quick Check? Shade those exercise numbers below.

① ② ③ ④ ⑤ ⑥ ⑦ ⑧

Numerical Expressions

Explore Order of Operations

Online Activity Use a real-world situation to complete the Explore.

> @ **INQUIRY** How can you evaluate a numerical expression? ×

Learn Writing Numerical Expressions

A **numerical expression** is a mathematical phrase that contains only numbers and mathematical operations. For example, $6 + 2 \div 1$ is a numerical expression.

Some numerical expressions contain multiplication. Multiplication can be represented in several ways, including a raised dot or parentheses. Here are some ways to represent the product of 2 and 3.

$2 \cdot 3$ $2(3)$ $(2)3$ $(2)(3)$

Example 1 Translate a Verbal Expression

Translate *one plus eight divided by three* into a numerical expression.

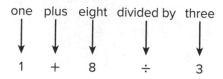

The translation is _____.

Check

Translate the verbal expression into a numerical expression. *seven times fifteen minus two times nine* _____

 Go Online You can complete an Extra Example online.

Today's Goals
- Write numerical expressions for verbal expressions.
- Evaluate numerical expressions.

Today's Vocabulary
numerical expression

exponent

base

evaluate

🗨 **Think About It!**

Identify the factors of the numerical expression 6(11).

🗨 **Think About It!**

What verbal phrase represents addition? subtraction? multiplication? division?

Example 2 Translate a Verbal Expression with Grouping Symbols

Translate *the sum of five and nine divided by two* into a numerical expression.

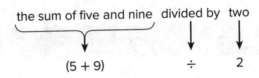

the sum of five and nine | divided by | two

$(5 + 9)$ | $\div$ | 2

The translation is _____.

🌐 Example 3 Write a Numerical Expression

BOWLING A person's handicap in bowling is usually found by subtracting the person's average from 200, multiplying by 2, and dividing by 3. Marley's bowling average is 170. Write a numerical expression for Marley's handicap in bowling.

The first step to find the handicap is to subtract the average from 200.

Marley's average is 170. $200 -$ _____

Next, multiply the difference by 2.   _____ $(200 - 170)$

Finally, divide the result by 3.

_____

Check

TEMPERATURE To convert a temperature in degrees Celsius to degrees Fahrenheit, multiply the temperature in degrees Celsius by 9, divide by 5, and add 32. Write a numerical expression to convert 37° Celsius to Fahrenheit. _____

A. $\dfrac{37 \cdot 9}{5} + 32$

B. $\dfrac{37 \cdot 9}{5 + 32}$

C. $\dfrac{37 \cdot 9}{5} \cdot 32$

D. $\dfrac{37 \cdot 5}{9} + 32$

🔵 **Go Online** You can complete an Extra Example online.

Learn Evaluating Numerical Expressions

An expression of the form x^n is read "x to the nth power." The word *power* is used to refer to the expression, the value, or the exponent of the expression.

When n is a positive integer, the **exponent** indicates the number of times a number is multiplied by itself. In a power, the **base** is the number being multiplied by itself.

To **evaluate** an expression means to find its value. If a numerical expression contains more than one operation, then the rule that lets you know which operation to perform first is called the **order of operations**.

Key Concept • Order of Operations
Step 1 Evaluate expressions inside grouping symbols.
Step 2 Evaluate all powers.
Step 3 Multiply and/or divide from left to right.
Step 4 Add and/or subtract from left to right.

Talk About It!

Explain how the order of operations applies when using the formula $\frac{1}{2}h(b_1 + b_2)$ to find the area of a trapezoid.

Example 4 Evaluate Expressions

Evaluate each expression.

a. 2^4

$2^4 = 2 \cdot 2 \cdot 2 \cdot 2$ Use 2 as a factor 4 times.

$= \underline{\hspace{1cm}}$ Multiply.

b. 4^5

$4^5 = 4 \cdot 4 \cdot 4 \cdot 4 \cdot 4$ Use 4 as a factor 5 times.

$= \underline{\hspace{1cm}}$ Multiply.

Example 5 Order of Operations

Evaluate $20 - 7 + 8^2 - 7 \cdot 11$.

$20 - 7 + 8^2 - 7 \cdot 11 = 20 - 7 + 64 - 7 \cdot 11$ $\underline{\hspace{2cm}}$

$= 20 - 7 + 64 - 77$ Multiply 7 and 11.

$= 13 + 64 - 77$ $\underline{\hspace{2cm}}$

$= 77 - 77$ $\underline{\hspace{2cm}}$

$= 0$ Subtract 77 from 77.

Think About It!

Write an expression that uses exponents and at least three different operations. Explain the steps you would take to evaluate the expression.

 Go Online You can complete an Extra Example online.

Copyright © McGraw-Hill Education

Check

Write the steps of the order of operations in the correct order.

Step 1 _____

Step 2 _____

Step 3 _____

Step 4 _____

🌐 Example 6 Write and Evaluate a Numerical Expression

ARCADE **Mellie is playing a bowling game at an arcade. She rolls two balls into the 30-point hole, four balls into the 20-point hole, and three balls into the 50-point hole. Write and evaluate an expression to find Mellie's total score.**

Part A Complete the table to write an expression for Mellie's total score.

To find Mellie's total score, find the number of points scored from each hole and add the products.

Words	two balls rolled into the 30-point hole	plus	four balls rolled into the 20-point hole	plus	three balls rolled into the 50-point hole
Expression	2 · _____	+	4 · _____	+	3 · _____

Part B Evaluate the expression.

$2 \cdot 30 + 4 \cdot 20 + 3 \cdot 50 =$ _____ + _____ + _____   Multiply.

= _____   Add.

Mellie scored _____ points.

Check

COMPUTERS A computer technician charges a flat fee of $50 plus $25 per hour. On Monday, he worked on Aika's computer for 2 hours. On Tuesday, he worked on Aika's computer for 3 hours.

Part A Which expression(s) represents Aika's bill? Select all that apply.

A. $50 + 25(2) + 25(3)$

B. $50 + 25(2 + 3)$

C. $25(2 + 3)$

D. $50 + 25(5)$

E. $25 + 50(2 + 3)$

Part B How much money does Aika owe the technician? _____

🔘 **Go Online** You can complete an Extra Example online.

Use a Source

Find data about the scoring in a game of interest to you where you can score different numbers of points for different plays. Write and evaluate an expression to represent a possible score.

Example 7 Expressions with Grouping Symbols

Evaluate each expression.

a. $\dfrac{(4+5)^2}{3(7-4)}$

$$\dfrac{(4+5)^2}{3(7-4)} = \dfrac{(9)^2}{3(3)}$$ Evaluate inside parentheses.

$$= \dfrac{81}{3(3)}$$ Evaluate the power in the numerator.

$$= \dfrac{81}{9}$$ Multiply 3 and 3 in the denominator.

$$= \underline{\hspace{1cm}}$$ Divide 81 by 9.

b. $15 - [10 + (3-2)^2] + 6$

$15 - [10 + (3-2)^2] + 6 = 15 - [10 + (\underline{\hspace{0.7cm}})^2] + 6$ Evaluate innermost parentheses.

$= 15 - [10 + \underline{\hspace{0.7cm}}] + 6$ Evaluate power.

$= 15 - [\underline{\hspace{0.7cm}}] + 6$ Add.

$= \underline{\hspace{0.7cm}} + 6$ Subtract.

$= \underline{\hspace{0.7cm}}$ Add.

Check

Evaluate $\dfrac{2^3 - 5}{15 + 9}$. _____

Learn Plan for Problem Solving

Using a **four-step problem-solving plan** can help you make sense of problems and persevere in solving them.

Key Concept • Four-Step Problem-Solving Plan
Step 1 Identify and understand the task.
Read the task carefully, and make sure you understand what question to answer or problem to solve.
Step 2 Plan your approach.
Choose a strategy. Plan the steps you will use to complete the task.
Step 3 Solve the problem.
Use the strategy you chose in Step 2 to solve the problem.
Step 4 Check the solution.
Make sure that your solution is reasonable and completes the task.

 Go Online You can complete an Extra Example online.

 Think About It!

Equivalent expressions have the same value. Are the expressions $(30 + 17) \times 10$ and $10 \times 30 + 10 \times 17$ equivalent? Why or why not?

Study Tip

Grouping Symbols
Grouping symbols such as parentheses ( ), brackets [ ], braces { }, and fraction bars are used to clarify or change the order of operations. So, evaluate expressions inside grouping symbols first. For fraction bars, evaluate the numerator and denominator before completing the division.

 Think About It!

How can using this four-step problem-solving plan help you effectively solve problems?

Example 8 Write and Evaluate Expressions

MONEY **Thursdays are Student Days at LSC Theaters. Student tickets are $5 and popcorn refills are free. You will buy a ticket and a 20-ounce bottle of water, and you will split the cost of a large tub of popcorn and large candy with a friend. How much money should you bring?**

Popcorn		Drinks		Snacks	
Mini Tub	$3.00	20 oz	$5.00	Fruit Snacks	$3.00
Medium Tub	$5.00	32 oz	$5.50	Hot Dog	$5.00
Large Tub	$7.50	44 oz	$6.00	Candy	$3.50

Understand We are given what you will buy, the cost of each item, and the cost of the popcorn and candy that will be split with a friend. We are asked to find the _____.

Plan

Step 1 Write an expression to represent the _____ of the ticket, drink, and food.

Step 2 _____ the expression to find the cost.

Solve

Step 1 Write the expression for the total cost in dollars.

ticket   drink   popcorn   candy

_____   +   _____ $+\frac{1}{2}\cdot$ _____ $+\frac{1}{2}\cdot$ _____

Step 2 Use the order of operations to evaluate the expression.

$5 + 5 + \frac{1}{2}\cdot 7.50 + \frac{1}{2}\cdot 3.50$ Original expression

$= 5 + 5 +$ _____ $+$ _____

$=$ _____

Check The cost of the ticket and water is $ _____. The total cost of the popcorn and candy is $ _____. So half of that cost is $ _____. The total is $10 + $5.50 or $ _____.

Check

PETS While she was on a 7-day vacation, Ms. Hernandez boarded her dog at a kennel. If her boarding budget is $250, can Ms. Hernandez afford a daily extra walk and one wash and nail trimming? Explain.

Service	Cost
Board	$24 per day
Wash	$45
Extra walk	$4 per day
Nail trimming	$12
Vitamins	$1 per day

_____

 Go Online You can complete an Extra Example online.

Math History Minute

Hungarian mathematician **George Pólya (1887–1985)** is known as the father of mathematical problem solving. One of his books, *How to Solve It*, was a bestseller that was translated into 17 languages. In it, Pólya describes a four-step plan for problem solving.

Practice

🔵 **Go Online** You can complete your homework online.

Example 1

Write a numerical expression for each verbal expression.

1. two plus twelve divided by four

2. eighteen more than five

3. seven more than three times eleven

4. twenty-five less than one hundred

5. six minus three minus one

6. fourteen decreased by three times four

7. twenty-four divided by six plus seven

8. eight times six divided by two minus nine

9. one hundred sixteen divided by four plus twenty-eight minus thirty-three

10. two hundred fifty-nine minus eighty-five plus sixty-two divided by two

Example 2

Write a numerical expression for each verbal expression.

11. the sum of three and seven divided by two

12. the difference of six and two divided by four

13. the sum of four and nine times three

14. eighteen divided by the sum of two and seven

15. ten divided by the product of four and five

16. the difference of eleven and four times five

17. the sum of one and two divided by twenty

18. the sum of two and four and six times eight

19. the sum of twelve and sixteen divided by the sum of three and four

20. the difference of twenty-two and six divided by the sum of five and three

21. the sum of thirty-six and fourteen divided by the product of two and five

22. the quotient of thirty-two and four divided by the sum of one and three

23. the sum of six and fifteen divided by the difference of thirteen and nine

24. the difference of thirty-one and seventeen divided by the product of ten and four

Example 3

25. SOLAR SYSTEM It takes Earth about 365 days to orbit the Sun. It takes Uranus about 85 times as long. Write a numerical expression to describe the number of days it takes Uranus to orbit the Sun.

26. TEST SCORES To find the average of a student's test scores, add the scores and divide by the number of tests. Suppose Ryan scored 85, 92, 88, and 98 on four tests. Write a numerical expression to describe Ryan's average test score.

27. HOMEWORK It took Carrie five less minutes than twice the amount of time as Hua to complete her homework. It took Hua thirty-five minutes to complete her homework. Write a numerical expression to describe the amount of time it took Carrie to complete her homework.

28. PERIMETER The perimeter of Stephanie's triangle is half the perimeter of Juan's triangle. Juan's triangle is shown. Write a numerical expression to describe the perimeter of Stephanie's triangle.

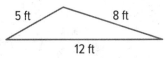

29. BEDROOM Shenandoah's rectangular bedroom is 12 feet long and 7 feet wide. Write a numerical expression to describe the area of Shenandoah's bedroom.

Examples 4, 5, and 7

Evaluate each expression.

30. 7^2

31. 14^3

32. 2^6

33. $35 - 3 \cdot 8$

34. $18 \div 9 + 2 \cdot 6$

35. $10 + 8^3 \div 16$

36. $[(6^3 - 9) \div 23]4$

37. $\dfrac{8 + 3^3}{12 - 7}$

38. $\dfrac{(1 + 6)9}{5^2 - 4}$

39. $4(16 \div 2 + 6)$

40. $13 - \dfrac{1}{3}(11 - 5)$

41. $(5 \cdot 2 - 9) + 2 \cdot \dfrac{1}{2}$

42. $62 - 3^2 \cdot 8 + 11$

43. $4^3 \div 8$

44. $20 + 3(8 - 5)$

45. $3[4 - 8 + 4^2(2 + 5)]$

46. $\dfrac{2 \cdot 8^2 - 2^2 \cdot 8}{2 \cdot 8}$

47. $25 + \left[(16 - 3 \cdot 5) + \dfrac{12 + 3}{5}\right]$

48. $7^3 - \dfrac{2}{3}(13 \cdot 6 + 9)4$

Example 6

49. BIOLOGY Lavania is studying the growth of a population of fruit flies in her laboratory. After 6 days, she had nine more than five times as many fruit flies as when she began the study. If she observes 20 fruit flies on the first day of the study, write and evaluate an expression to find the population of fruit flies Lavania observed after 6 days.

   a. Write an expression for the population of fruit flies Lavania observed after 6 days.

   b. Find the population of fruit flies Lavania observed after 6 days.

50. PRECISION The table shows how scores are calculated at diving competitions. Each of the five judges scores each dive from 1 to 10 in 0.5-point increments. Tyrell performs a dive with a degree of difficulty of 2.5. His scores from the judges are 8.0, 7.5, 6.5, 7.5, and 7.0.

Calculating a Diving Score	
Step 1	Drop the highest and lowest of the five judges' scores.
Step 2	Add the remaining scores to find the raw score.
Step 3	Multiply the raw score by the degree of difficulty.

   a. Write an expression to find Tyrell's score for the dive.

   b. What was Tyrell's score for the dive?

51. RAMP The side panel of a skateboard ramp is a trapezoid, as shown.

   a. Write an expression to find the amount of wood needed to build the two side panels of a skateboard ramp.

   b. How much wood is needed to build the two side panels of a skateboard ramp?

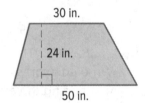

30 in.

24 in.

50 in.

Example 8

52. SKIING The cost of a ski trip is shown. The Sanchez family wants to purchase lift tickets for 2 adults and 3 children. They also need to rent 2 complete pairs of skis. If they also buy a 16-ounce hot chocolate for each person, find the total cost of the ski trip.

Lift Tickets		Rentals		Hot Chocolate	
Children	$34	Skis	$32	12 oz	$3
Seniors	$36	Poles Only	$10	16 oz	$4
Adults	$42	Snowboards	$29	20 oz	$5

Mixed Exercises

Write a numerical expression for each verbal expression.

53. eight to the fourth power increased by six

54. the sum of three and five to the third power times five plus one

55. CONSTRUCT ARGUMENTS Isabel wrote the expression $6 + 3 \times 5 - 6 + 8 \div 2$ and asked Tamara to evaluate it. When Tamara evaluated it, she got a value of 19. Isabel told Tamara that her value was incorrect and said that the value should have been 38. Who is correct? Justify your argument.

56. REASONING Write an expression that includes the numbers 2, 4, and 5, has a value of 50, and includes one set of parentheses.

57. CONSTRUCT ARGUMENTS Kelly buys 3 video games that cost $18.96 each. She also buys 2 pairs of earbuds that cost $11.50 each. She has a coupon for $2 off the price of each video game. Kelly uses a calculator, as shown, to find that the total cost of the items is $77.85. The cashier tells her that the total cost is $73.85. Who is correct? Justify your argument.

58. REASONING The expression $13.25 \times 5 + 6.5$ gives the total cost in dollars of renting a bicycle and helmet for 5 days. The fee for the helmet does not depend upon the number of days.

  a. What does 13.25 represent?

  b. How would the expression be different if the cost of the helmet were doubled?

59. PERSEVERE The figure shows a floor plan for a two-room apartment. Write an expression for the area of the apartment, in square feet, by first finding the area of each room and then adding. Then describe how you can write the expression in a different way.

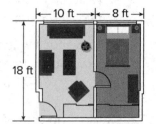

60. CREATE Describe a situation that could be represented by each expression.

  a. $9.95 + 0.75 \times 3$

  b. $15 \times 6 - 5 \times 6$

  c. $59 \times 5 - 25 \times 5 - 30$

61. FIND THE ERROR A student was asked to evaluate an expression. The student's work is shown. Describe any errors and find the correct value of the expression. Explain your reasoning.

$6^2 - 5 \times 2 + 2(9-7)$
$= 6^2 - 5 \times 2 + 2(2)$
$= 36 - 5 \times 2 + 2(2)$
$= 36 - 10 + 4$
$= 36 - 14$
$= 22$

62. PERSEVERE When is $4x < 4$? Use a drawing to justify your reasoning.

Algebraic Expressions

Learn Writing Algebraic Expressions

A **variable** is a letter used to represent an unspecified number or value.

An **algebraic expression** is an expression that contains at least one variable.

A **term** of an expression is a number, a variable, or a product or quotient of numbers and variables.

A **variable term** is a term that contains a variable.

A **constant term** is a term that does not contain a variable.

Today's Goals
- Write algebraic expressions for verbal expressions.
- Evaluate algebraic expressions.

Today's Vocabulary
variable

algebraic expression

term

variable term

constant term

define a variable

Example 1 Write a Verbal Expression

There are common verbal phrases associated with operations. These phrases can be used to interpret an algebraic expression.

Interpreting an Algebraic Expression	
Operation	Verbal Phrases
addition	more than, sum, plus, increased by, added to
subtraction	less than, subtracted from, difference, decreased by, minus
multiplication	product of, multiplied by, times
division	quotient of, divided by

Write a verbal expression for $5x^3 + 2$.

_____

😃 **Think About It!**

Write an algebraic expression that uses at least two variables, one constant, one product, and one power. Identify the terms, variables, constant, product, factors, and powers.

Check

Which verbal expression represents $\frac{2}{5}m^2$? _____

A. two fifths times the product of m and two

B. the quotient of two and five times the product of m and two

C. two fifths of m squared

D. two fifths times m cubed

🔘 **Go Online** You can complete an Extra Example online.

Go Online
An alternate method is available for this example.

Watch Out!

Quotient When a verbal expression refers to a quotient, the first term mentioned is the numerator and the second term mentioned is the denominator. For example, *the quotient of 6 and t* is written as $\frac{6}{t}$, while *the quotient of t and 6* is written as $\frac{t}{6}$.

Study Tip

Exponents The exponents 2 and 3 are frequently used in math and have special verbal phrases associated with them: x^2 is read "*x* squared," and x^3 is read "*x* cubed."

Example 2 Write a Verbal Expression with Grouping Symbols

Write a verbal expression for $a^4 + \frac{6b}{7}$.

$$a^4 \qquad + \qquad \frac{6b}{7}$$

a to the fourth power   plus   the quotient of 6 times *b* and 7

_____

Try an alternate method.

$$a^4 \qquad + \qquad \frac{6b}{7}$$

a to the fourth power   plus   6 times *b* divided by 7

_____

Compare and contrast the methods.

_____
_____
_____
_____

Example 3 Write an Algebraic Expression

Write an algebraic expression for each verbal expression.

a. 2 times the quantity *y* plus 11

The word *times* implies _____, *the quantity* implies _____, and *plus* implies _____.

So, the expression is written as _____.

b. *n* cubed increased by 5

The word *cubed* implies a power of ___, and *increased by* suggests _____.

The expression could be written as _____.

Check

Use the verbal phrase *18 increased by* the *product of* 3 and *d*.

Part A The key verbal phrases in the expression are italicized. Identify which operation each key phrase implies.

increased by implies _____

product of implies _____

Part B Write an algebraic expression for the verbal phrase.

_____

 Go Online You can complete an Extra Example online.

When writing an expression to represent a situation, choose a variable to represent each unknown value in the problem. This is called **defining a variable**.

🌐 Example 4 Write an Expression

SOCCER **In the group play stage of the FIFA World Cup, teams are placed in groups of 4, and they play each other. A team is awarded 3 points for a win, 1 point for a tie, and no points for a loss. Write an algebraic expression that represents the number of points accumulated by one team in the group play stage of the World Cup.**

Define variables for the unknown values.

Let w be the number of wins, t be the number of ties, and z be the number of losses for one team.

So, the number of points awarded for wins is _____, the number of points awarded for ties is _____, and the number of points awarded for losses is _____. The number of points accumulated is _____.

a. Write a verbal expression for the number of points accumulated and interpret the meaning of the variables in the context of the problem.

_____

b. What units are associated with the variables, the coefficients, and the expression?

_____

_____

_____

c. How would the expression change if a point were deducted for each loss?

_____

Check

MUSIC A music festival offers one-day and three-day passes. A one-day pass costs $100, and a three-day pass costs $250. Write an expression for the total ticket sales if n one-day passes and t three-day passes are sold. _____

🔵 **Go Online** You can complete an Extra Example online.

Modeling When writing an expression to model a situation, begin by identifying the important quantities and relationships.

Think About It!

Use the Substitution Property to replace each variable in the expressions with the appropriate value. Let $p = 6$, $q = 0.5$, and $r = 10$.

$3p =$ _____

$q^2 =$ _____

$4q + r =$ _____

$pr^3 =$ _____

 Online Activity Use a real-world situation to complete the Explore.

⊘ INQUIRY How are algebraic expressions useful in the real world?

Learn Evaluating Algebraic Expressions

Algebraic expressions can be evaluated for given values of the variables. The Substitution Property allows us to evaluate an algebraic expression by replacing the variables with their values.

Key Concept • Substitution Property	
Words	A quantity may be substituted for its equal in any expression.
Symbols	If $a = b$, then a may be replaced by b in any expression.
Example	If $m = 11$, then $4m = 4 \cdot 11$.

Example 5 Evaluate an Algebraic Expression

After applying the Substitution Property to an algebraic expression, you can find the value of the numerical expression by using the order of operations.

Evaluate $a^2(3b - a + 5) \div c$ if $a = 2$, $b = 6$, and $c = 4$.

$a^2(3b - a + 5) \div c =$

_____$(3 \cdot$ _____ $-$ _____ $+ 5) \div$ _____	$a = 2, b = 6, c = 4$
$= 2^2($_____ $- 2 + 5) \div 4$	Multiply 3 by 6.
$= 2^2($_____$) \div 4$	Subtract 2 from 18, add 5.
$=$ _____$(21) \div 4$	Evaluate 2^2.
$=$ _____ $\div 4$	Multiply 4 by 21.
$=$ _____	Divide 84 by 4.

Check

Evaluate $\dfrac{b(9 - c)}{a^2}$ if $a = 4$, $b = 6$, $c = 8$. _____

Talk About It!

How would you evaluate $a[(b - c) \div d] - f$ if you were given values of a, b, c, d, and f? How would you evaluate the expression differently if the expression were $a \cdot b - c \div d - f$?

⊘ Go Online You can complete an Extra Example online.

🌐 Example 6 Write and Evaluate an Algebraic Expression

WORLD RECORDS In 2004, Chad Fell set the record for the largest bubblegum bubble blown. Assume that the bubble was spherical. The surface area of a sphere is four times π multiplied by the radius squared.

Part A Complete the table to write an expression that represents the surface area of a sphere.

Words	four times π multiplied by radius squared
Variable	Let $r =$ _____ .
Expression	$4 \times \pi r^2$ or $4\pi r^2$

Part B The record-setting bubble had a radius of 25.4 centimeters. Find the surface area of this bubble.

$A = 4\pi r^2$ Surface area of a sphere

$= 4\pi(\underline{\hspace{1cm}})^2$ Replace r with 25.4.

$= 4\pi(\underline{\hspace{1cm}})$ Evaluate $25.4^2 = 645.16$.

$= \underline{\hspace{1cm}}\pi$ Multiply 4 by 645.16.

≈ 8107.32 Simplify.

The surface area of the bubble is approximately _____ cm².

Check

FOOTBALL The seating capacities of team stadiums in the AFC East Division of the National Football League are shown in the table.

Team	Number of Seats
Miami Dolphins	65,326
New England Patriots	66,829
Buffalo Bills	71,608
New York Jets	82,500

Part A Write an expression that represents the maximum number of attendees at Jets, Dolphins, and Patriots home games during the season. Let j be the number of Jets home games, d be the number of Dolphins home games, and p be the number of Patriots home games. _____ $j +$ _____ $d +$ _____ p

Part B Suppose that after the sixth week of the season, the Jets had played 4 home games, the Dolphins had played 3 home games, and the Patriots had played 2 home games. Based on your expression from Part A, find the maximum number of attendees at the Jets, Dolphins, and Patriots games after the sixth week of the season. _____

🌐 **Go Online** You can complete an Extra Example online.

Pause and Reflect

Did you struggle with anything in this lesson? If so, how did you deal with it?

Record your observations here.

Practice

Go Online You can complete your homework online.

Examples 1 and 2

Write a verbal expression for each algebraic expression.

1. $4q$

2. $\frac{1}{8}y$

3. $15 + r$

4. $w - 24$

5. $3x^2$

6. $\frac{r^4}{9}$

7. $2a + 6$

8. $r^4 \cdot t^3$

9. $25 + 6x^2$

10. $6f^2 + 5f$

11. $\frac{3a^5}{2}$

12. $9(a^2 - 1)$

13. $5g^6$

14. $(c - 2)d$

15. $4 - 5h$

16. $2b^2$

17. $7x^3 - 1$

18. $p^4 + 6r$

19. $3n^2 - x$

20. $(2 + 5)p$

21. $18(p + 5)$

Example 3

Write an algebraic expression for each verbal expression.

22. x more than 7

23. a number less 35

24. 5 times a number

25. one third of a number

26. f divided by 10

27. the quotient of 45 and r

28. three times a number plus 16

29. 18 decreased by 3 times d

30. k squared minus 11

31. 20 divided by t to the fifth power

32. the sum of a number and 10

33. 15 less than the sum of k and 2

34. the product of 18 and q

35. 6 more than twice m

Example 4

36. TECHNOLOGY There are 1024 bytes in a kilobyte. Write an expression that describes the number of bytes in a computer chip with n kilobytes.

37. THEATER H. Howard Hughes, Professor Emeritus of Texas Wesleyan College, and his wife Erin Connor Hughes attended a record 6136 theatrical shows. Write an expression for the average number of shows they attended per year if they accumulated the record over y years.

38. TIDES The difference between high and low tides along the Maine coast one week is 19 feet on Monday and x feet on Tuesday. Write an expression to show the average difference between the tides for Monday and Tuesday.

39. SALE The cost of a T-shirt is shown. Monica has a $10-off coupon. Write an expression that describes the cost of t T-shirts, not including sales tax.

40. GYM MEMBERSHIP Juliana wants to join a gym. The cost of a gym membership is a one-time $100 fee plus $30 per month. Write an expression that describes the cost of a gym membership after m months.

41. BOWLING The cost for bowling is $5 per player for shoe rentals and $45 per hour to book a lane. Suppose a group of f friends go bowling for h hours. Write an expression for the total cost for the group of friends to go bowling.

Example 5

Evaluate each expression if $g = 2$, $r = 3$, and $t = 11$.

42. $g + 6t$

43. $7 - gr$

44. $r^2 + (g^3 - 8)^5$

45. $(2t + 3g) \div 4$

46. $t^2 + 8rt + r^2$

47. $3g(g + r)^2 - 1$

Evaluate each expression if $a = 8$, $b = 4$, and $c = 16$.

48. $a^2bc - b^2$

49. $\dfrac{c^2}{b^2} + \dfrac{b^2}{a^2}$

50. $\dfrac{2b + 3c^2}{4a^2 - 2b}$

51. $\dfrac{3ab + c^2}{a}$

52. $\left(\dfrac{a}{b}\right)^2 - \dfrac{c}{a - b}$

53. $\dfrac{2a - b^2}{ab} + \dfrac{c - a}{b^2}$

Evaluate each expression if $x = 6$, $y = 8$, and $z = 3$.

54. $xy + z$

55. $yz - x$

56. $2x + 3y - z$

57. $2(x + z) - y$

58. $5z + (y - x)$

59. $5x - (y + 2z)$

60. $z^3 + (y^2 - 4x)$

61. $\dfrac{y + xz}{2}$

62. $\dfrac{3y + x^2}{z}$

Example 6

63. SCHOOLS Jefferson High School has 100 less than 5 times as many students as Taft High School.

   a. Write an expression to find the number of students at Jefferson High School if Taft High School has *t* students.

   b. How many students are at Jefferson High School if Taft High School has 300 students?

64. GEOGRAPHY Guadalupe Peak in Texas has an altitude that is 671 feet more than double the altitude of Mount Sunflower in Kansas.

   a. Write an expression for the altitude of Guadalupe Peak if Mount Sunflower has an altitude of *n* feet.

   b. What is the altitude of Guadalupe Peak if Mount Sunflower has an altitude of 4039 feet?

65. TRANSPORTATION The Plaid Taxi Cab Company charges a $1.75 base fee plus $3.45 per mile. Deangelo plans to take a Plaid taxi to the airport.

   a. Write an expression to find the cost for Deangelo to take a Plaid taxi *m* miles to the airport.

   b. How much will it cost for Deangelo to take a Plaid taxi 8 miles to the airport?

66. GEOMETRY The area of a circle is given by the product of π and the square of the radius.

   a. Write an expression for the area of a circle with radius *r*.

   b. What is the area of the circle shown at the right? Use 3.14 for π.

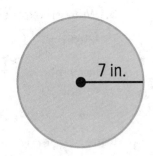

7 in.

Mixed Exercises

67. Consider the expression $\frac{5x}{2} + y^3$.

   a. Write two verbal expressions for $\frac{5x}{2} + y^3$.

   b. Evaluate the expression if $x = 4$ and $y = 2$.

68. Evaluate $\frac{7a + b}{b + c}$, if $a = 2$, $b = 6$, and $c = 4$.

69. Evaluate $x^2 + y^2 + z$, if $x = 7$, $y = 6$, and $z = 4$.

70. Evaluate $\frac{2b + c^2}{a}$, if $a = 2$, $b = 4$, and $c = 6$.

71. Evaluate $2 + x(2y + z)$, if $x = 5$, $y = 3$, and $z = 4$.

72. STRUCTURE Write an algebraic expression that includes a sum and a product. Write a verbal expression for your algebraic expression.

73. STRUCTURE Write a verbal expression that includes a difference and a quotient. Write an algebraic expression for your verbal expression.

74. USE A MODEL A toy manufacturer produces a set of blocks, with edge b, that can be used by children to build play structures. The production team is analyzing the amount of paint they need for a block.

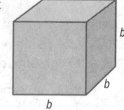

a. The production team decides to use one coat of paint for each block. Write an expression representing the minimum amount of paint needed for one block with edge b.

b. The production team decides one coat of paint is not enough, so they want to use two coats of paint for each block. Write an expression representing the minimum amount of paint needed for one block with edge b.

c. The production team purchases cans of paint that will cover 60 in^2. Write an inequality representing the maximum length of edge b, in inches, when the block is covered with the minimum amount of paint needed for two coats of paint.

75. REASONING During a long weekend, Devon paid a total of x dollars for a rental car so he could visit his family. He rented the car for 4 days at a rate of $36 per day. There was an additional charge of $0.20 per mile after the first 200 miles driven.

a. Write an expression to represent the amount Devon paid for additional mileage.

b. Write an expression to represent the number of miles over 200 miles that Devon drove.

c. How many miles did Devon drive overall if he paid a total of $174 for the car rental?

Higher-Order Thinking Skills

76. PERSEVERE For the cube, x represents a positive whole number. Find the value for x such that the volume of the cube and 6 times the area of one of its faces have the same value.

77. WRITE Describe how to write an algebraic expression from a real-world situation. Include a definition of algebraic expression in your own words.

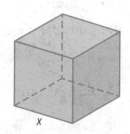

78. WRITE Explain the difference between an algebraic expression and a verbal expression.

79. CREATE Write a real-world situation that can be modeled by the inequality $10t + 5.25 \leq 50$.

Properties of Real Numbers

Learn Properties of Equality

There are several properties of equality that apply to the addition and multiplication of real numbers.

Key Concept • Properties of Equality	
Reflexive Property	
Words	Any quantity is equal to itself.
Symbols	For any number a, $a = a$.
Examples	$3 = 3$ $9 + 2 = 9 + 2$
Symmetric Property	
Words	If one quantity equals a second quantity, then the second quantity equals the first.
Symbols	For any numbers a and b, if $a = b$, then $b = a$.
Example	If $7 = 3 + 4$, then $3 + 4 = 7$.
Transitive Property	
Words	If one quantity equals a second quantity and the second quantity equals a third quantity, then the first quantity equals the third quantity.
Symbols	For any real numbers a, b, and c, if $a = b$ and $b = c$, then $a = c$.
Example	If $5 + 1 = 2 + 4$ and $2 + 4 = 6$, then $5 + 1 = 6$.

Example 1 Identify Properties of Equality

Identify the property of equality used to justify each statement. Explain your reasoning.

a. If 13 + 25 = 38, then 38 = 13 + 25.

_____ Property of Equality; $13 + 25 = 38$ and _____

b. $y + 4 = y + 4$

_____ Property of Equality; $y + 4$ is equal to _____.

Check

Identify the property of equality used to justify each statement.

a. $22 + 7 = 22 + 7$

b. If $36 = 17 + 19$, then $17 + 19 = 36$.

_____ _____

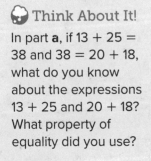 **Go Online** You can complete an Extra Example online.

Today's Goals
- Recognize the properties of equality and identiy.
- Evaluate numerical expressions by applying the Inverse and Identity Properties.
- Evaluate numerical expressions by applying the Commutative and Associative Properties.

Today's Vocabulary
additive identity

additive inverses

multiplicative identity

multiplicative inverses

Go Online
You may want to complete the Concept Check to check your understanding.

Think About It!
In part **a**, if $13 + 25 = 38$ and $38 = 20 + 18$, what do you know about the expressions $13 + 25$ and $20 + 18$? What property of equality did you use?

Copyright © McGraw-Hill Education

🌐 Example 2 Interpret Properties of Equality

COOKING **If the amount of sugar in a recipe is equal to the amount of flour plus 2 tablespoons, and the amount of flour plus 2 tablespoons is equal to the amount of milk, then the amount of sugar is equal to the amount of milk.**

a. Write a verbal expression in the spaces below.

☐ = ☐ , and ☐ = ☐ , then

☐ = ☐ .

Let a = the amount of sugar in the recipe, b = the amount of flour in the recipe plus 2 tablespoons, and c = the amount of milk in the recipe.

If we substitute a, b, and c for the verbal expressions, we can write the following algebraic expression.

If $a = b$ and $b = c$, then $a = c$.

b. This statement is an example of which property of equality?

This is an example of the _____ Property of Equality.

Check

WEIGHT The weight of a bag of oranges plus 13 ounces is equal to 5 pounds.

Part A Which quantity is equivalent to 5 pounds?

5 lb = _____

Part B This statement is an example of which property of equality?

_____

 Go Online You can complete an Extra Example online.

🗨 Talk About It!

Suppose the example stated, "If the amount of milk in a recipe is equal to the amount of sugar and the amount of sugar is equal to the amount of flour plus 2 tablespoons, then the amount of milk is equal to the amount of flour plus 2 tablespoons." Would this example still describe the Transitive Property of Equality? Explain your reasoning.

Watch Out!

Choosing a Variable Remember that any letter can be used to represent an algebraic expression, not just a, b, and c.

Example 3 Use Properties of Equality

Use the given property of equality to complete each statement.

a. $y - 21 = $ _____?_____; **Reflexive Property of Equality**

The Reflexive Property of Equality states that any quantity equals itself, so _____ = _____.

b. If 24 + 11 = 9 + 26 and 9 + 26 = z, then 24 + 11 = ____?____;
Transitive Property of Equality

The Transitive Property of Equality states that if one quantity _____ equals a second quantity _____ and the second quantity _____ equals a third quantity _____, then the first quantity _____ equals the third quantity _____. So, _____.

Check

Use the given property of equality to complete each statement.

a. If 43 + 9 = 10 + 42 and 10 + 42 = 52, then 43 + 9 = _____;
Transitive Property of Equality

b. $2m - 1 = $ _____; Reflexive Property of Equality

Learn Identities and Inverses

The sum of any number a and 0 is equal to a. Thus, 0 is called the **additive identity**. If the sum of two numbers is equal to the additive identity, like $4 + (-4) = 0$, then the two numbers are **additive inverses**.

Key Concept • Addition Properties	
Additive Identity Property	
Words	For any real number a, the sum of a and 0 is a.
Symbols	$a + 0 = 0 + a = a$
Examples	$5 + 0 = 5$
	$0 + 5 = 5$
Additive Inverse Property	
Words	A real number and its opposite are additive inverses of each other.
Symbols	$a + (-a) = 0$
Examples	$2 + (-2) = 0$
	$7 - 7 = 0$

Go Online You can complete an Extra Example online.

The product of any number a and 1 is equal to a. Thus, 1 is called the **multiplicative identity**.

The product of any number a and 0 is equal to 0. This is called the Multiplicative Property of Zero.

Two numbers with a product of 1 are called **multiplicative inverses** or **reciprocals**.

Key Concept • Multiplication Properties	
Multiplicative Identity Property	
Words	For any real number a, the product of a and 1 is a.
Symbols	$a \cdot 1 = a$ $1 \cdot a = a$
Examples	$4 \cdot 1 = 4$ $1 \cdot 4 = 4$
Multiplicative Property of Zero	
Words	For any real number a, the product of a and 0 is 0.
Symbols	$a \cdot 0 = 0$ $0 \cdot a = 0$
Examples	$12 \cdot 0 = 0$ $0 \cdot 12 = 0$
Multiplicative Inverse Property	
Words	For every real number $\frac{a}{b}$ where $a, b \neq 0$, there is exactly one number $\frac{b}{a}$ such that $\frac{a}{b} \cdot \frac{b}{a}$ is 1.
Symbols	$\frac{a}{b} \cdot \frac{b}{a} = 1$ $\frac{b}{a} \cdot \frac{a}{b} = 1$
Examples	$\frac{2}{3} \cdot \frac{3}{2} = 1$ $\frac{3}{2} \cdot \frac{2}{3} = 1$

Study Tip

Decimals as Fractions
When multiplying decimal values, consider the values as fractions to see if you can evaluate the expression using multiplicative inverses. For example, the value of $0.\overline{6} \cdot 1.5$ is not immediately clear, but if you rewrite the expression using fractions, then you can easily see that $\frac{2}{3} \cdot \frac{3}{2} = 1$.

Think About It!

If $x - y$ is an example of the Additive Inverse Property, then what must be true about the relationship between x and y?

Example 4 Evaluate Using the Addition Properties

Evaluate $4 - 2^2 + 8(2)$.

$4 - 2^2 + 8(2) = 4 - \underline{\hspace{1cm}} + 8(2)$ Simplify 2^2.

$= 4 - 4 + \underline{\hspace{1cm}}$ Multiply 8 by 2.

$= \underline{\hspace{1cm}} + 16$ Additive inverses: $4 - 4 = 0$

$= \underline{\hspace{1cm}}$ Additive identity: $0 + 16 = 16$

Check

Evaluate $13 + 2^2 + 0 = \underline{\hspace{0.5cm}}$.

Go Online You can complete an Extra Example online.

Example 5 Evaluate Using the Multiplicative Identity and Multiplicative Inverse

Evaluate $\frac{3}{2} \cdot \frac{2}{3} \cdot 7$.

$\frac{3}{2} \cdot \frac{2}{3} \cdot 7 = \underline{\hspace{1cm}} \cdot 7$ Multiplicative inverses: $\frac{3}{2} \cdot \frac{2}{3} = 1$

$\phantom{\frac{3}{2} \cdot \frac{2}{3} \cdot 7} = \underline{\hspace{1cm}}$ Multiplicative identity: $1 \cdot 7 = 7$

Think About It!

Why is the multiplicative identity 1 and not the same value as the additive identity, 0?

Check

Identify the property used in each step of the evaluation process.

$4 \cdot 1 + 0 - 4 + 3 = 4 + 0 - 4 + 3$ _____

$ = 4 - 4 + 3$ _____

$ = 0 + 3$ _____

$ = 3$ _____

Example 6 Evaluate Using the Multiplicative Property of Zero

Evaluate $[4 - 3(2) + 7] \cdot 0$.

Notice that this expression is the product of an expression and 0.

According to the Multiplicative Property of Zero, the product of any number and 0 is 0.

Therefore, $[4 - 3(2) + 7] \cdot 0 = \underline{\hspace{1cm}}$.

Check

Evaluate $(13 - 1) \cdot 0 + \frac{3}{4} \cdot \frac{4}{3} = \underline{\hspace{1cm}}$.

Watch Out!

Parentheses
Before evaluating the expression using the Multiplicative Property of Zero, pay close attention to parentheses to see if the entire expression or only part of it is being multiplied by 0. For example, $(5 + 9 - 2) \cdot 0 = 0$, but $5 + (9 - 2) \cdot 0 = 5 + 0 = 5$.

Go Online You can complete an Extra Example online.

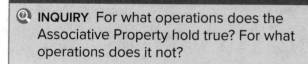

Online Activity Use a table to complete the Explore.

INQUIRY For what operations does the Associative Property hold true? For what operations does it not?

Learn Commutative and Associative Properties

An easy way to find the sum or product of numbers is to group, or associate, the numbers using the Associative Property.

For the addition and multiplication of real numbers, the order does not change their sum or product. This is called the Commutative Property.

Key Concept • Associative and Commutative Properties	
Associative Property	
Words	The way you group three or more numbers when adding or multiplying does not change their sum or product.
Symbols	For any numbers a, b, and c, $(a + b) + c = a + (b + c)$ and $(ab)c = a(bc)$.
Examples	$(2 + 9) + 4 = 2 + (9 + 4)$ $(3 \cdot 6) \cdot 5 = 3 \cdot (6 \cdot 5)$
Commutative Property	
Words	The order in which you add or multiply numbers does not change their sum or product.
Symbols	For any numbers a and b, $a + b = b + a$ and $a \cdot b = b \cdot a$.
Examples	$8 + 12 = 12 + 8$ $4 \cdot 9 = 9 \cdot 4$

🌐 Example 7 Evaluate Using the Associative Property

PERSONAL FINANCE Jalen wants to add up how much money he spent on gasoline in July. He grabs the four receipts he has for July and writes the expression that represents the total amount he spent. Evaluate the expression to determine how much money Jalen spent on gasoline in July.

☁️ **Think About It!**

Why did you use the Associative Property to evaluate this expression?

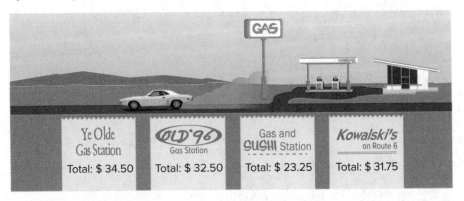

Ye Olde Gas Station — Total: $ 34.50

OLD 96 Gas Station — Total: $ 32.50

Gas and SUSHI Station — Total: $ 23.25

Kowalski's on Route 6 — Total: $ 31.75

34.50 + _____ + 23.25 + _____   Original expression

= (34.50 + 32.50) + (_____ + _____)   Associative (+)

= _____ + 55.00   Simplify.

= _____   Add.

Jalen spent _____ on gasoline in July.

Check

Determine the property used for each step of the evaluation process.

$114 + 71 + 19 + 26 = 114 + (71 + 19) + 26$ _____

$= 114 + 90 + 26$ Simplify.

$= 114 + 26 + 90$ _____

$= 140 + 90$ Simplify.

$= 230$ Simplify.

🔵 **Go Online** You can complete an Extra Example online.

Example 8 Evaluate Using the Commutative Property

Evaluate $\frac{5}{6} \cdot 9 \cdot \frac{6}{5}$.

$\frac{5}{6} \cdot 9 \cdot \frac{6}{5} = \frac{5}{6} \cdot$ _____ $\cdot 9$ Commutative ($\times$).

$= $ _____ $\cdot 9$ Multiplicative inverses: $\frac{5}{6} \cdot \frac{6}{5} = 1$

$= $ _____   Multiplicative Identity

Check

Josefina needs to evaluate $\frac{7}{3} \cdot 2 \cdot 3$. She wants to use the Commutative Property to more easily evaluate the expression. To do this, she should first multiply _____ and _____.

Example 9 Evaluate Using the Associative and Commutative Properties

GROCERIES Jade buys groceries once each week, and she wants to calculate how much she spent on groceries over a 4-week period. Evaluate the expression to find how much Jade spent at the grocery store.

Groceries	
Week	Amount Spent ($)
1	32
2	27
3	28
4	33

$32 + 27 + 28 + 33 = 32 + 28 +$ _____   Step 1

$= (32 + 28) + ($ _____ $)$ Step 2

$= 60 +$ _____   Step 3

$=$ _____   Step 4

Jade spent _____ on groceries over the 4-week period.

Check

Use the Commutative and Associative Properties to evaluate $\frac{5}{9} \cdot 9 \cdot \frac{9}{5} \cdot 4$. _____

Watch Out!

Associative and Commutative Properties

Before you apply the Associative or Commutative Property, make sure that only one operation is involved. For example, in $3 + 5 \cdot 2$, you cannot group $3 + 5$ or switch the 3 and 5 because that does not adhere to the order of operations.

Think About It!

For which step(s) did Jade use the Commutative Property? For which step(s) did Jade use the Associative Property?

Go Online

to learn about operations with rational numbers in Expand 1-3.

🖱 **Go Online** You can complete an Extra Example online.

Practice

Examples 1–3

Identify the property of equality used to justify each statement.

1. If $4 + 17 = 21$, then $21 = 4 + 17$.

2. $x + 3 = x + 3$

3. If $16 = 9 + 7$, then $9 + 7 = 16$.

4. If $6 + 2 = 4 + 4$ and $4 + 4 = 8$, then $6 + 2 = 8$.

Use the given property of equality to complete each statement.

5. If $23 + 14 = 37$, then $37 = 23 +$ _____;
Symmetric Property of Equality

6. If $a + 5 = b + 3$ and $a + 5 = 12$, then $b + 3 =$ _____;
Transitive Property of Equality

7. If $34 = 19 + 15$, then $19 + 15 =$ _____;
Symmetric Property of Equality

8. $b + 5 + 12 =$ _____;
Reflexive Property of Equality

9. TOLL ROADS Some toll highways assess tolls based on where a car entered and exited. The table shows the highway tolls for a car entering and exiting at a variety of exits. Assume that the toll for the reverse direction is the same.

Entered	Exited	Toll
Exit 8	Exit 10	$0.25
Exit 10	Exit 15	$1.00
Exit 15	Exit 18	$0.50
Exit 18	Exit 22	$0.75

a. Julio travels from Exit 8 to Exit 15. Which quantity is equivalent to Exit 8 to Exit 15?

b. What property would you use to determine the toll?

Examples 4–6

Evaluate each expression. Name the property used in each step.

10. $3(22 - 3 \cdot 7)$

11. $[3 \div (2 \cdot 1)]\frac{2}{3}$

12. $2(3 \cdot 2 - 5) + 3 \cdot \frac{1}{3}$

13. $2[5 - (15 \div 3)]$

14. $6 + 9[10 - 2(2 + 3)]$

15. $2(6 \div 3 - 1) \cdot \frac{1}{2}$

Examples 7–9

Evaluate each expression using properties of numbers. Name the property used in each step.

16. $25 + 14 + 15 + 36$

17. $4\frac{4}{9} + 7\frac{2}{9}$

18. $4.3 + 2.4 + 3.6 + 9.7$

19. $2 \cdot 8 \cdot 10 \cdot 2$

20. $1\frac{5}{6} \cdot 24 \cdot 3\frac{1}{11}$

21. $2\frac{3}{4} \cdot 1\frac{1}{8} \cdot 32$

22. $16 + 8 + 14 + 12$

23. $2 \cdot 4 \cdot 5 \cdot 3$

24. $6.4 + 2.7 + 1.6 + 5.3$

25. $\frac{4}{3} \cdot 7 \cdot 3 \cdot 10$

Evaluate each expression if $a = -1$, $b = 4$, and $c = 6$.

26. $4a + 9b - 2c$

27. $-10c + 3a + a$

28. $a - b + 5a - 2b$

29. $8a + 5b - 11a - 7b$

30. $3c^2 + 2c + 2c^2$

31. $3a - 4a^2 + 2a$

32. Name the property that is used in $5 \cdot n \cdot 2 = 0$. Then find the value of n.

33. Name two properties used to evaluate $7 \cdot 1 - 4 \cdot \frac{1}{4}$.

34. Evaluate $7 \cdot 2 \cdot 7 \cdot 5$ using properties of numbers. Name the property used in each step.

Mixed Exercises

Find the value of x. Then name the property used.

35. $8 = 8 + x$

36. $3.2 + x = 3.2$

37. $10x = 10$

38. $\frac{1}{2} \cdot x = \frac{1}{2} \cdot 7$

39. $x + 0 = 5$

40. $1 \cdot x = 3$

41. $\frac{4}{3} \cdot \frac{3}{4} = x$

42. $2 + 8 = 8 + x$

43. $x + \frac{3}{4} = 3 + \frac{3}{4}$

44. $\frac{1}{3} \cdot x = 1$

45. MENTAL MATH The triangular banner has a base of 9 centimeters and a height of 6 centimeters. Using the formula for area of a triangle, the banner's area can be expressed as $\frac{1}{2} \times 9 \times 6$. Gabrielle finds it easier to write and evaluate $\left(\frac{1}{2} \times 6\right) \times 9$ to find the area. Is Gabrielle's expression equivalent to the area formula? Explain.

46. FINANCE Felicity put down $800 on a used car. She took out a loan to pay off the balance of the cost of the car. Her monthly payment will be $175. After 9 months, how much will she have paid for the car?

47. ANATOMY The human body has 126 bones in the upper and lower extremities, 28 bones in the head, and 52 bones in the torso. Use the Associative Property to write and evaluate an expression that represents the total number of bones in the human body.

48. SCHOOL SUPPLIES At a local school supply store, a highlighter costs $1.25, a ballpoint pen costs $0.80, and a spiral notebook costs $2.75. Use mental math and the Associative Property of Addition to find the total cost if one of each item is purchased.

49. PERSEVERE Write two equations showing the Transitive Property of Equality. Justify your reasoning.

50. ANALYZE Determine whether the following statement is *sometimes*, *always*, or *never* true. Justify your argument. The Commutative Property holds for subtraction.

51. ANALYZE Provide examples to show that there is no Commutative Property or Associative Property for division. What is the relationship between the results when the order of division of two numbers is switched?

52. WRITE Explain why 0 has no multiplicative inverse.

53. ANALYZE The sum of any two whole numbers is always a whole number. So, the set of whole numbers {0, 1, 2, 3, 4, ...} is said to be closed under addition. This is an example of the **Closure Property**. State whether each statement is *true* or *false*. If false, justify your reasoning.

a. The set of whole numbers is closed under subtraction.

b. The set of whole numbers is closed under multiplication.

c. The set of whole numbers is closed under division.

54. ANALYZE Explain whether 1 can be an additive identity. Give an example to justify your reasoning.

55. WHICH ONE DOESN'T BELONG? Identify the equation that does not belong with the other three. Justify your conclusion.

$x + 12 = 12 + x$	$7h = h \cdot 7$	$1 + a = a + 1$	$(2j)k = 2(jk)$

56. CREATE Write an expression that simplifies to 160 using the Commutative and Associative Properties.

Distributive Property

Explore Using Rectangles with the Distributive Property

▶ **Online Activity** Use dynamic geometry software to complete the Explore.

> ⊘ **INQUIRY** What is the product of a and $(b + c)$? ✕

Explore Modeling the Distributive Property

▶ **Online Activity** Use algebra tiles to complete the Explore.

> ⊘ **INQUIRY** How can you use algebra tiles to find the product of two expressions? ✕

Learn Distributive Property with Numerical Expressions

The expressions $3(4 + 2)$ and $3 \cdot 4 + 3 \cdot 2$ are equivalent expressions because they have the same value, 18. This concept shows how the Distributive Property combines addition and multiplication. Multiplying a number by a sum of numbers is the same as doing each multiplication separately and then adding the products.

Key Concept • Distributive Property	
Symbols	For any numbers a, b, and c, $a(b + c) = ab + ac$ and $(b + c)a = ba + ca$ and $a(b - c) = ab - ac$ and $(b - c)a = ba - ca$.
Examples	$4(9 + 3) = 4 \cdot 9 + 4 \cdot 3$ $\quad\quad 4(12) = 36 + 12$ $\quad\quad\quad\quad 48 = 48$ $5(8 - 2) = 5 \cdot 8 - 5 \cdot 2$ $\quad\quad 5(6) = 40 - 10$ $\quad\quad\quad 30 = 30$

The Symmetric Property of Equality allows the Distributive Property to also be written in the reverse order $ab + ac = a(b + c)$.

▶ **Go Online** You can complete an Extra Example online.

Today's Goals
- Use the Distributive Property to evaluate expressions.
- Use the Distributive Property to simplify expressions.

Today's Vocabulary
coefficient

like terms

simplest form

equivalent expressions

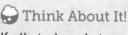 **Think About It!**
If $a(b + c) = ab + ac$, then what does $ab + ac$ equal?

Think About It!

Write an expression for the amount spent on back-to-school shopping in June and September. Then evaluate the expression.

Watch Out!

Decimals and Percents
Make sure to convert percents to decimals before performing calculations.

🌐 Apply Example 1 Use the Distributive Property

SHOPPING **Assume that high school students spend an average of $180 on back-to-school shopping during the months of June to September. Write an expression to represent the amount** of money spent in August if the amount of money spent in August is equal to the amount of money spent in July minus the amount of money spent in June. Evaluate your expression using the Distributive Property.

When are you likely to do the majority of back-to-school shopping?

24% June, 46% July, 22% Aug, 8% Sept

1. What is the task?

Describe the task in your own words. Then list any questions that you may have. How can you find answers to your questions?

_____

_____

2. How will you approach the task? What have you learned that you can use to help you complete the task?

_____

_____

_____

3. What is your solution?

Use your strategy to solve the problem.

What expression represents the amount spent in August?

_____

What is the total amount spent in August? _____

4. How can you know that your solution is reasonable?

✍ **Write About It!** Write an argument that can be used to defend your solution.

_____

_____

_____

🔎 Go Online You can complete an Extra Example online.

Check

SWIMMING Verdell's swim team practices 5 days a week. Each day they spend 15 minutes stretching, 45 minutes swimming laps, and 30 minutes lifting weights.

Part A Which expression(s) represent the number of minutes Verdell's team spends in practice each week? Select all that apply. _____

    A. $5(15 + 45 + 30)$ B. $5(15) + 45 + 30$

    C. $5(15) + 5(45) + 5(30)$ D. $5(15) + 5(45) + 30$

    E. $5 + 15 + 45 + 30$

Part B How much time does Verdell's team spend in practice each week? ____

    A. 90 minutes   B. 150 minutes

    C. 330 minutes   D. 450 minutes

Example 2 Mental Math

Use the Distributive Property to rewrite and evaluate each expression.

a. 5 · 99

$5 \cdot 99 = 5(100 - 1)$	Think: $99 = 100 - 1$
$= 5(100) - 5(1)$	Distributive Property
$= \underline{\hspace{1cm}} - \underline{\hspace{1cm}}$	Multiply.
$= \underline{\hspace{1cm}}$	Subtract.

b. 4 · 1002

$4 \cdot 1002 = 4(1000 + 2)$	Think: $1002 = 1000 + 2$
$= 4(1000) + 4(2)$	Distributive Property
$= \underline{\hspace{1cm}} + \underline{\hspace{1cm}}$	Multiply.
$= \underline{\hspace{1cm}}$	Add.

Check

Part A Estimate the value of the expression $7(51)$.

    $7(\underline{\hspace{1cm}}) = 350$, so $7(51)$ will be a little _____ than 350.

Part B Which expression(s) use(s) the Distributive Property to rewrite and find the exact value of the expression $7(51)$? _____

    A. $51(7 - 3); 204$ B. $7(50 + 1); 357$ C. $51(7 + 3); 510$

    D. $51(7) - 51(3); 204$ E. $7(50) + 7(1); 357$ F. $51(7) + 51(3); 510$

Go Online You can complete an Extra Example online.

Think About It!

How can you use the Distributive Property to rewrite and evaluate the expression $8(1100)$?

Learn Distributive Property with Algebraic Expressions

The **coefficient** is the numerical factor of a term.

Like terms are terms with the same variables, with corresponding variables having the same exponent.

An expression is in **simplest form** when it is replaced by an equivalent expression having no like terms or parentheses.

The Distributive Property and the properties of equality can be used to show that $6x + 2x = 8x$. In this expression, $6x$ and $2x$ are like terms.

$$6x + 2x = (6 + 2)x \qquad \text{Distributive Property}$$
$$= 8x \qquad \text{Substitution}$$

The expressions $6x + 2x$ and $8x$ are called **equivalent expressions** because they represent the same value for any value of the variable.

Example 3 Distribute an Algebraic Expression from the Left

Rewrite $4(5x - 7)$ using the Distributive Property. Then simplify.

$$4(5x - 7) = 4 \cdot 5x - 4 \cdot 7 \qquad \text{Distributive Property}$$
$$= 20x - 28 \qquad \text{Multiply.}$$

Problem-Solving Tip

Make a Model
It can be helpful to visualize a problem using algebra tiles or folded paper.

Check

Simplify the expression. ____

$-6(r + 3g - t)$

A. $-6r - 18g + 6t$

B. $-6r + 18g - 6t$

C. $6r - 18g + 6t$

D. $6r + 18g - 6t$

Go Online You can complete an Extra Example online.

Example 4 Distribute an Algebraic Expression from the Right

Rewrite $(3y^2 + y - 8)6$ using the Distributive Property. Then simplify.

$(3y^2 + y - 8)6 = 6(3y^2) + 6(y) + 6(-8)$ Distributive Property

$ = 18y^2 + 6y - 48$ Multiply.

🗨 **Talk About It!**

Emilio says you can add $18y^2$ and $6y$ to get $24y^3$. Do you agree or disagree? Justify your answer.

Example 5 Combine Like Terms

Simplify each expression.

a. $14a + 18a$

$14a + 18a = (\underline{\hspace{1cm}} + \underline{\hspace{1cm}})a$ Distributive Property

$ = \underline{\hspace{1.5cm}}$ Substitution

b. $4b^2 + 9b - 3b$

$4b^2 + 9b - 3b = 4b^2 + (\underline{\hspace{1cm}} - \underline{\hspace{1cm}})b$ Distributive Property

$ = \underline{\hspace{1.5cm}}$ Substitution

💭 **Think About It!**

What are the like terms in part **a** and part **b**?

Check

Simplify the expression. If not possible, choose *simplified*. _____

$b^2 + 13b + 13$

A. $13b^2 + 13b$

B. $26b^2b$

C. $26b^3$

D. simplified

 Go Online You can complete an Extra Example online.

Watch Out!

Like Terms
$4b^2$ and $6b$ are not like terms because they have different exponents.

Example 6 Write and Simplify Expressions

Part A Complete the table to write an algebraic expression for *three times the sum of 2x and 3y decreased by twice the difference of 4x and y.*

Words	three times the sum of 2x and 3y	decreased by	twice the difference of 4x and y
Expression			

Part B Simplify the expression and indicate the properties used.

$3(2x + 3y) - 2(4x - y)$

$= 3(2x) + 3(3y) - 2(4x) - 2(-y)$ Distributive Property

$= \underline{}x + \underline{}y - \underline{}x + \underline{}y$ Multiply.

$= 6x - 8x + 9y + 2y$ Commutative (+)

$= (\underline{} - \underline{})x + (\underline{} + \underline{})y$ Distributive Property

$= \underline{}x + \underline{}y$ Simplify.

Check

Which expressions are equivalent to *4 times the sum of 2 times x and 6*? Write each expression in the appropriate box.

- $8(x - 3)$
- $4(2x + 6)$
- $(4 + 2x)6$
- $4(2 + x + 6)$
- $8x + 24$
- $4(2x) + 4(6)$

Equivalent	Not Equivalent

🔵 **Go Online** You can complete an Extra Example online.

Practice

🔾 **Go Online** You can complete your homework online.

Example 1

Use the Distributive Property to rewrite each expression. Then evaluate.

1. (4 + 5)6

2. 7(13 + 12)

3. 6(6 − 1)

4. (3 + 8)15

5. 14(8 − 5)

6. (9 − 4)19

7. OPERA Aran's drama class is planning a field trip to see Mozart's famous opera *Don Giovanni*. Tickets cost $39 each, and there are 23 students and 2 teachers going on the field trip.

    a. Write an expression to find the group's total ticket cost.

    b. What is the group's total ticket cost?

8. SALARY In a recent year, the median salary for an engineer in the United States was $55,000 and the median salary for a computer programmer was $52,000.

    a. Write an expression to estimate the total cost for a business to employ an engineer and a programmer for 5 years.

    b. Estimate the total cost for a business to employ an engineer and a programmer for 5 years.

9. COSTUMES Isabella's ballet class is performing a spring recital for which they need butterfly costumes. Each butterfly costume is made from $3\frac{3}{5}$ yards of fabric.

    a. Write an expression to find the number of yards of fabric needed for 10 costumes.

    b. Use the Distributive Property to find the number of yards of fabric needed for 10 costumes. Show your work. (Hint: A mixed number can be written as the sum of an integer and a fraction.)

10. REASONING Letisha and Noelle each opened a checking account, a savings account, and a college fund. The chart shows the amounts that they deposit into each account every month.

	Checking	Savings	College
Letisha	$125	$75	$50
Noelle	$250	$50	$50

    a. Write an expression to find the amount in Letisha's checking, savings, and college accounts after 12 months.

    b. How much is in Letisha's checking, savings, and college accounts after 12 months?

Example 2

Use the Distributive Property to rewrite and evaluate each expression.

11. $7 \cdot 497$

12. $6(525)$

13. $36 \cdot 3\frac{1}{4}$

14. $\left(4\frac{2}{7}\right)21$

15. $5 \cdot 89$

16. $9 \cdot 99$

17. $15 \cdot 104$

18. $15\left(2\frac{1}{3}\right)$

19. $12 \cdot 98$

20. $8 \cdot 1.5$

21. $3 \cdot 10.2$

22. $5\left(4\frac{1}{5}\right)$

Examples 3 and 4

Rewrite each expression using the Distributive Property. Then simplify.

23. $2(x + 4)$

24. $(5 + n)3$

25. $(4 - 3m)8$

26. $-3(2x - 6)$

27. $(2 - 4n)17$

28. $11(4d + 6)$

29. $\left(\frac{1}{3} - 2b\right)27$

30. $4(8p + 16q - 7r)$

31. $6(2c - cd^2 + d)$

32. $7(h - 10)$

33. $3(m + n)$

34. $2(x - y + 1)$

35. $\left(\frac{1}{2} + 6a\right)14$

36. $-2(7m - 8n - 5p)$

37. $(0.3 - 6x)9$

38. $-4(4a + 2b - \frac{1}{2}c)$

Example 5

Simplify each expression. If not possible, write *simplified*.

39. $13r + 5r$

40. $3x^3 - 2x^2$

41. $7m + 7 - 5m$

42. $5z^2 + 3z + 8z^2$

43. $7m + 2m + 5p + 4m$

44. $6x + 4y + 5x$

45. $3m + 5g + 6g + 11m$

46. $4a + 5a^2 + 2a^2 + a^2$

47. $5k + 3k^3 + 7k + 9k^3$

48. $6x^2 + 14x - 9x$

49. $17g + g$

50. $2x^2 + 6x^2$

51. $7a^2 - 2a^2$

52. $3y^2 - 2y + 9$

53. $3q^2 + q - q^2$

Example 6

Consider each verbal expression.

a. Write an algebraic expression to represent the verbal expression.

b. Simplify the expression and indicate the properties used.

54. *The product of 9 and t squared, increased by 3 times the sum of 2 and t squared*

55. *The product of 3 and a, plus 5 times the difference of a and b*

56. *3 times the sum of r and d squared increased by 2 times the sum of r and d squared*

Mixed Exercises

Simplify each expression.

57. $3x + 7(3x + 4)$

58. $4(fg + 3g) + 5g$

59. $6d + 4(3d + 5)$

60. $2(6x + 4) + 7x$

61. $4y^3 + 3y^3 + y^4$

62. $a + \frac{a}{5} + \frac{2}{5}a$

63. $4(2b - b)$

64. $2(n + 2n - 5)$

65. $7(2x + y) + 6(x + 5y)$

66. REASONING A theater has m seats per row on the left side of the aisle and n seats per row on the right side of the aisle. There are r rows of seats.

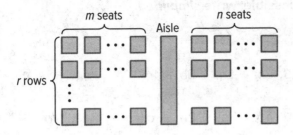

a. Explain how you can use the Distributive Property to write two different expressions that represent the total number of seats in the theater.

b. Suppose you double the number of seats in each row on the left side of the aisle. Does this double the number of seats in the theater? Use one of the expressions you wrote in part **a** to justify your answer.

67. FENCES Demonstrate the Distributive Property by writing two equivalent expressions to represent the perimeter of the fenced dog park.

m

Dog Park n

68. PERSEVERE Use the Distributive Property to simplify $6x^2[(3x - 4) + (4x + 2)]$.

69. FIND THE ERROR Ariana is shipping 8 bags of granola to a customer. Each bag weighs 22 ounces and the maximum weight she can ship in one box is 10 pounds 5 ounces. She makes the calculations at the right and decides that she can ship the bags in one box. Do you agree? Explain your reasoning.

$$8(22) = 8(20 + 2)$$
$$= 8(20) + 2$$
$$= 160 + 2$$
$$= 162 \text{ ounces}$$

70. ANALYZE Determine whether the following statement is *true or false*. Justify your argument. The Distributive Property is a property of both addition and multiplication.

71. WRITE Why is it helpful to represent verbal expressions algebraically?

72. CREATE Write an expression that simplifies to $2a + 14$ using the Distributive Property.

Expressions Involving Absolute Value

Explore Distance Between Points on a Number Line

Online Activity Use graphing technology to complete the Explore.

> **INQUIRY** How can you find the distance between any two values x and y on a number line?

Learn Evaluating Expressions Involving Absolute Value

The **absolute value** of a number is its distance from 0 on a number line.

Key Concept • Absolute Value of Variables	
Words	For any real number x, if x is positive or zero, then the absolute value of x is x. If x is negative, then the absolute value of x is the opposite of x.
Symbols	For any real number x, $\|x\| = x$ if $x \geq 0$, and $\|x\| = -x$ if $x < 0$.
Examples	If $x = 12$, then $\|x\| = 12$. If $x = -7$, then $\|x\| = -(-7) = 7$.

Example 1 Write an Absolute Value Expression

THERMOMETERS The accuracy of a meat thermometer is the positive difference between the temperature reading on the thermometer t and the actual temperature of the meat m. Complete the statements about which expressions are and are not equivalent to the accuracy of a meat thermometer.

$|m - t|$ This expression _____ equivalent to the accuracy of a meat thermometer because it represents the difference between the measurements, and it is always positive because it is the _____ of the difference.

$|m| - |t|$ This expression _____ equivalent to the accuracy of a meat thermometer in all cases because if $t > m$, then $|m| - |t|$ is

_____.

$|t - m|$ This expression _____ equivalent to the accuracy of a meat thermometer because it represents the difference between the measurements, and it is _____ because it is the _____ of the difference.

$|t| - |m|$ This expression _____ equivalent to the accuracy of a meat thermometer in all cases because if m _____ t, then $|t| - |m|$ is _____.

Today's Goal
• Evaluate absolute value expressions.

Today's Vocabulary
absolute value

Study Tip

Interpreting −x
Do not read −x as *negative x* but as *the opposite of x.* Because x represents a negative number whenever $x < 0$, the expression −x represents a positive number.

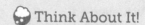 **Think About It!**

Is the expression −x always a negative value? If not, for what values of x is −x positive?

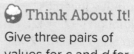 **Think About It!**

Give three pairs of values for c and d for which $|c + d|$ is not equal to $|c| + |d|$.

Study Tip

Opposites If *a* and *b* are real numbers and *a* ≠ *b*, the difference *a* − *b* is the opposite of *b* − *a*. The absolute values of opposites are always equal.

Think About It!

How would you evaluate the expression

$|-2(xy + 5y)|$?

Study Tip

Order of Operations Apply the order of operations when evaluating an expression inside grouping symbols. Perform all the multiplication from left to right and then do the addition and subtraction from left to right.

Watch Out!

Additive Inverses Although $|-2| = |2|$, replacing *x* with −2 will not give the same value as replacing *x* with 2. The evaluation of the expression will be the same only if 3 + 4*x* equals −11 for some other value of *x*.

Check

AMUSEMENT PARKS At an amusement park, a vendor attempts to guess a person's weight. If they are not within 3 pounds, the person will win a prize. If a person weighs 168 pounds and the vendor makes a guess of *x* pounds, which expression represents the number of pounds a winning guess is away from the actual weight? _____

A. $|3 - x|$ **B.** $|x + 3|$

C. $|168 - x|$ **D.** $|3 - 168|$

Example 2 Evaluate the Absolute Value of an Algebraic Expression

Evaluate $|-2xy + 5y|$ if $x = 6$ and $y = -3$.

$$|-2xy + 5y| = |-2(6)(-3) + 5(-3)|$$ Replace *x* with 6 and *y* with −3.

$$= |\underline{\quad} - 15|$$ Multiply.

$$= |\underline{\quad}|$$ Subtract 15 from 36.

$$= \underline{\quad}$$ $|21| = 21$

Check

Evaluate $-|5a + 3(2ab - 1)|$ if $a = -2$ and $b = -3$. _____

Example 3 Evaluate an Expression Involving Absolute Value

When evaluating algebraic expressions, absolute value bars act as a grouping symbol. Perform any operations inside the absolute bars first.

Evaluate $23 - |3 + 4x|$ if $x = 2$. Select a statement below to justify each step.

Replace *x* with 2.	Multiply.	$3 + 8 = 11$
$|11| = 11$	Simplify.	

$$23 - |3 + 4x| = 23 - |3 + 4(2)|$$ _____

$$= 23 - |3 + 8|$$ _____

$$= 23 - |11|$$ _____

$$= 23 - 11$$ _____

$$= 12$$ _____

Check

Evaluate $1.4 - 2.5|5y + 0.6|$ if $y = -3$. __

A. −34.6   **B.** −28.6

C. −15.84   **D.** 37.4

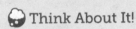 **Go Online** You can complete an Extra Example online.

Practice

Example 1

1. **POOLS** The accuracy of a pool thermometer is the positive difference between the temperature reading on the thermometer t and the actual temperature of the pool p. Write two absolute value expressions equivalent to the accuracy of a pool thermometer.

2. **ROLLERCOASTER** At a theme park, a person must be a certain height to ride a rollercoaster. A person must be h inches tall, plus or minus 1.5 inches. Anoki says the absolute value expression $|1.5 - h|$ represents an acceptable height. David says the absolute value expression $|h - 1.5|$ represents an acceptable height. Who is correct? Explain.

3. **WATER DEPTH** An *echo sounder* is a device used to determine the depth of water by measuring the time it takes a sound produced just below the water surface to return, or echo, from the bottom of the body of water. The accuracy of an echo sounder is the positive difference between the depth of water reading on the echo sounder r and the actual depth of water w. Write two absolute value expressions equivalent to the accuracy of an echo sounder.

4. **GOLF** A certain company designs and ships boxes of golf balls. Each box must weigh 540 grams. Write two absolute value expressions that represent the number of grams a box weighing g grams is away from the desired weight.

Example 2

Evaluate each expression if $m = -4$, $n = 1$, $p = 2$, $q = -6$, $r = 5$, and $t = -2$.

5. $|-n - 2mp|$

6. $|12 + 2t|$

7. $|q - 2mt|$

8. $|3r + 6m|$

9. $|p + 4q - 3r|$

10. $|16 + 4(3q + p)|$

11. $|2m + 6(q - t)|$

12. $-|10 - 7r + 8m + 2p|$

13. $-|14 - 6n + 7(q + 2t)|$

Example 3

Evaluate each expression if $a = 2$, $b = -3$, $c = -4$, $h = 6$, $y = 4$, and $z = -1$.

14. $|2b - 3y| + 5z$

15. $15 - |2 - 3a|$

16. $|a - 5| - 1$

17. $|b + 1| + 8$

18. $5 - |c + 1|$

19. $|a + b| - c$

20. $5 + |2b|$

21. $|4 - h| - b$

22. $|2 - b - h| - h$

Evaluate each expression if $a = -2$, $b = -3$, $c = 2$, $x = 2.1$, $y = 3$, and $z = -4.2$.

23. $|2x + z| + 2y$

24. $4a - |3b + 2c|$

25. $-|5a + c| + |3y + 2z|$

26. $-a + |2x - a|$

27. $|y - 2z| - 3$

28. $3|3b - 8c| - 3$

Evaluate each expression if $a = -\frac{1}{2}$, $b = \frac{3}{4}$, and $c = -\frac{2}{3}$.

29. $-|6c - 16b| + 1$

30. $14 + 2|3c + 10a|$

31. $|-2a - 20b| - 12c$

32. $12a - |-16b|$

33. $|5 - 15c| + a$

34. $|2 - (a - 6b)| + 18c$

Mixed Exercises

35. GPS A golf GPS is a device that can be used to determine the distance a golf ball is from a pin. The accuracy of a golf GPS is the positive difference between the distance a golf ball is from a pin on the golf GPS g and the actual distance a golf ball is from a pin d.

    a. Write two absolute value expressions equivalent to the accuracy of a golf GPS.

    b. Evaluate the expression if the actual distance from the golf ball to the pin is 70 meters and the distance the golf ball is from the pin on the golf GPS is 75 meters.

36. STRUCTURE The students in Mrs. Mangione's class attempt to guess the number of marbles in a jar to earn 2 extra credit points on their next exam. Suppose there are 1206 marbles in a jar and a student makes a guess of m marbles.

    a. Write two absolute value expressions that represent the difference between the guess and the actual number of marbles in the jar.

    b. Evaluate the expression if a student guesses there are 1100 marbles in the jar.

37. CREATE Describe a real-world situation that could be represented by the absolute value expression $|x - 89|$.

38. FIND THE ERROR The accuracy of a rain gauge is the positive difference between the amount of rain in the rain gauge g and the actual amount of rain r. Sam says the absolute value expression $|g| - |r|$ is equivalent to the accuracy of a rain gauge. Is Sam correct? Explain your reasoning.

39. ANALYZE Diaz claims that if a and b are real numbers, then $|a + b|$ is always equal to $|a| + |b|$. Determine whether his claim is *true* or *false*. Justify your argument.

Descriptive Modeling and Accuracy

Learn Descriptive Modeling

Descriptive modeling is a way to mathematically describe real-world situations and the factors that cause them. A **metric** is a rule for assigning a number to some characteristic or attribute.

Metrics can be used to make comparisons. In sports, earned run average is used to compare baseball pitchers, and the quarterback rating compares the performance of football quarterbacks. In banking, a person's debt-to-income ratio can determine whether the person qualifies for a loan. In a good metric, factors that are important in the situation are considered and included in the metric. For example, the quarterback rating includes measures of passing, running, penalties, and other factors that make a quarterback effective.

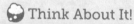 Example 1 Use Descriptive Modeling

COLLEGE ATHLETICS Some universities use a metric called the Academic Index to qualify high school athletic recruits. A student athlete with a 3.1 G.P.A. received scores of 610 in reading, 640 in writing, and 700 in math on the SAT. Use the expression to determine whether the student athlete qualifies at a university that requires an Academic Index of 186 or greater.

G.P.A.	G.P.A. Value
4.0	80
3.9	79
3.8	78
3.7	77
3.6	75
3.5	73
3.4	71
3.3	70
3.2	69
3.1	68
3.0	67

$$2\left[\frac{\left(\frac{\text{Reading Score} + \text{Writing Score}}{2}\right) + \text{Math Score}}{20}\right] + \text{G.P.A. Value}$$

Step 1 Find all values for the metric.

Use the table to determine the G.P.A. value. For a 3.1 G.P.A., the value is 68.

(continued on the next page)

Today's Goals
- Define and use appropriate quantities for the purpose of descriptive modeling.
- Choose a level of accuracy appropriate to limitations on measurements when reporting quantities.

Today's Vocabulary
descriptive modeling

metric

accuracy

Think About It!
Some universities require student athletes to take the SAT twice and then use both scores when determining their Academic Index. How do you think this could affect the score?

Copyright © McGraw-Hill Education

Think About It!

What other attributes of high school recruits do you think universities might consider when creating metrics to determine qualification?

Step 2 Substitute values in the metric.

$$2\left[\dfrac{\left(\dfrac{\text{Reading Score} + \text{Writing Score}}{2}\right) + \text{Math Score}}{20}\right] + \text{G.P.A. Value}$$

$$= 2\left[\dfrac{\left(\dfrac{610 + 640}{2}\right) + 700}{20}\right] + \underline{\quad}$$ 　Reading 610, writing 640, math 700, and G.P.A. 68

$$= 2\left[\dfrac{\left(\dfrac{1250}{2}\right) + 700}{20}\right] + 68$$ 　Add 610 and 640.

$$= 2\left[\dfrac{625 + 700}{20}\right] + 68$$ 　Divide by 2.

$$= 2\left[\dfrac{1325}{20}\right] + 68$$ 　Add 625 and 700.

$$= 2(\underline{\quad}) + 68$$ 　Divide by 20.

$$= \underline{\quad} + 68$$ 　Multiply by 2.

$$= \underline{\quad}$$ 　Simplify.

The Academic Index is _____.

Step 3 Evaluate by using the metric.

Because the Academic Index is greater than 186, this student _____ qualified to attend the university as an athlete.

Check

PARKS Rachelle wants to determine the best state park for hiking and fishing.

Part A Use the metric to calculate a score for each park. Round to the nearest tenth.

$$\text{Park Score} = 100\left[0.2\left(\dfrac{\text{online rating}}{5}\right) + 0.4\left(\dfrac{\text{miles of trails}}{25}\right) + 0.4\left(\dfrac{\text{fish weight}}{10}\right)\right]$$

State Park	Online Star Rating	Hiking Trails (mi)	Best Fish (lb)	Park Score
Gooseberry Falls	4.8	20	9.8	
Lake Maria	4.3	14	8.2	
Maplewood	4.9	25	7.3	
Camden	4.3	15.8	10.1	

Part B How might someone who enjoys hiking much more than fishing change this metric?

_____

_____

_____

🐺 **Go Online** You can complete an Extra Example online.

Example 2 Compare Metrics

HEIGHT A child's adult height can be predicted using several metrics. Given the height of the mother, father, and their son at 2 years old, use the metrics to predict the son's height as an adult.

72"   65"   35"

Use a Source

Find information to create a metric to measure something that is important to you. Explain how your metric includes the factors that you think are important to measure.

Method 1 The Gray Method

The Gray Method uses the average heights of the parents, adjusted by the gender of the child. For a boy, the mother's height is multiplied by $\frac{13}{12}$, and for a girl, the father's height is multiplied by $\frac{12}{13}$.

$$\frac{\frac{13}{12} \cdot \text{mother's height} + \text{father's height}}{2}$$

$$= \frac{\frac{13}{12} \cdot 65 + 72}{2} \qquad \text{mother's height 65, father's height 72}$$

$$= \frac{70.42 + 72}{2} \qquad \text{Multiply } \frac{13}{12} \text{ and 65.}$$

$$= \frac{142.42}{2} \qquad \text{Add 70.42 and 72.}$$

$$= \underline{\hspace{2cm}} \qquad \text{Simplify.}$$

Using the Gray Method, the boy will be about _____ inches tall as an adult.

Method 2 The Doubling Method

The Doubling Method multiplies the height of a child by 2 at a specific age to predict the child's height as an adult. Height at 24 months is used for boys, and height at 18 months is used for girls.

$2 \cdot$ height of child

$= 2 \cdot \underline{\hspace{1.5cm}} \qquad$ Boy's height of 35 inches

$= \underline{\hspace{1cm}} \qquad$ Simplify.

Using the Doubling Method, the boy will be _____ inches tall as an adult.

Go Online You can complete an Extra Example online.

Copyright © McGraw-Hill Education

Check

LOANS Elan is applying for a home loan. At National Road Bank, Elan's debt-to-income ratio must be 0.36 or less to qualify for a loan, and at New Savings Bank his mortgage-to-income ratio must be 0.28 or less.

Part A Use the two metrics and the information provided to determine whether Elan qualifies for a home loan at each bank. Round to the nearest hundredth.

Monthly Income	Monthly Debt	Monthly Mortgage
$3650	$1165	$1068

National Road Bank: Debt-to-Income Ratio

$$\frac{\text{Monthly Debt}}{\text{Monthly Income}} = \underline{\hspace{2cm}}$$

New Savings Bank: Mortgage-to-Income Ratio

$$\frac{\text{Monthly Mortgage}}{\text{Monthly Income}} = \underline{\hspace{2cm}}$$

Part B Compare the results of the two metrics. How effective are each of the metrics as measures of whether Elan can afford to buy a house?

_____

_____

_____

_____

Learn Accuracy

All measurements are approximations. When you measure something, you are limited by the measurement tool that you are using. **Accuracy** is the nearness of a measurement to the true value of the measure.

The accuracy needed for baking cookies, timing the final seconds of a basketball game, and determining the gold medalist of a 100-meter dash are very different.

Whether measurements should be rounded depends on how the measurement will be used and the limitations of the units in which the measurement is taken.

 Go Online You can complete an Extra Example online.

🌐 Example 3 Decide Where to Round

ROAD TRIP **Damien and two of his friends are taking a road trip. They plan to share the responsibility of driving and will each drive an equal distance. Damien's GPS shows that the total distance is 172 miles. Determine the exact distance that each person should drive. Then determine a more appropriate driving distance for each person given the limitations of the situation.**

To determine the exact distance each person should drive, divide the total distance by _____.

$$172 \text{ miles} \div \text{_____} = \text{_____ miles}$$

Because the distance given by the GPS is accurate to the nearest mile, the distance each driver will drive should be rounded to the nearest mile. Each driver will drive about _____ miles.

Check

VACATION Inchiro has saved $400.00 to spend on his 7-day vacation. He plans to budget his $400.00 by spending the same amount each day of the vacation. Determine the appropriate amount he should spend each day.

$_____

Think About It!
The total cost of fuel for the trip was $20. If they split the cost equally, how much should each person pay? What unit of measure limits the accuracy of the solution?

🌐 Example 4 Find an Appropriate Level of Accuracy

SPACE SHUTTLE **In 2012, NASA's space shuttle *Endeavor* traveled approximately 897 miles from the Kennedy Space Center to Houston, Texas, by a shuttle carrier aircraft. Then it traveled about 1381 miles to the Los Angeles International Airport. Finally, a truck pulled *Endeavor* 12 miles through the streets of Los Angeles to the California Science Center.**

If the shuttle carrier aircraft flew at an average speed of 287 miles per hour and the truck pulled *Endeavor* at an average speed of 1.3 miles per hour, determine the total amount of time it took *Endeavor* to travel from the Kennedy Space Center to the California Science Center with a reasonable level of accuracy.

Because the parts of the space shuttle's journey from the Kennedy Space Center to Houston and then from Houston to Los Angeles are at the same speed, add those two distances, _____ + _____ or _____.

$$\frac{2278 \text{ miles}}{287 \frac{\text{miles}}{\text{hours}}} + \frac{12 \text{ miles}}{1.3 \frac{\text{miles}}{\text{hours}}} \approx \text{_____ hours} + \text{_____ hours}$$

$$\approx \text{_____ hours}$$

The total travel time for *Endeavor* from the Kennedy Space Center to the California Science Center was about _____ hours.

🌐 **Go Online** You can complete an Extra Example online.

Talk About It!
Why is it unreasonable to say that it took 17.168 hours for Endeavor to reach the science center?

Copyright © McGraw-Hill Education

Check

POSTAGE A school is hosting a marching band competition and plans to mail postcards, fliers, and large information packets to other schools. The school will mail 150 postcards, which cost $0.34 in postage, and between 75 to 100 fliers, which cost $0.49 in postage. Forty-three information packets will be mailed. The cost of mailing the information packets varies, but the average is $1.59. Select the total mailing cost that represents the most reasonable level of accuracy. _____

A. $156.15

B. $160

C. $162.245

D. $168.37

🌐 Example 5 Determine Accuracy

POPULATION The U.S. Census Bureau Web site shows a counter that displays the population of the United States on a certain day as 329,158,023. How accurate is the reported population? Explain your reasoning.

United States Population
329,158,023

COMPONENTS OF POPULATION CHANGE
One birth every **7 seconds**
One death every **13 seconds**
One international migrant (net) every **29 seconds**
Net gain of one person every **11 seconds**

Because there is no way to count every person in the United States at any given moment, giving an exact population does not make sense. The number of births, deaths, and immigrations varies, so the population does not increase at a steady rate. The Web site uses averages to estimate the population at a specific time.

It would be more appropriate for the Web site to report the population as _____.

Check

BIOLOGY A science magazine reported that there are, on average, 37 trillion cells that make up the human body. Select the option that best describes the accuracy of the magazine. _____

A. The magazine is accurate because scientists can count every cell.

B. The magazine is probably accurate because the number is not very specific.

C. The magazine is not accurate because there is no way to count all of the cells of a person.

D. The magazine is not accurate because the number of cells is always changing.

🌐 **Go Online** You can complete an Extra Example online.

Practice

Go Online You can complete your homework online.

Example 1

1. **TEST SCORES** A teacher compares the ratio of the number of questions answered correctly to the total number of questions on a test as a metric. For a student to earn an A or B on a test, the ratio must be greater than or equal to 0.8. The last test given by the teacher had a total of 40 questions. Using this metric, what is the least number of questions a student can answer correctly to earn an A or B on the test?

$$\frac{\text{number of questions answered correctly}}{\text{total number of questions}}$$

2. **DRIVER'S TEST** The Department of Motor Vehicles, DMV, uses a metric to determine whether a person earns a driver's license. In one state, the total number of possible points on the written portion of the driver's exam is 46. A person will pass the written portion of the driver's exam by scoring 84% or greater. The table shows the number of points different people earned on the written portion of the driver's exam. How many people passed the written portion of the driver's exam?

38	40	41	45	39
35	40	46	43	37
41	42	44	41	46
40	38	32	44	45
39	40	30	43	45

3. **TRACK** A college track coach compares the ratio of time it takes a runner to run 100 meters to 12 seconds. For a runner to be on the team, the ratio must be less than or equal to 0.95. What is the slowest time 100 meters can be run to make the team?

Example 2

4. **DEBT-TO-INCOME RATIO** Find Jada's debt-to-income ratio if her monthly expenses are $1850 and her monthly salary is $2500.

5. **DEBT-TO-INCOME RATIO** Find Victoria's debt-to-income ratio if her monthly expenses are $1280 and her monthly salary is $2500.

6. **PLUMBING** Raven is deciding between two plumbing services. Service Provider A multiplies the average number of hours spent at a residence by $50, where the average number of hours spent at a residence is 1.5 hours for a new house and 3.75 hours for an old house. Service Provider B multiplies the exact number of hours spent at a residence by $60. Suppose Raven has a new house and needs a plumber for 1.75 hours. Find the cost of service charge using both methods.

7. **INVESTING** Hector is deciding how much he should invest each year. The Automatic Method multiplies the average income by 10%, where the average income is $50,000 for an employee that has been at the same company for 10 years or less and $60,000 for an employee that has been at the same company for more than 10 years. The Exact Method multiplies the exact income by 7.5%. Suppose Hector has been at the same company for 12 years and his income last year was $75,000. Find the amount Hector should invest using both methods.

Example 3

8. **MONEY** Jordan has $20 to share among 3 people. Jordan types 20 ÷ 3 into his calculator and gets 6.666666667. How much should he give to each person?

9. **SNACKS** Ms. Miller has 14 snack bars to share among 6 students. Ms. Miller types 14 ÷ 6 into her calculator and gets 2.333333333. How many snack bars should she give each student?

10. **EVENT PLANNING** Max is planning a banquet for the National Honors Society. Approximately 60 people will be attending the banquet. If 8 people can fit comfortably at a table, how many tables should he have?

11. **GARDEN** Emily wants to plant flowers in a narrow rectangular plot that is 1 foot by 4.5 feet. The flowers she wants to plant need to be spaced at least 8 inches apart. How many plants should she buy for the garden?

12. **LEMONADE** Justin has 64 ounces of lemonade to divide among 9 people. When he types 64 ÷ 9 into his calculator, the number that appears is 7.1111111. How much lemonade should he give to each person?

13. **MONEY** Darnell has $1000 he wants to divide among his 3 children. How much should Darnell give each of his children?

Example 4

14. **FUNDRAISING** At a bake sale, the golf team sold all 50 cupcakes for $1.50 each. They sold almost all of the 100 cookies for $1.00 each. The team also received donations from 7 people averaging $4.25 per donation. Determine the total amount of money the golf team collected from the bake sale with a reasonable level of accuracy.

15. **EXPENSES** Santino spends an average of $125 per month on clothing, not including sales tax. He also spends about $180 per month going out to eat and an average of $130 per week on groceries, including sales tax. If the tax rate is 7.25%, find the total amount he spends on food and clothing in a year, including sales tax, with a reasonable level of accuracy.

16. **TIME MANAGEMENT** Ava spends about 45 minutes studying each day. She also practices the piano for an average of 20 minutes per day, and she practices soccer 1.5 hours three times a week. If Ava decides to reduce the time she does each activity by $\frac{1}{6}$, find the total number of hours she spends studying and practicing in a year with a reasonable level of accuracy.

Example 5

17. SCHOOLS The superintendent at Hartgrove High School says there are 3103 students enrolled at the school. How accurate is the reported enrollment? Explain your reasoning.

18. POPULATION The U.S. Census Bureau Web site shows that the population of Texas on July 1, 2016 was 27,862,596. How accurate is the reported population? Explain your reasoning.

19. TRAFFIC LIGHTS A map maker reported that there were about 12,000 traffic lights in New York City. How accurate is the report? Explain your reasoning.

20. SAND A mathematician reported that there are 1,578,932 grains of sand in one cubic foot. How accurate is the report? Explain your reasoning.

Mixed Exercises

21. USE A SOURCE A coach compares the ratio of the number of free throws made to the total number of attempted free throws as a metric. For a player to be selected as a free throw shooter when the other team is given a technical foul, the ratio must be greater than or equal to 0.82. Find the number of free throws made and the number of free throws attempted for three former NBA players. Using the metric, which players would and would not be selected as a free throw shooter when the other team is given a technical foul?

22. REASONING A carpenter is measuring the length of a living room. Should the carpenter measure the length in feet, inches, meters, or kilometers to be most accurate? Explain.

23. SPACE Which unit of measure is the most appropriate for measuring the distance from Earth to the star Polaris: feet, kilometers, or light-years? Explain.

24. POPULATION Juanita and Trevor are doing research about the deer population in Ohio. Juanita says there are over 750,000 deer in Ohio. Trevor says there are 734,928 deer in Ohio. Who is more accurate? Explain your reasoning.

The graph shows how the number of visitors at a local zoo is related to the average daily temperature. The line shown is the line that most closely approximates the data in the scatter plot. Use the graph for Exercises 25–27.

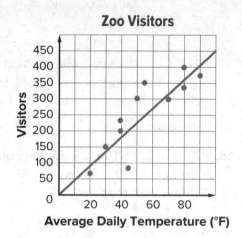

Zoo Visitors

Average Daily Temperature (°F)

25. STRUCTURE Describe the line in terms of accuracy.

26. USE ESTIMATION Use the line to approximate the number of visitors at the zoo for an average daily temperature of 50°F. Compare this to the actual number of visitors given by the point on the graph for an average daily temperature of 50°F.

27. REASONING Explain why some points are above the line and some points are below the line.

28. METRICS Suppose two mortgage companies compare the ratio of the monthly mortgage payment to the total monthly income as their metric. Suppose 0.3 is the ideal metric for the debt-to-income ratio for Company A, and 0.28 is the ideal metric for the debt-to-income ratio for Company B. Provided the target mortgage payment is the same for either company, then which of these mortgage companies requires a greater monthly income? Explain.

🧁 Higher-Order Thinking Skills

29. WRITE Suppose you start your own company. When hiring employees, you want to set certain metrics, such as typing speed. What other attributes of employees do you think you might consider when creating metrics to determine hiring qualifications?

30. FIND THE ERROR Mr. Moreno's students are weighing materials for a chemistry experiment. Four students weigh the same sample using different scales:

          100 g   104 g   105 g   103.5 g

Mr. Moreno tells the students that they each weighed the amount correctly. Explain how this is possible.

31. WRITE Lamont stops at a gas station that sells gasoline at 3.29\frac{9}{10}$ per gallon. He pumps 8.618 gallons of gasoline into the tank. How much will Lamont pay for gas? How much accuracy is possible? How much accuracy is necessary? Explain.

Essential Question

How can mathematical expressions be represented and evaluated?

Module Summary

Lessons 1-1 and 1-2

Numerical and Algebraic Expressions

- A numerical expression contains only numbers and mathematical operations.

- To evaluate an expression means to find its value. If a numerical expression contains more than one operation, the rule that lets you know which operation to perform first is called the *order of operations*.

- The Substitution Property allows you to evaluate an algebraic expression by replacing the variables with their values.

Lessons 1-3 and 1-4

Properties of Real Numbers and Distributive Property

- The Reflexive Property states that any quantity is equal to itself.

- The Symmetric Property states that if one quantity equals a second quantity, then the second quantity equals the first.

- The Transitive Property states that if one quantity equals a second quantity and the second quantity equals a third quantity, then the first quantity equals the third quantity.

- The Associative Property states that the way you group three or more numbers when adding or multiplying does not change their sum or product.

- The Commutative Property states that the order in which you add or multiply numbers does not change their sum or product.

- The Distributive Property states that multiplying a number by a sum of numbers is the same as doing each multiplication separately and then adding the products.

Lesson 1-5

Expressions Involving Absolute Value

- The absolute value of a number is its distance from 0 on the number line.

- Absolute value is always greater than or equal to zero.

Lesson 1-6

Descriptive Modeling and Accuracy

- Descriptive modeling is a way to mathematically describe real-world situations and the factors that cause them.

- A metric is a rule for assigning a number to a characteristic or attribute.

- Accuracy is the nearness of a measurement to the true value of the measure.

Study Organizer

 Foldables

Use your Foldable to review the module. Working with a partner can be helpful. Ask for clarification of concepts as needed.

Test Practice

1. MULTIPLE CHOICE Which is equivalent to 2^5? (Lesson 1-1)

Ⓐ 10

Ⓑ 16

Ⓒ 24

Ⓓ 32

2. MULTI-SELECT The table shows the prices of several items at a movie theater. Which expressions represent the total cost of 4 movie tickets, 2 popcorns, and 1 bottled water? Select all that apply. (Lesson 1-1)

Item	Cost
Ticket	$9.75
Popcorn	$6.25
Soda	$5.50
Water	$4.75
Box of Candy	$3.50

Ⓐ 4(9.75) + 2(6.25) + 4.75

Ⓑ 4(9.75) + 2(5.50) + 4.75

Ⓒ 39.00 + 12.50 + 4.75

Ⓓ 54.75

Ⓔ 56.25

3. OPEN RESPONSE Write an algebraic expression that represents *five times the quantity x increased by seven, minus four cubed.* (Lesson 1-2)

4. MULTIPLE CHOICE What is the value of the expression $9x^2 + 4x - 11$ when $x = 3.2$? Express your answer as a decimal, rounded to the nearest hundredth. (Lesson 1-2)

Ⓐ 20.232

Ⓑ 30.6

Ⓒ 59.4

Ⓓ 93.96

5. MULTIPLE CHOICE Which algebraic expression represents the verbal expression *the product of five and a number, decreased by eleven*? (Lesson 1-2)

Ⓐ $5n - 11$

Ⓑ $11 - 5n$

Ⓒ $5(n - 11)$

Ⓓ $11 - (n + 5)$

6. OPEN RESPONSE Evaluate the expression $5[13 - (3^2 + 2^2)]$. (Lesson 1-3)

7. MULTIPLE CHOICE Which expression is NOT a way to represent $2 \cdot 3\frac{5}{6} + 2 \cdot 12 + 2 \cdot 1\frac{1}{6}$? (Lesson 1-3)

Ⓐ $2\left(3\frac{5}{6} + 12 + 1\frac{1}{6}\right)$

Ⓑ $2 \cdot 5 + 12$

Ⓒ $2(5 + 12)$

Ⓓ 34

8. OPEN RESPONSE Ayumi, a chef, wants to determine how many meals she cooked in one evening. The table shows the four meals she made and the number of people that were served each meal. (Lesson 1-3)

Meal	Number of People
Lasagna	27
Spaghetti & Meatballs	21
Steak & Potatoes	19
Shrimp Scampi	13

She uses the following steps to determine how many total meals she cooked.

Step 1: $27 + 21 + 19 + 13$

Step 2: $27 + 13 + 21 + 19$

Step 3: $(27 + 13) + (21 + 19)$

Step 4: $40 + 40$

Step 5: 80

Which property did Ayumi use in Step 3?

[   ]

9. TABLE ITEM Indicate whether each of the statements is *true* or *false*. (Lesson 1-3)

Statement	True	False
$4(6 - 2 \times 3) = 0$		
$11(3^2 - 9) + 2\left(\frac{1}{2}\right) = 0$		
$4 \cdot 0 + 4^2 - 2^3 - (2 + 2 \cdot 3) = 0$		

10. MULTI-SELECT Which expressions could be used to evaluate $418(27)$? (Lesson 1-4)

Ⓐ $418(20 - 7)$

Ⓑ $(420 - 2)(27)$

Ⓒ $(400 - 18)(27)$

Ⓓ $(418)(20 + 7)$

Ⓔ $(418)(30 - 3)$

11. TABLE ITEM Indicate whether each expression represents the verbal expression *negative seven times the quantity triple m minus eleven.* (Lesson 1-4)

Expression	Yes	No
$-7(m^3 - 11)$		
$-7(3m) - 7(-11)$		
$-21m - 77$		
$-21m + 77$		
$-21m - 11$		
$-7m^3 + 77$		

12. MULTIPLE CHOICE Which is the simplified expression of $-8(2m + 9k - 13)$? (Lesson 1-4)

Ⓐ $-16m + 9k - 13$

Ⓑ $-16m - 72k + 104$

Ⓒ $-16m - 72k - 104$

Ⓓ $16m - 72k - 104$

13. MULTI-SELECT A group of 8 artists plans to attend a quilting class and purchase lunch. (Lesson 1-4)

Which expression(s) represents the total cost for all 8 artists?

(A) 8(10) + 14

(B) 10(8 + 14)

(C) 8(10 + 14)

(D) 80 + 140

(E) 80 + 112

14. MULTI-SELECT If x and y are both integers, which expression(s) are equivalent to $|x - y|$? (Lesson 1-5)

(A) $|y - x|$

(B) $|y + x|$

(C) $|y| - |x|$ if $x \leq y$ and $x \geq 0$

(D) $|x| - |y|$ if $y > x$

(E) $|y| + |x|$ if $y > x$

15. OPEN RESPONSE What is the value of the expression $4^2 + 3|4x - 9|$ when $x = -2$? (Lesson 1-5)

16. OPEN RESPONSE A player's secondary average (SecA) is a way to look at the extra bases gained without regard to batting average. The formula for SecA is SecA $= \frac{T - H + B + S - C}{A}$, where T is total bases, H is hits, B is bases from balls or walks, S is stolen bases, C is number of times caught stealing, and A is times at bat.

Player	T	H	B	S	C	A
Altuve	186	116	39	22	3	330
Murphy	183	110	17	2	3	315
Ortiz	187	94	45	2	0	279
Ramos	139	84	23	0	0	251

Find each player's SecA. Round to the nearest hundredth if necessary. (Lesson 1-6)

17. OPEN RESPONSE A marketing manager bought 4 advertisements for $1345 each. She reported to her supervisor that she spent about $4000 out of her budget. Did the marketing manager report her spending to a reasonable level of accuracy? Explain. (Lesson 1-6)

Equations in One Variable

What Will You Learn?

Place a check mark (✓) in each row that corresponds with how much you already know about each topic **before** starting this module.

KEY

👎 — I don't know.   👌 — I've heard of it.   👍 — I know it!

	Before			After		
	👎	👌	👍	👎	👌	👍
write equations to represent relationships						
interpret equations that represent relationships						
solve one-step equations by using addition and subtraction						
solve one-step equations by using multiplication and division						
solve multi-step equations						
solve equations with variables on each side						
solve equations by applying the Distributive Property						
solve equations that involve absolute value						
solve proportions						
solve an equation with more than one variable for a specific variable						
convert units of measure by using dimensional analysis						

📖 Foldables Make this Foldable to help you organize your notes about equations. Begin with four sheets of grid paper.

1. **Fold** four sheets of grid paper in half along the width.

2. **Unfold** each sheet and tape to form one long piece.

3. **Label** each piece with the lesson number as shown. Label the last piece for vocabulary. Refold to form a booklet.

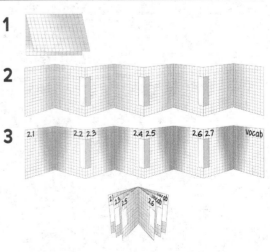

What Vocabulary Will You Learn?

Check the box next to each vocabulary term that you may already know.

☐ constraint

☐ dimensional analysis

☐ equation

☐ equivalent equations

☐ formula

☐ identity

☐ literal equation

☐ multi-step equation

☐ proportion

☐ solution

☐ solve an equation

Are You Ready?

Complete the Quick Review to see if you are ready to start this module.
Then complete the Quick Check.

Quick Review

Example 1

Write an algebraic expression for the phrase *the quotient of five and w decreased by eight.*

the quotient of five and w decreased by eight

$\frac{5}{w}$ $\qquad$ $-$ $\qquad$ 8

The expression is $\frac{5}{w} - 8$.

Example 2

Evaluate $9 + \frac{4^2}{2} - 2(5 \times 2 - 8)$.

$9 + \frac{4^2}{2} - 2(5 \times 2 - 8)$ Original expression

$= 9 + \frac{4^2}{2} - 2(2)$ Evaluate inside the parentheses.

$= 9 + 8 - 2(2)$ Evaluate the power and divide.

$= 9 + 8 - 4$ Multiply.

$= 13$ Add and subtract.

Quick Check

Write an algebraic expression for each verbal expression.

1. six times a number n increased by two

2. a number d squared minus three

3. the sum of four times b and nine

Evaluate each expression.

4. $(7 + 3)^2 - 4$

5. $4(11 - 5) \div 3$

6. $\frac{1}{3}(21) + \frac{1}{8}(32)$

7. $3 \cdot 2^3 + 64 \div 8$

8. $\frac{11 - 3}{2} + 9$

9. $6[(5 - 3)^2 + 8] \div 2$

How did you do?

Which exercises did you answer correctly in the Quick Check? Shade those exercise numbers below.

Writing and Interpreting Equations

Copyright © McGraw-Hill Education

Explore Writing Equations by Modeling a Real-World Situation

Online Activity Use a real-world situation to complete the Explore.

×

INQUIRY What steps can you use to write equations to represent a real-world situation?

Today's Goals
- Translate sentences into equations.
- Translate equations into sentences.

Today's Vocabulary
equation

constraint

Learn Writing Equations

A mathematical statement that contains two expressions and an equal sign, =, is an **equation**.

Key Concept • Writing Equations	
Step 1	Identify each unknown and assign a variable to it.
Step 2	Identify the givens and their relationships.
Step 3	Write the sentence as an equation.

Example 1 Write an Equation for a Sentence

Write an equation for the sentence.

Twenty minus the quotient of 7 and x is the same as twice x.

Recall that a quotient is the result of division.

Twenty minus the quotient of 7 and x is the same as **twice x.**

$$20 - \qquad \frac{7}{x} \qquad = \qquad 2x$$

The equation is _____.

Check

Write an equation for the sentence.

Four times a number less 10 is equal to 16. ___

    A. $4x - 10 = 16$ B. $4(x - 10) = 16$

    C. $10 - 4x = 16$ D. $4x + 10 = 16$

 **Go Online** You can complete an Extra Example online.

Think About It!

What distinguishes an expression from an equation?

Study Tip

Verbal Phrases When writing the verbal form of an equation, *is* and *equals* can be used interchangeably.

Copyright © McGraw-Hill Education

🌐 **Example 2** Write an Equation

LIFE ONLINE **Of 799 teens surveyed about what they do online, some use a social network. Of those on a social network, 430 say people their age are "mostly kind" online and the remaining 193 do not. Write an equation to find the number of teens surveyed who are not on a social network.**

Social Teens Mostly Kind Online

Not "Mostly Kind" 193

"Mostly Kind" 430

Step 1 Identify each unknown and assign a variable to it.

Let n = the number of teens surveyed who _____ on a social network.

Step 2 Identify the givens and their relationship.

The givens are:

- _____ teens were surveyed.

- Some number of the teens use a social network.

- _____ of those on a social network say people their age are "mostly kind" online. The other _____ do not.

The 430 and 193 make up the group on a social network. The rest of the 799 surveyed are not on a social network.

Step 3 Write the sentence as an equation.

The sum of the teens on a social network and those not on a social network is 799.

_____

Check

READING Etu has read 12 of the 32 chapters in his assigned book. He plans to finish the book by reading c chapters each for 8 days until the book is due. Which equation best represents the situation? _____

A. $12 - 8c = 32$

B. $12 + \dfrac{c}{8} = 32$

C. $12 + 8c = 32$

D. $12 - \dfrac{c}{8} = 32$

🌐 **Go Online** You can complete an Extra Example online.

💭 Think About It!

Is there only one equation that represents the situation? Justify your argument.

Example 3 Write an Equation with Multiple Variables

GEOMETRY **Translate the sentence into a formula.**

The perimeter of a rectangle is twice the sum of the length and the width.

Step 1 Identify unknowns.

_____

Step 2 Assign variables.

Let P = perimeter, ℓ = _____, and w = _____.

Step 3 Identify the givens and their relationships.

Twice means _____.

Twice the sum means you _____ first, then _____.

Step 4 Write an equation.

The formula for the perimeter of a rectangle is _____.

Check

Translate the sentence into a formula.

MOTORS The horsepower of a motor is the product of the motor speed and the torque divided by 5252. _____

A. $H = \dfrac{M}{5252T}$

B. $H = \dfrac{MT}{5252}$

C. $H = \dfrac{5252}{MT}$

D. $H = \dfrac{5252M}{T}$

BAGELS Plain and cinnamon raisin bagels are the most popular flavors. Each year, 24 million more than twice as many packages of plain bagels are sold as cinnamon raisin. There were 136 million packages of plain bagels sold last year. Create an equation that can be used to find the number of millions of packages of cinnamon raisin bagels, c, sold last year.

_____

_____

Learn Interpreting Equations

Look for the relationships in an equation by interpreting each part of the expressions in the equation.

As you interpret an equation that represents a real-life situation, consider that the equation may be viewed as a constraint in the situation. In mathematics, a **constraint** is a condition that a solution must satisfy. These conditions limit the number of possible solutions. The solutions of the equation meet the constraints of the problem.

Go Online You can complete an Extra Example online.

Think About It!

Why is it helpful to identify all the unknowns before writing an equation?

Talk About It!

What is an example of a real-life constraint? Explain.

Example 4 Write a Sentence for an Equation

Write a sentence for the equation.

$$2z - 1 = 5$$

Check

Write a sentence for the equation.

a. $6z - 15 = 45$ _____

    A. Six times a number z, minus fifteen equals forty-five.

    B. Six times the quantity of a number z minus fifteen equals forty-five.

    C. Six times the difference of a number z and fifteen equals forty-five.

    D. Six plus z minus fifteen equals forty-five.

b. $(y + 3)^2 = 25$ _____

    A. y plus 3 squared is 25.

    B. y squared plus 3 is 25.

    C. The quantity y plus 3 squared is 25.

    D. y plus 3 is 25.

Example 5 Write a Sentence for an Equation with Grouping Symbols

Write a sentence for the equation.

$$3(y + 1) = 12$$

The parentheses tell us that 3 is *three times* the expression in the parentheses.

The parentheses can be written as *the quantity*.

Write $y + 1$ as y plus 1 or *the sum of y and one*.

Write $= 12$ as *equals twelve* or *is twelve*.

One of the sentences that represents the equation $3(y + 1) = 12$ is:

_____

 Go Online You can complete an Extra Example online.

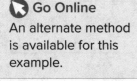

Watch Out!

Parentheses When writing a sentence for an equation with parentheses, the phrase *the quantity* should be written immediately before the terms that are contained in the parentheses, not at the beginning of the sentence.

Go Online

An alternate method is available for this example.

Check

Select the sentence(s) that represent(s) the equation. _____

$7(p + 23) = 102$

A. Seven times the sum of p and twenty-three is the same as one hundred two.

B. Seven times p plus twenty-three equals one hundred two.

C. Seven times the quantity p plus twenty-three equals one hundred two.

D. The quantity seven times p plus twenty-three is the same as one hundred two.

E. Seven times the sum of p and twenty-three is one hundred two.

Example 6 Interpret an Equation

GEOMETRY Write a sentence for the formula for the surface area of a rectangular prism $S = 2\ell w + 2\ell h + 2wh$. Then interpret the equation in the context of the situation.

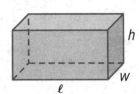

From the equation, we see that the surface area of a rectangular prism depends on the length, width, and height.

S =	2ℓw +	2ℓh +	2wh
Surface area equals		Two times length times height plus	

Think About It!
What did you already know that helped you interpret the formula?

- The first term, $2\ell w$, is two times the area of a _____. In the prism above, the area of the bottom face is _____. The top is the same shape, so it has the same area. This term represents the sum of the areas of the bottom and top faces.

- The second term, $2\ell h$, is the sum of the areas of the front and _____ faces.

- The third term, $2wh$, is the sum of the areas of the _____ and right faces.

So, the surface area of the rectangular prism is the _____ _____.

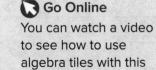
Go Online
You can watch a video to see how to use algebra tiles with this example.

Go Online You can complete an Extra Example online.

Check

FINANCE The formula for a loan balance is $b = p\left(1 + \frac{r}{12}\right) - d$ if the previous balance is p, the annual interest rate is r, and a payment of d is made.

Part A Write a sentence for the formula. _____

A. The balance equals the previous balance multiplied by the quantity one plus the annual interest rate divided by 12 minus the payment.

B. The balance equals the previous balance multiplied by the annual interest rate divided by 12 minus the payment.

C. The balance equals the previous balance multiplied by the sum of one and the annual interest rate minus the payment.

D. The balance equals the previous balance minus the payment.

Part B Select each sentence that is a correct interpretation of the equation in the context of the situation. Select all that apply. _____

A. The expression $\frac{r}{12}$ represents the monthly interest rate.

B. The expression $p\left(1 + \frac{r}{12}\right)$ represents the previous balance plus interest.

C. If no payments are made, then $d = 0$ and the balance is the same as the previous balance.

D. The expression $\left(1 + \frac{r}{12}\right)$ represents the monthly interest rate.

E. If no payment is made, then $d = 0$ and the balance is the previous balance plus interest.

Pause and Reflect

Did you struggle with anything in this lesson? If so, how did you deal with it?

Record your observations here.

Go Online You can complete an Extra Example online.

Practice

📎 **Go Online** You can complete your homework online.

Example 1

Write an equation for each sentence.

1. Two added to three times a number m is the same as 18.

2. The product of five and the sum of a number x and three is twelve.

3. The quotient of 24 and x equals 14 minus 2 times x.

4. Nine times a number y subtracted from 85 is seven times the sum of four and y.

Example 2

5. **WALKING** Lily has walked 2 miles. Her goal is to walk 6 miles. Lily plans to reach her goal by walking 3 miles each hour h for the rest of her walk. Write an equation to find the number of hours it will take Lily to reach her goal.

6. **MATH** Paulina has completed 24 of the 42 math problems she was assigned for homework. She plans to finish her homework by completing 9 math problems each hour h. Write an equation to find the number of hours it will take Paulina to complete her math homework assignment.

7. **ATHLETICS** Of 107 athletes surveyed about what sport they play, some play basketball. Of those that play basketball, 48 play baseball and the remaining 33 do not play baseball. Write an equation to find the number of athletes surveyed who do not play basketball.

8. **SALES** Cars and trucks are the most popular vehicles. Last year, the number of cars sold was 39,000 more than three times the number of trucks sold. There were 216,000 cars sold last year. Write an equation that can be used to find the number of trucks, t, sold last year.

Example 3

Translate each sentence into an equation or formula.

9. Twice a increased by the cube of a equals b.

10. Seven less than the sum of p and t is as much as 6.

11. The sum of x and its square is equal to y times z.

12. Four times the sum of f and g is identical to six times g.

13. The area A of a square is the length of a side ℓ squared.

14. The perimeter P of a triangle is equal to the sum of the lengths of sides a, b, and c.

15. The perimeter of a rectangle is equal to 2 times the length plus twice the width.

16. The density of an object is the quotient of its mass and its volume.

17. Simple interest is computed by finding the product of the principal amount p, the interest rate r, and the time t.

18. The surface area of a rectangular prism is 2 times the sum of the width, w, times height, h, and length, l, times width and length times height.

Examples 4 and 5

Write a sentence for each equation.

19. $j + 16 = 35$

20. $4m = 52$

21. $7(p + 23) = 102$

22. $r^2 - 15 = t + 19$

23. $\frac{2}{5}v + \frac{3}{4} = \frac{2}{3}x^2$

24. $\frac{1}{3} - \frac{4}{5}z = \frac{4}{3}y^3$

25. $g + 10 = 3g$

26. $2(t + 4q) = 2q + 4t$

27. $4(a + b) = 9a$

28. $8(2y - 6x) = 4 + 2x$

29. $\frac{1}{2}(f + y) = f - 5$

30. $k^2 - n^2 = 2b$

Example 6

Write a sentence for each formula. Then interpret the equation in the context of the situation.

31. GEOMETRY The formula for the volume of a cylinder is $V = \pi r^2 h$, where V is the volume, r is the length of the radius of the base, and h is the height of the cylinder.

32. GEOMETRY The formula for the volume of a cube is $V = s^3$, where V is the volume and s is the side length.

33. FINANCE The simple interest formula is given by $I = Prt$, where $I =$ interest, $P =$ principal, $r =$ rate, and $t =$ time.

34. FINANCE The compound interest formula is given by $A = P\left(1 + \frac{r}{n}\right)^{nt}$, where A is the amount, P is the principal, r is the rate, n is the number of times interest is compounded per year, and t is the time in years.

35. SCIENCE Newton's second law of motion is $F = ma$, where F is the force acting on an object, m is the mass of the object and a is the acceleration of the object.

36. SCIENCE The formula $d = rt$ relates the distance traveled d, the rate of travel r, and the time spent traveling t.

Mixed Exercises

For Exercises 37–40, match each sentence with an equation.

A. $g^2 = 2(g - 10)$ **B.** $\frac{1}{2}g + 32 = 15 + 6g$ **C.** $g^3 = 24g + 4$ **D.** $3g^2 = 30 + 9g$

37. One half of g plus thirty-two is as much as the sum of fifteen and six times g.

38. A number g to the third power is the same as the product of 24 and g plus 4.

39. The square of g is the same as two times the difference of g and 10.

40. The product of 3 and the square of g equals the sum of thirty and the product of nine and g.

Translate each sentence into an equation.

41. The difference of the square of y and twelve is the same as the product of five and x.

42. The difference of f and five times g is the same as 25 minus f.

43. Three times b less than 100 is equal to the product of 6 and b.

44. Four times the sum of 14 and c is a squared.

Translate each equation into a sentence.

45. $4n = x(5 - n)$ **46.** $2b - 10 = 4$ **47.** $y + 3x^2 = 5x$

Translate each sentence into a formula.

48. The area A of a circle is pi times the radius r squared.

49. The volume V of a rectangular prism equals the product of the length ℓ, the width w, and the height h.

50. REASONING The area of a kitchen is 182 square feet. This is 20% of the area of the first floor of the house. Let F represent the area of the first floor. Write an equation to represent the situation.

51. REASONING Katie is twice as old as her sister Mara. The sum of their ages is 24. Write a one-variable equation to represent the situation.

52. GEOMETRY The formula $F + V = E + 2$ shows the relationship between the number of faces F, edges E, and vertices V of a polyhedron, such as a pyramid. Write the formula in words.

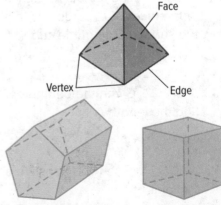

53. STRUCTURE A recycling company charges business owners $10 for each cubic yard of waste removed from their facility plus a 10% fuel charge based on the total monthly bill. Let w represent the number of cubic yards of waste removed during the month. Write an equation to describe the total cost c of the recycling service per month.

 **Higher Order Thinking Skills**

54. WRITE Determine whether the two sentences describe the same equation. Explain.

The product of x and y plus z equals w.

The product of x and the sum of y and z equals w.

55. ANALYZE Determine whether the equation is an accurate translation of the sentence. Explain.

a. The square of the product of 4 and a number is equal to 8 times the sum of the number and 6. $(4n)^2 = 8(n + 6)$

b. Three more than one-half a number is equal to 2 less than the number.
$\frac{n}{\frac{1}{2}} + 3 = n - 2$

56. PERSEVERE Translate the formula $A = \frac{b_1 + b_2}{2} \cdot h$ into words. Let A represent the area. List any constraints on the variables.

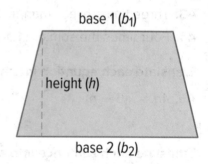

base 1 (b_1)

height (h)

base 2 (b_2)

57. CREATE Write a scenario for the equation $12a + 10(a - 1) = 188$.

58. CREATE Write a problem about your favorite television show that uses the equation $x + 8 = 30$.

59. ANALYZE The surface area of a three-dimensional object is the sum of the area of the faces. If ℓ represents the length of the side of a cube, write a formula for the surface area of the cube.

60. ANALYZE Given the perimeter P and width w of a rectangle, write a formula to find the length ℓ.

61. WRITE How can you translate a verbal sentence into an algebraic equation? Explain.

Solving One-Step Equations

Explore Using Algebra Tiles to Solve One-Step Equations Involving Addition or Subtraction

 Online Activity Use algebra tiles to complete the Explore.

> @ **INQUIRY** How can you model and solve addition and subtraction equations?

Explore Using Algebra Tiles to Solve One-Step Equations Involving Multiplication

 Online Activity Use algebra tiles to complete the Explore.

> @ **INQUIRY** How can you use algebra tiles to solve multiplication equations?

Learn Solving One-Step Equations Involving Addition or Subtraction

To **solve an equation** means to find all values of the variable that make the equation true. Each value that makes an equation true is a **solution**. **Equivalent equations** have the same solution.

Key Concept • Addition Property of Equality	
Words	If a number is added to each side of a true equation, the resulting equivalent equation is also true.
Symbols	For any real numbers a, b, and c, if $a = b$, then $a + c = b + c$.

Key Concept • Subtraction Property of Equality	
Words	If a number is subtracted from each side of a true equation, the resulting equivalent equation is also true.
Symbols	For any real numbers a, b, and c, if $a = b$, then $a - c = b - c$.

Today's Goals
- Solve equations by using addition and subtraction.
- Solve equations by using multiplication and division.

Today's Vocabulary
solve an equation

solution

equivalent equations

Think About It!
What happens if you add 5 to each side of $x - 5 = 15$? Which Property of Equality are you using?

Go Online
You may want to complete the Concept Check to check your understanding.

Example 1 Solve by Adding

Use the Addition Property of Equality to solve $g - 25 = 113$.

Horizontal Method		*Vertical Method*
$g - 25 = 113$	Original equation	$g - 25 = 113$
$g - 25 + 25 = 113$ _____	Add 25 to each side.	$\underline{\quad + 25 \quad}$ _____
$g =$ ____	Simplify.	$g =$ ____

CHECK

$$g - 25 = 113 \qquad \text{Original equation}$$
$$\underline{\quad} - 25 \overset{?}{=} 113 \qquad \text{Substitute 138 for } g.$$
$$113 = 113 \qquad \text{True}$$

Check

Solve $\frac{2}{3} + w = 1\frac{1}{2}$. State which property of equality you used. __

A. $\frac{5}{6}$; Subtraction Property of Equality

B. $\frac{5}{6}$; Addition Property of Equality

C. $\frac{13}{6}$; Addition Property of Equality

D. $\frac{13}{6}$; Subtraction Property of Equality

Example 2 Solve by Subtracting

Use the Subtraction Property of Equality to solve $27 + k = 30$.

Horizontal Method		*Vertical Method*
$27 + k = 30$	Original equation	$27 + k = \;\; 30$
27 ____ $+ k = 30$ ____	Subtract 27 from each side.	____ ____
$k =$ ____	Simplify.	$k =$ ____

CHECK

$$27 + k = 30 \qquad \text{Original equation}$$
$$27 + \underline{\quad} \overset{?}{=} 30 \qquad \text{Substitute 3 for } k.$$
$$30 = 30 \qquad \text{True}$$

Check

Solve $a + 26 = 35$.

$a =$ ____

 **Go Online** You can complete an Extra Example online.

🌎 Example 3 Write a One-Step Equation

TENNIS In tennis, the Grand Slam tournaments are the four most prestigious annual events. At one point in his career, Roger Federer had won three more Grand Slam singles titles than Rafael Nadal. If at that time Roger Federer held the record for the most Grand Slam singles titles won with 17, how many Grand Slam singles titles had Rafael Nadal won?

Complete the table to write an equation that represents the number of Grand Slam singles titles Rafael Nadal won.

Words	Roger Federer	won	three	more than	Rafael Nadal
Variable	Let n = the number of singles _____ won.				
Equation	____		____	____	____

$17 = n + 3$ Original equation

$17 \underline{\quad} = n + 3 \underline{\quad}$ Subtract 3 from each side.

$\underline{\quad} = n$ Simplify.

Rafael Nadal had won ____ Grand Slam singles titles.

Check

DOGS On average, a male bulldog weighs 15 pounds less than a male golden retriever. If the average male bulldog weighs 50 pounds, write and solve an equation to find the average weight of a male golden retriever. ____

A. $50 = w - 15$; 65 pounds

B. $50 = w + 15$; 35 pounds

C. $50 = w - 15$; 35 pounds

D. $50 = 15 - w$; 65 pounds

Use a Source

Choose another men's singles tennis player and research the number of Grand Slam singles titles he has won. Write your own equation relating the number of Grand Slam singles titles he has won to the 17 titles of Roger Federer.

🌐 **Go Online** You can complete an Extra Example online.

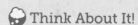 Think About It!

What happens if you divide each side of $8x = 32$ by 8? Which property of equality does this demonstrate?

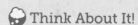 Think About It!

How could you use the Division Property of Equality to simplify $ax = 32$ to $x = \frac{32}{a}$? How does this relate to using the Division Property of Equality to solve $8x = 32$?

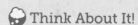 Think About It!

Describe a method you could use to check your solution for **part a.**

Learn Solving One-Step Equations Involving Multiplication or Division

You can also use the Multiplication Property of Equality and the Division Property of Equality to solve equations.

Key Concept • Multiplication Property of Equality	
Words	If an equation is true and each side is multiplied by the same nonzero number, then the resulting equation is equivalent.
Symbols	For any real numbers a, b, and c, if $a = b$, then $ac = bc$.
Example	If $x = 3$, then $8x = 24$.

Key Concept • Division Property of Equality	
Words	If an equation is true and each side is divided by the same nonzero number, the resulting equation is equivalent.
Symbols	For any real numbers a, b, and c, $c \neq 0$, if $a = b$, then $= \frac{a}{c} = \frac{b}{c}$.
Example	If $x = -35$, then $\frac{x}{7} = \frac{-35}{7}$ or -5.

Example 4 Solve Equations by Multiplying or Dividing

Solve each equation.

a. $\frac{3}{8}x = \frac{9}{4}$

$$\frac{3}{8}x = \frac{9}{4}$$ Original equation

$$\left(\frac{8}{3}\right)\frac{3}{8}x = \underline{\hspace{1cm}} \frac{9}{4}$$ Multiply each side by $\frac{8}{3}$, the reciprocal of $\frac{3}{8}$.

$$x = \underline{\hspace{1cm}}$$ Simplify.

b. $42 = -14y$

$$42 = -14y$$ Original equation

$$\frac{42}{-14} = \frac{-14y}{-14}$$ Divide each side by -14.

$$-3 = y$$ Simplify.

Check

Solve the equation $6y = 54$.

$y = \underline{\hspace{1cm}}$

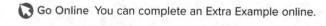 **Go Online** You can complete an Extra Example online.

🌐 Apply Example 5 Solve by Multiplying

SURVEY Kenji took a survey of the sophomore class. If 96 sophomores, or two-thirds of the class, said they were going to the football game on Saturday, how many sophomores were in the survey?

1. What is the task?

Describe the task in your own words. Then list any questions that you may have. How can you find the answers to your questions?

_____

_____

_____

2. How will you approach the task? What have you learned that you can use to help you complete the task?

_____

_____

_____

_____

3. What is your solution?

What equation represents the number of sophomores surveyed?

_____

How many sophomores were in the survey?

_____

4. How can you know that your solution is reasonable?

⚡ **Write About It!** Write an argument that can be used to defend your solution.

_____

_____

_____

_____

_____

🌐 **Go Online** You can complete an Extra Example online.

👓 **Think About It!**

How would your equation change if the 96 sophomores planning to attend the game represented three-fourths of the class? What would you multiply each side of the equation by to solve the new equation?

Check

FASHION Imani is making costumes for a play. She spent $146.58 on 21 yards of fabric. Write and solve an equation to find how much Imani paid for each yard of fabric. _____

A. $21p = 146.58$; $6.98 per yard

B. $146.58p = 21$; $0.14 per yard

C. $146.58(21) = p$; $3078.18 per yard

D. $21p = 146.58$; $3078.18 per yard

Pause and Reflect

Did you struggle with anything in this lesson? If so, how did you deal with it?

 Record your observations here.

Go Online You can complete an Extra Example online.

Practice

Examples 1, 2, and 4

Solve each equation.

1. $v - 9 = 14$

2. $44 = t - 72$

3. $-61 = d + (-18)$

4. $18 + z = 40$

5. $-4a = 48$

6. $12t = -132$

7. $18 - (-f) = 91$

8. $-16 - (-t) = -45$

9. $\frac{1}{3}v = -5$

10. $\frac{u}{8} = -4$

11. $\frac{a}{6} = -9$

12. $-\frac{k}{5} = \frac{7}{5}$

13. $\frac{3}{4} = w + \frac{2}{5}$

14. $-\frac{1}{2} + a = \frac{5}{8}$

15. $-\frac{t}{7} = \frac{1}{15}$

16. $-\frac{5}{7} = y - 2$

17. $v + 914 = -23$

18. $447 + x = -261$

19. $-\frac{1}{7}c = 21$

20. $-\frac{2}{3}v = -22$

21. $\frac{3}{5}q = -15$

22. $\frac{n}{8} = -\frac{1}{4}$

23. $\frac{c}{4} = -\frac{9}{8}$

24. $\frac{2}{3} + r = -\frac{4}{9}$

25. $y - 7 = 8$

26. $w + 14 = -8$

27. $p - 4 = 6$

28. $-13 = 5 + x$

29. $98 = b + 34$

30. $y - 32 = -1$

31. $n + (-28) = 0$

32. $y + (-10) = 6$

33. $-1 = t + (-19)$

34. $j - (-17) = 36$

35. $14 = d + (-10)$

36. $u + (-5) = -15$

37. $11 = -16 + y$

38. $c - (-3) = 100$

39. $47 = w - (-8)$

40. $x - (-74) = -22$

41. $4 - (-h) = 68$

42. $-56 = 20 - (-j)$

43. $12z = 108$

44. $-7t = 49$

45. $18f = -216$

46. $-22 = 11v$

47. $-6d = -42$

48. $96 = -24a$

49. $\frac{c}{4} = 16$

50. $\frac{a}{16} = 9$

51. $-84 = \frac{d}{3}$

52. $-\frac{d}{7} = -13$

53. $\frac{t}{4} = -13$

54. $31 = -\frac{1}{6}n$

55. **SUPREME COURT** Chief Justice William Rehnquist served on the Supreme Court for 33 years until his death in 2005. Write and solve an equation to determine the year he was confirmed as a justice on the Supreme Court.

56. **SALARY** In a recent year, the annual salary of the Governor of New York was $179, 000. During the same year, the annual salary of the Governor of Tennessee was $94,000 less than that. Write and solve an equation to find the annual salary of the Governor of Tennessee in that year.

57. **WEATHER** On a cold January day, Kiara noticed that the temperature dropped 21 degrees over the course of the day to −9°C. Write and solve an equation to determine what the temperature was at the beginning of the day.

58. **FARMING** The Rolling Hills Farm is 126 acres. This is $\frac{1}{4}$ the size of the Briarwood Farm. Write and solve an equation to determine the number of acres of the Briarwood Farm.

59. **SOCCER** During the season, 13% of the players who signed up for the soccer league dropped out. A total of 174 players finished the season.

   a. Assign a variable. Write an expression for the number of players who finished the season. Explain your reasoning.

   b. Write an equation to find the number of players who signed up for the soccer league.

   c. Solve the equation to find the number of players who signed up for the soccer league.

Mixed Exercises

Write an equation for each sentence. Then solve the equation.

60. Six times a number is 132.

61. Two thirds equals negative eight times a number.

62. Five elevenths times a number is 55.

63. Four fifths is equal to ten sixteenths of a number.

64. Three and two thirds times a number equals two ninths.

65. Four and four fifths times a number is one and one fifth.

Solve each equation. Check your solution.

66. $\frac{x}{9} = 10$

67. $\frac{b}{7} = -11$

68. $\frac{3}{4} = \frac{c}{24}$

69. $\frac{2}{3} = \frac{1}{8}y$

70. $\frac{2}{3}n = 14$

71. $\frac{3}{5}g = -6$

72. $4\frac{1}{5} = 3p$

73. $-5 = 3\frac{1}{2}x$

74. $6 = -\frac{1}{2}n$

75. $-\frac{2}{5} = -\frac{z}{45}$

76. $-\frac{g}{24} = \frac{5}{12}$

77. $-\frac{v}{5} = -45$

78. $-6 = \frac{2}{3}z$

79. $\frac{2}{7}q = -4$

80. $\frac{5}{9}p = -10$

81. $\frac{a}{10} = \frac{2}{5}$

82. $d - 8 = 6$

83. $-28 = p + 21$

84. $-7x = 63$

85. $-\frac{t}{5} = -8$

86. $y + (-16) = -12$

87. $\frac{3}{5}y = -9$

88. $-8d = -64$

89. $-\frac{3}{4}y = \frac{8}{20}$

90. VACATION The Lopez family is on vacation in Tennessee. They drove 210 miles from Memphis to Nashville and continued driving to Knoxville. By the time they reached Knoxville, they had traveled a total of 390 miles.

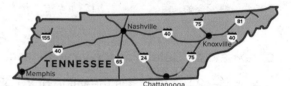

   a. Define the variable and write an equation that represents the distance from Nashville to Knoxville.

   b. Which property of equality could you use to isolate the variable in your equation? Explain your reasoning.

   c. How far is Knoxville from Nashville? How can you verify that your solution is accurate?

   d. If the Lopez family drives from Nashville to Chattanooga instead of Nashville to Knoxville, they will drive 47 fewer miles. Write an equation that represents the distance from Memphis to Chattanooga through Nashville. How far is Chattanooga from Nashville?

91. TICKETS Julian and Makayla order season tickets for the local soccer team. The ticket package they choose costs $780 and includes tickets to 12 games.

   a. Write and solve an equation that represents the cost per game.

   b. Single game tickets cost $85. How much do they save per game by using season tickets?

92. TACOS Orlando spent $18 at a taco truck. He ordered 4 tacos. Write and solve an equation to find the cost of each taco.

STRUCTURE Solve each equation. State the Property of Equality used.

93. $\frac{x}{9} = 24$

94. $m - 183 = -79$

95. $972 + y = 748$

96. $-\frac{4}{5}p = 32$

97. $135 = 9b$

98. $45 = \frac{3}{2}z$

Higher Order Thinking Skills

99. WHICH ONE DOESN'T BELONG Identify the equation that does not belong with the other three. Justify your conclusion.

| $n + 14 = 27$ | $12 + n = 25$ | $n - 16 = 29$ | $n - 4 = 9$ |

100. PERSEVERE Determine the value for each statement below.

    a. If $x - 9 = 12$, what is the value of $x + 1$?

    b. If $n + 7 = -4$, what is the value of $n + 1$?

101. CREATE Write an equation that you would use the Addition Property of Equality to solve.

102. ANALYZE Determine whether each sentence is *sometimes*, *always*, or *never* true. Justify your argument.

    a. $x + x = x$ **b.** $x + 0 = x$

103. ANALYZE How would you solve $5x = 35$? How would you solve $5 + x = 35$? How are the methods similar and how are they different?

104. WRITE Consider the Multiplication Property of Equality and the Division Property of Equality. Explain why they can be considered the same property. Which one do you think is easier to use?

Solving Multi-Step Equations

Explore Using Algebra Tiles to Model Multi-Step Equations

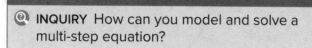 **Online Activity** Use algebra tiles to complete the Explore.

② INQUIRY How can you model and solve a multi-step equation?

Learn Solving Multi-Step Equations

A **multi-step equation** is an equation that uses more than one property of equality to solve it. To solve this type of equation, you can undo each operation using properties of equality. Working backward in the order of operations makes this process simpler. Each step in this process results in equivalent equations.

Operation	Opposite Operation
Addition	Subtraction
Subtraction	Addition
Multiplication	Division
Division	Multiplication

Example 1 Solve Multi-Step Equations

Use properties of equality to solve each equation. Check your solutions.

a. $2a - 6 = 4$

$$2a - 6 = 4$$ Original equation.

$$2a - 6 \underline{\hspace{1cm}} = 4 \underline{\hspace{1cm}}$$ Add 6 to each side.

$$2a = 10$$ Simplify.

$$\frac{2a}{2} = \frac{10}{2}$$ Divide each side by 2.

$$a = 5$$ Simplify.

(continued on the next page)

Today's Goal
• Solve equations involving more than one operation.

Today's Vocabulary
multi-step equation

🐷 **Think About It!**

In $4x - 2 = 5$, which two operations are being used? Which operations would you use to work backward in the order of operations to solve the equation?

CHECK

Check your solution by substituting the result back into the original equation.

$2a - 6 = 4$	Original equation
$2(5) - 6 \overset{?}{=} 4$	Substitute 5 for a.
$10 - 6 \overset{?}{=} 4$	Simplify.
$4 = 4$	True

b. $\dfrac{n+1}{-2} = 15$

$\dfrac{n+1}{-2} = 15$	Original equation.
$\underline{\hspace{1cm}}\left(\dfrac{n+1}{-2}\right) = \underline{\hspace{1cm}}(15)$	Multiply each side by -2.
$n + 1 = \underline{\hspace{1cm}}$	Simplify.
$n + 1 \underline{\hspace{1cm}} = -30 \underline{\hspace{1cm}}$	Subtract 1 from each side.
$n = \underline{\hspace{1cm}}$	Simplify.

Check your solution by substituting the result back into the original equation.

Check

Solve $3m + 4 = -11$.

$m = \underline{\hspace{0.8cm}}$

Solve $8 = \dfrac{x-5}{7}$.

$x = \underline{\hspace{0.8cm}}$

🌐 **Example 2** Write and Solve a Multi-Step Equation

FUNDRAISING **The student council raised $\dfrac{2}{5}$ of the money they need to cover the cost of the school dance with a bake sale. They raised an additional $150 selling raffle tickets. If the student council has raised $630, what is the cost of the dance? Write an equation for the problem. Then solve the equation.**

🔘 **Go Online** You can complete an Extra Example online.

Study Tip

Assumptions To solve an equation, you must assume that the original equation has a solution.

🔘 **Go Online**

You can watch a video to see how to use a graphing calculator with this example.

Study Tip

Multiplicative Inverse A number multiplied by its reciprocal is 1.

💬 **Talk About It!**

Would it work to first multiply each side by $\dfrac{5}{2}$? Explain your reasoning.

Complete the table to write an equation for the cost of the dance.

Words	Two fifths of the cost	plus 150	is 630.
Variable	Let c = the cost of _____ .		
Equation		+ 150	

$$\frac{2}{5}c + 150 = 630 \qquad \text{Original equation}$$

$$\frac{2}{5}c + 150 \underline{} = 630 \underline{} \qquad \text{Subtract 150 from each side.}$$

$$\frac{2}{5}c = 480 \qquad \text{Simplify.}$$

$$\underline{}\left(\frac{2}{5}\right)c = \underline{}(480) \qquad \text{Multiply each side by } \frac{5}{2}.$$

$$c = 1200 \qquad \text{Simplify.}$$

The dance costs $ _____ .

Check

BASKETBALL A sporting goods store sold $\frac{2}{3}$ of its basketballs, but 8 were returned. Now the store has 38 basketballs. How many were there originally? Write an equation for the problem. Then solve the equation. __

A. $\frac{2}{3}b + 8 = 38$; 45

B. $\frac{2}{3}b - 8 = 38$; 69

C. $\frac{1}{3}b + 8 = 38$; 90

D. $\frac{1}{3}b - 8 = 38$; 138

Example 3 Solve Multi-Step Equations with Letter Coefficients

Some equations have coefficients that are represented by letters. To solve these equations, apply the process of solving equations to isolate the variable.

Solve $ax + 7 = 5$ for x. Assume that $a \neq 0$.

$$ax + 7 = 5 \qquad \text{Original equation}$$

$$ax + 7 \underline{} = 5 \underline{} \qquad \text{Subtract 7 from each side.}$$

$$ax = -2 \qquad \text{Simplify.}$$

$$\frac{ax}{a} = \frac{-2}{a} \qquad \text{Divide each side by } a.$$

$$x = \underline{} \qquad \text{Simplify.}$$

Think About It!

Why do you have to assume that $a \neq 0$ when solving the equation?

Go Online You can complete an Extra Example online.

Check

Solve $2 - ax = -8$ for x. Assume $a \neq 0$. _____

A. $x = -\dfrac{10}{a}$

B. $x = -\dfrac{6}{a}$

C. $x = \dfrac{6}{a}$

D. $x = \dfrac{10}{a}$

Pause and Reflect

Did you struggle with anything in this lesson? If so, how did you deal with it?

Record your observations here.

 Go Online You can complete an Extra Example online.

Practice

Go Online You can complete your homework online.

Examples 1

Use properties of equality to solve each equation. Check your solution.

1. $3t + 7 = -8$

2. $8 = 16 + 8n$

3. $-34 = 6m - 4$

4. $9x + 27 = -72$

5. $\frac{y}{5} - 6 = 8$

6. $\frac{f}{-7} - 8 = 2$

7. $1 + \frac{r}{9} = 4$

8. $\frac{k}{3} + 4 = -16$

9. $\frac{n - 2}{7} = 2$

10. $14 = \frac{6 + z}{-2}$

11. $-11 = \frac{a - 5}{6}$

12. $\frac{22 - w}{3} = -7$

Example 2

13. SHOPPING Ricardo spent half of his allowance on school supplies. Then he bought a snack for $5.25. When he arrived home, he had $22.50 left. Write and solve an equation to find the amount of Ricardo's allowance a.

14. SHOPPING Liza earned some money by taking care of her neighbor's pet. She bought a drink for $1.95, and a concert ticket for $30. She bought a ring for $7.20, and then spent two-thirds of the remaining money on a wireless speaker. If Liza has $38.50 left, write and solve an equation to find the amount of money m Lisa earned by taking care of her neighbor's pet.

15. PET SHELTERS Henry works at a pet shelter after school. He purchases a large package of dog treats. He sets aside 10 treats and distributes the rest equally among the 15 dogs in the shelter. If each dog received 4 treats, write and solve an equation to find the number of treats t that were in the original package.

16. BASKETBALL The average number of points a basketball team scored for three games was 63 points. In the first two games, they scored the same number of points, which was 6 points more than they scored in the third game. Write and solve an equation to find the number of points the team scored in each game.

17. HUMAN HEIGHT Micah's adult height is one less than twice his height at age 2. Micah's adult height is 71 inches. Write and solve an equation to find Micah's height h at age 2.

Example 3

Solve each equation for x. Assume $a \neq 0$.

18. $ax + 3 = 23$

19. $4 = ax - 14$

20. $ax - 5 = 19$

21. $6 + ax = -29$

22. $\frac{8}{ax} - 5 = -3$

23. $18 - ax = 42$

24. $5 = \frac{5}{ax} + 1$

25. $-3 = ax + 11$

26. $-7 = -ax - 16$

Mixed Exercises

Solve each equation. Check your solution.

27. $3x + 8 = 29$

28. $\frac{a}{6} - 5 = 9$

29. $\frac{5r}{2} - 6 = 19$

30. $\frac{n}{3} - 8 = -2$

31. $5 + \frac{x}{4} = 1$

32. $-\frac{h}{3} - 4 = 13$

33. $5(1 + n) = -5$

34. $-27 = -6 - 3p$

35. $-\frac{a}{6} + 5 = 2$

REASONING **Write and solve an equation to find each number.**

36. A number is divided by 2, and then the quotient is increased by 8. The result is 33.

37. Two is subtracted from a number, and then the difference is divided by 3. The result is 30.

38. **PERSEVERE** The sum of 4 consecutive odd integers is equal to zero.

   a. Write an equation to model the sentence.

   b. Solve the equation to find the numbers. Check your solution.

39. **FIND THE ERROR** Kadija and Jorge are solving $\frac{1}{2}n + 5 = \frac{17}{2}$. Jorge uses the Subtraction Property of Equality followed by the Multiplication Property of Equality. Kadija also uses the Subtraction Property of Equality, but because n is multiplied by $\frac{1}{2}$, Kadija claims that the Division Property of Equality can be used to isolate the variable. Which student is correct? Explain your reasoning.

40. **CREATE** Write a problem that can be represented by the equation $11.9p + 23.1 = 273$. Define the variable and solve the equation.

41. **ANALYZE** Solve each equation for x. Assume that $a \neq 0$.
   a. $ax + 7 = 5$
   b. $\frac{1}{a}x - 4 = 9$
   c. $2 - ax = -8$

42. **ANALYZE** Determine whether each equation has a solution. Justify your answer.
   a. $\frac{a+4}{5+a} = 1$
   b. $\frac{1+b}{1-b} = 1$
   c. $\frac{c-5}{5-c} = 1$

43. **ANALYZE** Determine whether the following statement is *sometimes*, *always*, or *never* true. Justify your argument.

   The sum of three consecutive odd integers equals an even integer.

44. **WRITE** Write a paragraph explaining the order of the steps that you would take to solve a multi-step equation.

Solving Equations with the Variable on Each Side

Explore Modeling Equations with the Variable on Each Side

 Online Activity Use graphing technology to complete the Explore.

> ⊘ **INQUIRY** How can you solve an equation with the variable on each side?

Learn Solving Equations with the Variable on Each Side

Sometimes, the variable will appear on each side of an equation. To solve these equations, use the Addition or Subtraction Property of Equality to write an equivalent equation with the variable terms on one side and the numbers without variables, or constants, on the other side.

Example 1 Solve an Equation with the Variable on Each Side

Solve $5 + 7a = 4a - 13$. Check your solution.

$5 + 7a = 4a - 13$	Original equation
$\underline{} = -\underline{4a}$	Subtract $4a$ from each side.
$5 + 3a = -13$	Simplify.
$\underline{-5} = \underline{}$	Subtract 5 from each side.
$3a = -18$	Simplify.
$\dfrac{3a}{3} = \dfrac{-18}{3}$	Divide each side by 3.
$a = -6$	Simplify.

CHECK $\quad 5 + 7a = 4a - 13$	Original equation
$5 + 7(-6) \overset{?}{=} 4(-6) - 13$	Substitution, $a = -6$
$5 + -42 \overset{?}{=} -24 - 13$	Multiply.
$-37 = -37$	True

 Go Online You can complete an Extra Example online.

Today's Goals
- Solve equations with the variable on each side.
- Solve equations by applying the Distributive Property.
- Prove that equations are identities or have no solution.

Today's Vocabulary
identity

💭 **Think About It!**

Leon says that when you solve $5 + 7a = 4a - 13$, you can just combine $7a$ and $4a$ because they are like terms. Explain whether Leon is correct.

Study Tip

Solving an Equation
You may want to combine the terms with a variable on one side before isolating a constant.

🌐 Example 2 Write an Equation with the Variable on Each Side

CONTEST **The results of the 2015 Nathan's Hot Dog Eating Contest are shown.**

Suppose the men's and women's winners, Matt and Miki, decide to compete against each other. To make the competition more interesting, Matt will not start until Miki has eaten 20 hot dogs. **Assume that Matt and Miki eat at a constant rate throughout the competition. Based on the number of hot dogs eaten in 10 minutes by Matt and Miki, how many minutes after Matt starts eating will they have eaten the same number of hot dogs?**

Hot Dog Eating Contest

Winner	Hotdogs Eaten (Including Buns)
Matt Stonie (Men)	62
Miki Sudo (Women)	38

Contest Duration 10 minutes

Read the Problem

We want to find the number of minutes m for which Matt and Miki have eaten the same number of hot dogs.

Matt eats _____ hot dogs in 10 minutes, which is a rate of _____ hot dogs per minute. The number of hot dogs he has eaten m minutes after starting is _____.

Miki eats _____ hot dogs in 10 minutes, which is a rate of _____ hot dogs per minute. She is given a 20-hot dog head start, so the number of hot dogs she has eaten m minutes after Matt starts is $3.8m +$ _____.

The equation _____ represents this situation.

Solve the Problem

$$6.2m = 3.8m + 20 \qquad \text{Original equation}$$

$$6.2m \text{\_\_\_\_} = 3.8m + 20 \text{\_\_\_\_} \qquad \text{Subtract } 3.8m \text{ from each side.}$$

$$2.4m = 20 \qquad \text{Simplify.}$$

$$\frac{2.4m}{2.4} = \frac{20}{2.4} \qquad \text{Divide each side by 2.4.}$$

$$m \approx \text{\_\_\_\_} \qquad \text{Simplify.}$$

After approximately _____ minutes, Matt and Miki will have eaten the same number of hot dogs.

 Go Online You can complete an Extra Example online.

Check

BASKETBALL Nolan and Victor were two of the top scoring freshman players in a college basketball conference last season. The table shows how many points Nolan and Victor scored and how many games they played last season. The points Nolan and Victor score this season will be combined with their points from last season to give their total career points. This season, Nolan is hoping to catch up to Victor and have the same number of career points. Assume that Nolan and Victor play every game and score at the same constant rate as last season.

Player	Games Played	Points
Nolan	30	750
Victor	34	782

Part A

Based on each player's average scoring rate, write an equation that represents the number of games it will take Nolan to accumulate the same number of career points as Victor.

$$25p + \underline{\hspace{1cm}} = \underline{\hspace{1cm}} p + \underline{\hspace{1cm}}$$

Part B

Based on your equation in Part A, after how many games this season will Nolan and Victor have scored the same number of career points?

_____ games

Learn Solving Equations Involving the Distributive Property

Some equations contain grouping symbols. Grouping symbols can include parentheses ( ), brackets [ ], and fraction bars.

The steps for solving an equation can be summarized as follows.

Step 1 Simplify the expressions on each side. Remove any grouping symbols. Use the Distributive Property as needed.

Step 2 Use the Addition and/or Subtraction Properties of Equality to get the variable terms on one side of the equation and the constant terms on the other side. Simplify.

Step 3 Use the Multiplication and Division Properties of Equality to solve.

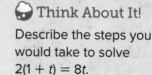 **Go Online** You can complete an Extra Example online.

Think About It!
Describe the steps you would take to solve $2(1 + t) = 8t$.

Study Tip

Grouping Symbols
Some expressions, like $2 - [11 + 5(p - 8)]$, contain grouping symbols inside of grouping symbols. To simplify these expressions, work from the inside out by first simplifying the expression within the innermost grouping symbol.

Example 3 Solve an Equation with Grouping Symbols

Solve $7(n - 1) = -2(3 + n)$.

$7(n - 1) = -2(3 + n)$	Original equation
$7n - 7 = $ _____	Distributive Property
$7n - 7$ _____ $= -6 - 2n$ _____	Add $2n$ to each side.
$9n - 7 = -6$	Simplify.
$9n - 7$ _____ $= -6$ _____	Add 7 to each side.
$9n = 1$	Simplify.
$\frac{9n}{9} = \frac{1}{9}$	Divide each side by 9.
$n = $ _____	Simplify.

Check

Solve $7(n - 2) + 8 = 3(n - 4) - 2$.

$n = $ _____

Copyright © McGraw-Hill Education

Math History Minute

During the short life of Indian mathematician **Srinivasa Ramanujan (1887–1920)**, he compiled nearly 3900 results, which included proofs of theorems, equations, and identities, nearly all of which have been proven correct. Ramanujan was known as a genius and an *autodidact*, which is a person who is self-taught.

Example 4 Solve an Equation with a Fraction Bar

Solve $5y = \frac{12y + 16}{4}$.

$5y = \frac{12y + 16}{4}$	Original equation
_____ $(5y) = $ _____ $\left(\frac{12y + 16}{4}\right)$	Multiply each side by 4.
$20y = 12y + 16$	Simplify.
$20y$ _____ $= 12y + 16$ _____	Subtract $12y$ from each side.
$8y = 16$	Simplify.
$\frac{8y}{8} = \frac{16}{8}$	Divide each side by 8.
$y = $ _____	Simplify.

🔊 **Go Online** You can complete an Extra Example online.

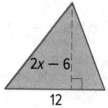

🌐 **Example 5** Write an Equation with Grouping Symbols

GEOMETRY **Find the value of x so that the figures have the same area.**

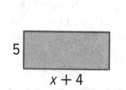

5

$x + 4$

$2x - 6$

12

The area of the rectangle is $5(x + 4)$, and the area of the triangle is $\frac{1}{2}(12)(2x - 6)$. The equation ___$(x + 4) = \frac{1}{2}(12)($___$)$ represents the situation where the areas of the figures are the same.

$5(x + 4) = \frac{1}{2}(12)(2x - 6)$	Original equation
$5(x + 4) =$ ___$(2x - 6)$	Multiply $\frac{1}{2}$ and 12.
___ $=$ ___	Distributive Property
$5x + 20$ ___ $= 12x - 36$ ___	Add 36 to each side.
$5x + 56 = 12x$	Simplify.
$5x + 56$ ___ $= 12x$ ___	Subtract $5x$ from each side.
$56 = 7x$	Simplify.
$\frac{56}{7} = \frac{7x}{7}$	Divide each side by 7.
$8 = x$	Simplify.

Check

GEOMETRY Find the value of x so that the figures have the same area.

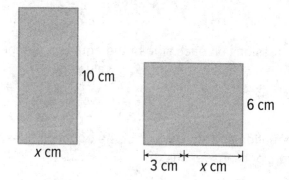

10 cm

x cm

6 cm

3 cm x cm

$x =$ ___

🐾 **Go Online** You can complete an Extra Example online.

Learn Identities and Equations with No Solutions

One solution	No solution	Identity
Words		
An equation has one solution if exactly one value of the variable makes the equation true.	An equation has no solution if there is no value of the variable that makes the equation true.	An **identity** is an equation that is true for all values of its variables.

Example 6 Solve an Equation with No Solution

Solve $6(y - 5) = 2(10 + 3y)$.

$$6(y - 5) = 2(10 + 3y)$$ Original equation

$$6y - \rule{1cm}{0.4pt} = 20 + \rule{1cm}{0.4pt}$$ Distributive Property

$$6y - 30 - 6y = 20 + 6y - 6y$$ Subtract $6y$ from each side.

$$\rule{1cm}{0.4pt} \neq \rule{1cm}{0.4pt}$$ Simplify.

Since -30 _____ 20, this equation has _____.

Example 7 Solve an Identity

Solve $7x + 5(x - 1) = 12x - 5$.

$$7x + 5(x - 1) = 12x - 5$$ Original equation

$$7x + \rule{1cm}{0.4pt} = 12x - 5$$ Distributive Property

$$\rule{1cm}{0.4pt} = 12x - 5$$ Simplify.

$$0 = 0$$ Subtract $12x - 5$ from each side.

Since the expressions on each side of the equation are the same, this equation is an _____. It is true for _____ values of x.

Check

Solve each equation and state whether the equation has *one solution*, has *no solution*, or is an *identity*.

a. $8(g + 6) = 5g + 3(g + 16)$ _____

b. $5x + 5 = 3(5x - 4) - 10x$ _____

c. $3w + 2 = 7w$ _____

d. $3(2b - 1) - 7 = 6b - 10$ _____

Go Online You can complete an Extra Example online.

Talk About It!

Could you tell that the equation was an identity before the final step? Explain your reasoning.

Go Online to

practice what you've learned about solving linear equations in the Put It All Together over Lessons 2-1 through 2-4.

Practice

Go Online You can complete your homework online.

Examples 1, 3, and 4

Solve each equation. Check your solution.

1. $7c + 12 = -4c + 78$

2. $2m - 13 = -8m + 27$

3. $9x - 4 = 2x + 3$

4. $6 + 3t = 8t - 14$

5. $\dfrac{b - 4}{6} = \dfrac{b}{2}$

6. $\dfrac{3v + 12}{6} = \dfrac{4v}{3}$

7. $2(r + 6) = 4(r + 4)$

8. $6(n + 5) = 3(n + 16)$

9. $5(g + 8) - 7 = 117 - g$

10. $12 - \dfrac{4}{5}(x + 15) = \dfrac{2}{5}x + 6$

11. $3(3m - 2) = 2(3m + 3)$

12. $6(3a + 1) - 30 = 3(2a - 4)$

13. $7n + 6 = 4n - 9$

14. $-6(2r + 8) = -10(r - 3)$

15. $5 - 3(w + 4) = w - 7$

16. $2x - 5(x - 3) = 2(x - 10)$

Example 2

17. OLYMPICS In the 2010 Winter Olympic Games, the United States won 1 more than 4 times the number of gold medals France won. The United States won 7 more gold metals than France. Write and solve an equation to find the number of gold medals each country won.

18. REASONING Diego's sister is twice his age minus 9 years. She is also as old as half the sum of the ages of Diego and both of his 12-year-old twin brothers. Write and solve an equation to find the ages of Diego and his sister.

19. NATURE The table shows the current heights and average growth rates of two different species of trees. Write and solve an equation to find how long it will take for the two trees to be the same height.

Tree Species	Current Height	Annual growth
A	38 inches	4 inches
B	45.5 inches	2.5 inches

20. WEIGHT A dog weighs two pounds less than three times the weight of a cat. The dog also weighs twenty-two more pounds than the cat. Write and solve an equation to find the weights of the dog and the cat.

Example 5

21. GEOMETRY Supplementary angles are two angles with measures that have a sum of 180°. Complementary angles are two angles with measures that have a sum of 90°. The measure of the supplement of an angle is 10° more than twice the measure of the complement of the angle. Let $90 - x$ equal the degree measure of the complement angle and $180 - x$ equal the degree measure of the supplement angle. Write and solve an equation to find the measure of the angle.

22. GEOMETRY Write and solve an equation to find the value of x so that the figures have the same area.

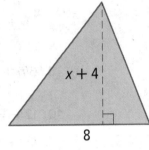
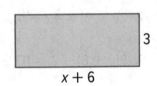

23. GEOMETRY Write and solve an equation to find the value of x so that the figures have the same area.

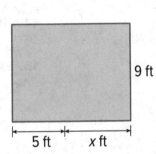
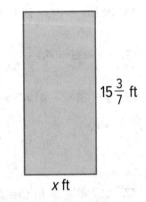

24. GEOMETRY Write and solve an equation to find the value of x so that the figures have the same area. The area of a trapezoid is $\frac{1}{2}h(b_1 + b_2)$.

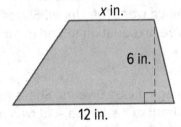
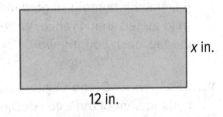

Solve each equation and state whether the equation has *one solution*, *no solution*, or is an *identity*.

25. $-6y - 3 = 3 - 6y$

26. $\frac{1}{2}(x + 6) = \frac{1}{2}x - 9$

27. $8q + 12 = 4(3 + 2q)$

28. $21(x + 1) - 6x = 15x + 21$

29. $12y + 48 - 4y = 8(y - 6)$

30. $8(z + 6) = 4(2z + 12)$

31. $2a + 2 = 3(a + 2)$

32. $\frac{1}{4}x + 5 = \frac{1}{4}x$

33. $7(c + 9) = 7c + 63$

34. $4k + 3 = \frac{1}{4}(8k + 16)$

35. $3b - 13 + 4b = 7b + 1$

36. $\frac{1}{2}(\frac{1}{2}m - 8) = \frac{1}{4}(m - 16)$

Mixed Exercises

Solve each equation. Check your solution.

37. $2x = 2(x - 3)$

38. $\frac{2}{5}h - 7 = \frac{12}{5}h - 2h + 3$

39. $-5(3 - q) + 4 = 5q - 11$

40. $2(4r + 6) = \frac{2}{3}(12r + 18)$

41. $\frac{3}{5}f + 24 = 4 - \frac{1}{5}f$

42. $\frac{1}{12} + \frac{3}{8}y = \frac{5}{12} + \frac{5}{8}y$

43. $6.78j - 5.2 = 4.33j + 2.15$

44. $14.2t - 25.2 = 3.8t + 26.8$

45. $3.2k - 4.3 = 12.6k + 14.5$

46. $5[2p - 4(p + 5)] = 25$

47. $m - 9 = \frac{2m - 12}{3}$

48. $\frac{3d - 2}{8} = -d + 16\frac{1}{4}$

49. Twice the greater integer of two consecutive odd integers is 13 less than three times the lesser integer.

    a. Write an equation to find the two consecutive odd integers.

    b. What are the integers in ascending order?

50. Two times the quantity of eight times a number plus two is equal to three times the quantity of two times the same number minus seven.

    a. Write an equation to find the number.

    b. Solve the equation to find the number.

51. USE A MODEL The perimeter of Figure 1 is four times a number minus three. The perimeter of Figure 2 is two times the same number plus five. The perimeters of Figure 1 and Figure 2 are the same.

    a. Write an equation to find the number.

    b. Solve the equation to find the number.

    c. What is the perimeter of Figure 2? Explain.

    d. Find the perimeter of Figure 1. Compare the perimeter of Figure 1 to the perimeter you found for Figure 2 to justify the value of k is correct. Show your work.

52. STRUCTURE Find two consecutive even integers such that twice the lesser of two integers is 4 less than two times the greater integer.

    a. Write and solve an equation to find the integers.

    b. Does the equation have one solution, no solution, or is it an identity? Explain.

53. FIND THE ERROR Anthony and Patty are solving the equation $y - m = m - y + 1$ for y. Is either correct? Explain why or why not.

Anthony	Patty
$y - m = m - y + 1$	$y - m = m - y + 1$
$2y - m = m + 1$	$2y - m = m + 1$
$2y = 2m + 1$	$2y = 1$
$y = m + \frac{1}{2}$	$y = \frac{1}{2}$

54. PERSEVERE Write an equation with variables on each side of the equal sign, at least one fractional coefficient, and a solution of -6. Discuss the steps you used.

55. CREATE Create an equation with at least two grouping symbols for which there is no solution.

56. WRITE Compare and contrast solving equations with variables on both sides of the equation to solving one-step or multi-step equations with a variable on one side of the equation.

57. ANALYZE Determine whether each solution is correct. If it is incorrect; find the correct solution. Justify your argument.

a.
$$2(g + 5) = 22$$
$$2g + 5 = 22$$
$$2g + 5 - 5 = 22$$
$$2g = 17$$
$$g = 8.5$$

b.
$$5d = 2d - 18$$
$$5d - 2d = 2d - 18 - 2d$$
$$3d = -18$$
$$d = -6$$

c.
$$-6z + 13 = 7z$$
$$-6z + 13 - 6z = 7z - 6z$$
$$13 = z$$

58. PERSEVERE Find the value of k for which each equation is an identity.

a. $k(3x - 2) = 4 - 6x$

b. $15y - 10 + k = 2(ky - 1) - y$

59. CREATE Write an equivalent equation to $x = 8$ that has the variable x on both sides.

60. ANALYZE Solve $5x + 2 = ax - 1$ for x. Assume $a \neq 5$. Describe each step.

Solving Equations Involving Absolute Value

Explore Modeling Absolute Value

Today's Goal
- Solve absolute value expressions.

🔄 **Online Activity** Use an infographic to complete the Explore.

> ⊗
>
> ⊘ **INQUIRY** How is margin of error related to absolute value?

Learn Solving Equations Involving Absolute Value

Absolute value equations contain at least one absolute value expression. The simplest form of an absolute value equation is $|x| = n$. Since absolute value represents distance, you must consider the case where the solution is x units from zero in the negative direction and the case where the solution is x units from zero in the positive direction.

Key Concept • Solving Absolute Value Equations	
Words	When solving equations that involve absolute values, there are two cases to consider.
	Case 1 The expression inside the absolute value symbol is positive or zero.
	Case 2 The expression inside the absolute value symbol is negative.
Symbols	For any real numbers a and b, if $|a| = b$ and $b \geq 0$, then $a = b$ or $a = -b$.
Example	$|d| = 3$, so $d = 3$ or $d = -3$.

Consider the equation $|x| = 3$. This means that the two points on the number line where the distance between 0 and x is 3 are solutions to the equation. The distance between 0 and -3 is 3, so -3 is a solution to the equation. The distance between 0 and 3 is 3, so 3 is also a solution to the equation.

💭 **Think About It!**

Clark says that the solution set for $|x| = 8$ is 8. Is he correct? Why or why not?

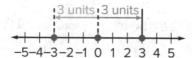

If $|x| = 3$, then $x = -3$ or $x = 3$. Thus the solution set is $\{-3, 3\}$. You can graph the solution set by graphing each solution on the number line.

For each absolute value equation, you must consider both cases. To solve an absolute value equation, first isolate the absolute value on one side of the equal sign if it is not already by itself.

Example 1 Solve an Absolute Value Equation When $n > 0$

Solve $|y + 2| = 4$. Then graph the solution set.

Case 1

If y is nonnegative, then $|y| = y$.

$y + 2 = 4$	Original equation
$y + 2 - 2 = 4 - 2$	Subtract 2 from each side.
$y = \underline{\quad}$	Simplify.

The solution set is $\{\underline{\quad}, 2\}$.

Graph points at -6 and 2 on the number line.

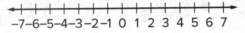

$-7-6-5-4-3-2-1\ 0\ 1\ 2\ 3\ 4\ 5\ 6\ 7$

Case 2

If y is negative, then $|y| = -y$.

$y + 2 = -4$

$y + 2 - 2 = -4 - 2$

$y = \underline{\quad}$

CHECK

Substitute -6 and 2 into the original equation.

$|y + 2| = 4$	Original equation
$|-6 + 2| \overset{?}{=} 4$	Substitute.
$|-4| \overset{?}{=} 4$	Simplify.
$4 = 4 \checkmark$	Take the absolute value.

$|y + 2| = 4$

$|2 + 2| \overset{?}{=} 4$

$|4| \overset{?}{=} 4$

$4 = 4 \checkmark$

Check

Graph the solution set of $|2t - 4| = 8$.

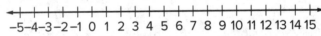

$-5-4-3-2-1\ 0\ 1\ 2\ 3\ 4\ 5\ 6\ 7\ 8\ 9\ 10\ 11\ 12\ 13\ 14\ 15$

Example 2 Solve an Absolute Value Equation When $n < 0$

Solve $|3x - 4| = -1$.

$|3x - 4| = -1$ means that the distance between $3x$ and 4 is $\underline{\quad}$. Since distance cannot be $\underline{\quad}$, the solution is the $\underline{\quad}$. The solution set is $\underline{\quad}$.

Check

Which statement must be true for the solution of $|ax + b| = c$ to be $\varnothing$?

_____

A. If a is negative, the solution will be $\varnothing$.

B. If b is negative, the solution will be $\varnothing$.

C. If c is negative, the solution will be $\varnothing$.

D. If c is positive, the solution will be $\varnothing$.

▶ **Go Online** You can complete an Extra Example online.

💬 Talk About It!

Would the solution change if the equation were changed to $|ax - 4| = -n$, where $n > 0$? Explain your reasoning.

🌐 **Example 3** Solve an Absolute Value Equation

MUSIC Depending on the size of each song, Luna's phone holds an average of 2000 songs, give or take 250 songs. Write and solve an equation involving absolute value to find the maximum and minimum number of songs that Luna's phone can hold.

Your Music
Library
is at
FULL CAPACITY
2000 songs

Red–Yellow Blues
Corso Girl
Ghostways   3:25
Orbital
Warp Spasm
Titans of Mars
17 Years
Where Did You Go?
~~Buried Under the Subway~~

Songs | Playlists | Artists

The maximum and minimum number of songs will differ from the average by _____ songs. Complete the table to write an equation that represents the maximum and minimum number of songs.

Words	The difference between the number of songs and _____ is 250.
Variable	Let x = the number of songs on Luna's phone.
Equation	_____

💭 Think About It!
Does the solution set mean that Luna's phone can hold only 1750 or 2250 songs? If not, how many songs can it hold?

Case 1

$x - 2000 = 250$ Original
                                        equation

$\underline{+\ 2000}$ $\underline{+\ 2000}$ Add
                                        2000.

$x = 2250$ Simplify.

Case 2

$x - 2000 = -250$

$\underline{+\ 2000}$ $\underline{+\ 2000}$

$x = 1750$

The solution set is {_____, _____}. The maximum and minimum number of songs are 2250 and 1750, respectively.

Check

SKYDIVING It takes approximately 6 minutes for a skydiver to land after she jumps out of a plane, give or take 30 seconds. What is the range of time, in seconds, it could take the skydiver to land?

[_____, _____]

🌐 **Go Online** You can complete an Extra Example online.

Example 4 Write an Absolute Value Equation

Write an equation involving absolute value for the graph.

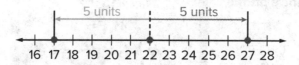

Find the point that is the same distance from 17 and from 27 on the number line. This is the midpoint between 17 and 27, which is _____.

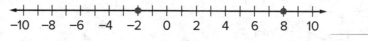

So an equation is $|x$ _____ $| = $ _____.

Check

Label each graph with the correct equation.

$|x + 3| = 6$ $|x - 1| = 3$ $|x - 1| = 6$ $|x - 3| = 5$

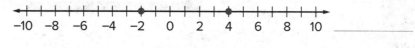

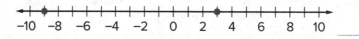

Pause and Reflect

Did you struggle with anything in this lesson? If so, how did you deal with it?

> Record your observations here.

🔵 **Go Online** You can complete an Extra Example online.

💭 **Think About It!**
Write a general rule that can be used for exercises similar to Example 4.

Practice

◥ **Go Online** You can complete your homework online.

Examples 1 and 2

Solve each equation. Then graph the solution set.

1. $|n - 3| = 5$

2. $|f + 10| = 1$

3. $|v - 2| = -5$

4. $|4t - 8| = 20$

5. $|8w + 5| = 21$

6. $|6y - 7| = -1$

7. $|x + 5| = -3$

8. $|-2y + 6| = 6$

9. $\left|\frac{3}{4}a - 3\right| = 9$

10. $|2x - 3| = 7$

Solve each equation.

11. $|7 - 2q| = 3$

12. $|4x - 2| = 26$

13. $|w + 1| = 5$

14. $|n + 2| = -1$

15. $|m - 2| = 2$

16. $|5c - 3| = 1$

17. $|2t + 6| = 4$

18. $|8k - 5| = -4$

Example 3

19. ENGINEERING *Tolerance* is an allowance made for imperfections in a manufactured object. The manufacturer of an oven specifies a temperature tolerance of ±15°F. This means that the temperature inside the oven will be within 15°F of the temperature to which it is set. Write and solve an absolute value equation to find the maximum and minimum temperatures inside the oven when the thermostat is set to 400°F.

20. POLLS Candidate A and Candidate B are running for mayor. A poll was taken to determine which candidate would likely win the election. The poll is accurate within ±5%. Write and solve an absolute value equation to find the maximum and minimum percent of voters who will vote for Candidate A if 38% of the voters in the poll voted for Candidate A.

21. STATISTICS The most familiar statistical measure is the arithmetic mean, or average. A second important statistical measure is the standard deviation, which is a measure of how far the data are from the mean. For example, the mean score on the Wechsler IQ test is 100 and the standard deviation is 15. This means that people within one standard deviation of the mean have IQ scores that are 15 points higher or lower than the mean.

a. One year, the mean mathematics score on the ACT test was 20.9 with a standard deviation of 5.3. Write an absolute value equation to find the maximum and minimum scores within one standard deviation of the mean.

b. What is the range of ACT mathematics scores within one standard deviation of the mean? within two standard deviations of the mean?

22. AVIATION The graph shows the results of a survey that asked 4300 students ages 7 to 18 what they thought would be the most important benefit of air travel in the future. There are about 40 million students in the United States. If the margin of error is ±3%, what is the range of the number of students ages 7 to 18 who would likely say that "finding new resources for Earth" is the most important benefit of future flight?

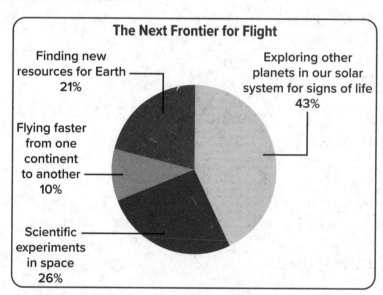

The Next Frontier for Flight

Finding new resources for Earth
21%

Exploring other planets in our solar system for signs of life
43%

Flying faster from one continent to another
10%

Scientific experiments in space
26%

Source: *The World Almanac*

23. MANUFACTURING A hardware store sells bags of rock salt that are labeled as weighing 35 pounds. The equipment used to package the salt produces bags with a weight that is within 8 ounces of the label weight. Write and solve an absolute value equation to determine the maximum and minimum weights for the bag of rock salt. Justify each step in the solution.

Example 4

Write an equation involving absolute value for each graph.

24.

```
-5-4-3-2-1 0 1 2 3 4 5
```

25.

```
-10 -8 -6 -4 -2 0 2 4 6 8 10
```

26.

```
-5-4-3-2-1 0 1 2 3 4 5
```

27.

```
-7-6-5-4-3-2-1 0 1 2 3
```

28.

```
-5-4-3-2-1 0 1 2 3 4 5
```

29.

```
-7-6-5-4-3-2-1 0 1 2 3
```

30.

```
-5-4-3-2-1 0 1 2 3 4 5
```

31.

```
-5-4-3-2-1 0 1 2 3 4 5
```

Mixed Exercises

Solve each equation. Then graph the solution set.

32. $\left|-\frac{1}{2}b - 2\right| = 10$

33. $|-4d + 6| = 12$

34. $|5f - 3| = 12$

35. $2|h| - 3 = 8$

36. $4 - 3|q| = 10$

37. $\frac{4}{|p|} + 12 = 14$

Write an equation involving absolute value for each graph.

38.

```
-5-4-3-2-1 0 1 2 3 4 5
```

39.

```
-5-4-3-2-1 0 1 2 3 4 5
```

40.

```
-5-4-3-2-1 0 1 2 3 4 5
```

41.

```
-2.0 -1.5 -1.0 -0.5 0 0.5 1.0 1.5 2.0
```

42.

```
-2 -1 0 1 2
```

43.

```
-3 -2 -1 0 1 2 3
```

44. REGULARITY For tropical fish, aquarium water should be set to 76°F with an allowance of 4°.

  a. Explain how to write an absolute value equation to represent this situation.

  b. Explain the steps to solve the absolute value equation. What do the solutions represent?

45. REASONING The temperature of a refrigerator is 38°F give or take 2°.

    a. Write an equation to find the maximum and minimum temperatures of the refrigerator.

    b. Solve the equation to find the maximum and minimum temperatures of the refrigerator.

46. STRUCTURE A quality control inspector at a bolt factory examines random bolts that come off the assembly line. All bolts being made must be a tolerance of 0.04 mm. The inspector is examining bolts that are to have a diameter of 6.5 mm. Write and solve an absolute value equation to find the maximum and minimum diameters of bolts that will pass his inspection.

47. SWIMMING POOL Chlorine is added to a swimming pool to sanitize the water and make it safe for swimming. The chlorine should be in the range of 2–4 ppm (parts per million). Write an equation that represents the maximum and minimum chlorine concentration.

48. FISH TANK Tom has a 10 gallon fish tank that he wants to fill with neon tetra fish. Tom calculates the number of fish that will fit in the tank using three different methods. Write an equation to represent the maximum and minimum number of fish that will fit in the tank.

Method	Number of Fish
Method 1	5
Method 2	9
Method 3	8

49. PERSEVERE If three points a, b, and c lie on the same line, then b is between a and c if and only if the distance from a to c is equal to the sum of the distances from a to b and from b to c. Write an absolute value equation to represent the definition of betweenness.

50. ANALYZE Translate the sentence $x = 5 \pm 2.3$ into an equation involving absolute value. Explain.

51. FIND THE ERROR Chris and Cami are solving $|x + 3| = -6$. Is either of them correct? Explain your reasoning.

Chris
$|x + 3| = 6$ or $|x + 3| = -6$
$x + 3 = 6$ or $x + 3 = -6$
$x = 3$ or $x = -9$

Cami
$|x + 3| = -6$
The solution is $\emptyset$.

52. WRITE Explain why an absolute value can never be negative.

53. CREATE Describe a real-world situation that could be represented by the absolute value equation $|x - 4| = 10$.

Solving Proportions

Explore Comparing Two Quantities

Online Activity Use graphing technology to complete the Explore.

> **INQUIRY** How can you solve for an unknown value if two quantities have a proportional relationship?

Learn Solving Proportions

A **proportion** is an equation stating that two ratios are equivalent.

Example 1 Solve a Proportion

Solve the proportion. If necessary, round to the nearest hundredth.

$$\frac{x}{45} = \frac{15}{25}$$

$\dfrac{x}{45} = \dfrac{15}{25}$	Original proportion
$45\left(\dfrac{x}{45}\right) = 45\left(\dfrac{15}{25}\right)$	Multiply each side by 45.
$x = \dfrac{45(15)}{25}$	Simplify.
$x = \dfrac{675}{25}$	Multiply.
$x = 27$	Divide.

CHECK

Check your solution by substituting into the original proportion and check to see if the fractions are equal.

$\dfrac{x}{45} = \dfrac{15}{25}$	Original proportion
$\dfrac{27}{45} \overset{?}{=} \dfrac{15}{25}$	Substitute.
$\dfrac{3}{5} = \dfrac{3}{5}$	True.

Go Online You can complete an Extra Example online.

Today's Goal
- Solve proportions.

Today's Vocabulary
proportion

💭 Think About It!
What is another equation that you could write to solve the proportion? Explain your reasoning.

Check

Solve $\frac{n-4}{8} = \frac{3}{2}$. If necessary, round to the nearest hundredth. _____

A. 16

B. 9.33

C. 8

D. 4.75

Example 2 Solve a Proportion with Two Missing Quantities

Solve $\frac{x}{9} = \frac{2x-3}{24}$. If necessary, round to the nearest tenth.

$\frac{x}{9} = \frac{2x-3}{24}$	Original proportion
$9\left(\frac{x}{9}\right) = 9\left(\frac{2x-3}{24}\right)$	Multiply each side by 9.
$x = \frac{9(2x-3)}{24}$	Simplify.
$x = \frac{18x-27}{24}$	Distributive Property.
$24x = 24\left(\frac{18x-27}{24}\right)$	Multiply each side by 24.
$24x = 18x - 27$	Simplify.
$24x - 18x = 18x - 18x - 27$	Subtract 18x from each side.
$6x = -27$	Simplify.
$\frac{6x}{6} = \frac{-27}{6}$	Divide each side by 6.
$x = -4.5$	Simplify.

> ☁ **Think About It!**
>
> How would the problem differ if the second ratio were 24 over 2x − 3?

Check

Solve $\frac{x}{12} = \frac{2x-5}{18}$. If necessary, round to the nearest hundredth. _____

A. −10

B. −3.75

C. 0.83

D. 10

🔁 **Go Online** You can complete an Extra Example online.

🌐 Example 3 Solve a Proportion by Using a Constant Rate

GEOGRAPHY **Parts of Mexico City are sinking at a rate of 140 centimeters every 5 years. If this rate remains constant, how many centimeters will the city sink in the next 12 years?**

Step 1 Estimate the solution.

In 10 years, Mexico City will sink 140(_____) or _____ centimeters. Because 10 years is slightly _____ 12 years, Mexico City will sink _____ 280 centimeters in 12 years.

Step 2 Write a proportion.

Let c represent _____.

$$\frac{\text{city sinks 140 cm}}{\text{in 5 years}} = \frac{\text{city sinks } c \text{ cm}}{\text{in 12 years}}$$

Step 3 Solve the proportion.

$\dfrac{140}{5} = \dfrac{c}{12}$	Original proportion
$12\left(\dfrac{140}{5}\right) = 12\left(\dfrac{c}{12}\right)$	Multiply each side by 12.
$\dfrac{12(140)}{5} = c$	Simplify.
$\dfrac{1680}{5} = c$	Simplify.
$336 = c$	Divide.

CHECK
How do you know your solution is reasonable?

_____

_____

_____

_____

Check

MIXTURE Oscar makes fruit punch to sell from his food truck by mixing 8 parts cranberry juice to 3 parts pineapple juice. How many cups of pineapple juice would Oscar need to mix with 48 cups of cranberry juice to make his punch? _____ cups

🌐 **Go Online** You can complete an Extra Example online.

💬 Talk About It!
Would you really expect the rate of sinking to remain constant over the entire time period? Explain.

🌐 Example 4 Solve a Percent Problem by Using a Proportion

MIXTURES A guide company makes a trail mix of raisins and mixed nuts. How many pounds of raisins does the guide company need to mix with 14 pounds of mixed nuts to make the trail mix 30% raisins?

METHOD 1 : raisins : trail mix

Let r represent the number of pounds of raisins.

Let $r + 14$ represent the number of pounds of the trail mix.

Write and solve a proportion.

$$\frac{\text{raisins}}{\text{trail mix}} = \frac{30}{100}$$ 30% of the trail mix is raisins.

$$\frac{r}{r + 14} = \frac{30}{100}$$ Substitute $r + 14$ for the amount of the trail mix.

$$(r + 14)\left(\frac{r}{r + 14}\right) = (r + 14)\frac{30}{100}$$ Multiply each side by $r + 14$.

$$r = (r + 14)\frac{3}{10}$$ Simplify.

$$r = \frac{3}{10}r + \frac{14 \cdot 3}{10}$$ Distributive Property

$$r - \frac{3}{10}r = \frac{3}{10}r - \frac{3}{10}r + \frac{42}{10}$$ Subtract $\frac{3}{10}r$ from each side.

$$\frac{7}{10}r = \frac{42}{10}$$ Simplify.

$$\frac{10}{7}\left(\frac{7}{10}r\right) = \frac{10}{7}\left(\frac{42}{10}\right)$$ Multiply each side by $\frac{10}{7}$.

$$r = \frac{42}{7} \text{ or } 6$$ Simplify.

METHOD 2 : raisins : mixed nuts

Let r represent the number of pounds of raisins, when the number of pounds of mixed nuts is 14.

Write and solve a proportion.

$$\frac{\text{raisins}}{\text{mixed nuts}} = \frac{30}{70}$$ The ratio of raisins to mixed nuts is 30 : 70.

$$\frac{r}{14} = \frac{30}{70}$$ Substitute 14 for the amount of mixed nuts.

$$14\left(\frac{r}{14}\right) = 14\left(\frac{30}{70}\right)$$ Multiply each side by 14.

$$r = \frac{14 \cdot 30}{70}$$ Simplify.

$$r = \frac{420}{70} \text{ or } 6$$ Simplify.

Check

MIXTURE Ayita is making a plant food mixture to use in her garden. The mixture is to be 20% plant food and 80% water. She needs to make 12 gallons of the mixture to cover her entire garden. Which proportions can be used to find the amount of plant food p she will need? Select all that apply. _____

A. $\frac{20}{80} = \frac{p}{12}$ B. $\frac{20}{100} = \frac{p}{12}$ C. $\frac{80}{100} = \frac{p}{12}$

D. $\frac{12 - p}{12} = \frac{80}{100}$ E. $\frac{20}{p} = \frac{80}{12}$

🐦 **Go Online** You can complete an Extra Example online.

Study Tip

Setting Up Ratios

It is a good idea to write the ratio in words to start the problem. Then read the problem to find the numbers or expressions to write each of the two ratios in the proportion.

☁ Think About It!

After multiplying each side by $r + 14$ in Method 1, Raja's resulting equation was $r = \frac{3}{10}r + 14$. What error did Raja make?

Practice

⊙ **Go Online** You can complete your homework online.

Example 1

Solve each proportion. If necessary, round to the nearest hundredth.

1. $\frac{3}{8} = \frac{15}{a}$

2. $\frac{t}{2} = \frac{6}{12}$

3. $\frac{4}{9} = \frac{13}{q}$

4. $\frac{15}{35} = \frac{g}{7}$

5. $\frac{7}{10} = \frac{m}{14}$

6. $\frac{8}{13} = \frac{v}{21}$

7. $\frac{w}{2} = \frac{4.5}{6.8}$

8. $\frac{1}{0.19} = \frac{12}{n}$

9. $\frac{2}{0.21} = \frac{8}{n}$

10. $\frac{2.4}{3.6} = \frac{k}{1.8}$

11. $\frac{t}{0.3} = \frac{1.7}{0.9}$

12. $\frac{7}{1.066} = \frac{z}{9.65}$

13. $\frac{x-3}{5} = \frac{6}{10}$

14. $\frac{7}{x+9} = \frac{21}{36}$

15. $\frac{10}{15} = \frac{4}{x-5}$

16. $\frac{6}{14} = \frac{7}{x-3}$

17. $\frac{7}{4} = \frac{f-4}{8}$

18. $\frac{3-y}{4} = \frac{1}{9}$

Example 2

Solve each proportion. If necessary, round to the nearest hundredth.

19. $\frac{4v+7}{15} = \frac{6v+2}{10}$

20. $\frac{9b-3}{9} = \frac{5b+5}{3}$

21. $\frac{2n-4}{5} = \frac{3n+3}{10}$

22. $\frac{2}{g+6} = \frac{4}{5g+10}$

23. $\frac{x}{3} = \frac{3x+2}{6}$

24. $\frac{w+3}{7} = \frac{w-1}{8}$

25. $\frac{4q-3}{5} = \frac{2q+1}{7}$

26. $\frac{5}{7k+4} = \frac{2}{2k-3}$

27. $\frac{m+1}{9} = \frac{m+2}{2}$

28. $\frac{j-5}{2} = \frac{j+8}{7}$

29. $\frac{9f+3}{10} = \frac{2f-4}{5}$

30. $\frac{2c-1}{3} = \frac{c+2}{4}$

31. $\frac{5n-2}{8} = \frac{n+8}{3}$

32. $\frac{h-7}{4} = \frac{2h+1}{3}$

33. $\frac{14}{3y+5} = \frac{3}{y}$

34. $\frac{p+10}{8} = \frac{2p-7}{4}$

35. $\frac{7}{14-d} = \frac{3}{18+d}$

36. $\frac{2z-4}{5} = \frac{3z+3}{10}$

Example 3

37. BOATING Dedra's boat used 5 gallons of gasoline in 4 hours. At this rate, how many gallons of gasoline will the boat use in 10 hours?

38. WATER A dripping faucet wastes 3 cups of water every 24 hours. How much water is wasted in a week?

39. PRECISION In November 2010 the average cost of 5 gallons of regular unleaded gasoline in the United States was $14.46. What was the average cost for 16 gallons of gasoline?

40. SHOPPING Stevenson's Market is selling 3 packs of stylus pens for $5.00. How much will 10 packs of stylus pens cost at this price?

41. STATE YOUR ASSUMPTION During basketball practice, Brent made 36 free throws in 3 minutes.

  a. How many free throws will Brent make in 5 minutes?

  b. What assumption did you make in part **a**? Explain.

42. NAILS Human fingernails grow at an average rate of 3.47 millimeters per month. How much will they grow in 20 months?

43. PICTURE Jasmine enlarged the size of a picture to a height of 15 inches. What is the new width of the picture if it was originally 6 inches wide by 4 inches tall?

44. TRAVEL Roscoe is exchanging $121 for Euros for his upcoming trip to Germany. If $2 can be exchanged for 1.78 Euros, how many Euros will Roscoe have?

Example 4

45. FUNDRAISER Owen is organizing a fundraiser. The proceeds will be split between a charity and the expenses from the fundraiser. Owen would like the cost of the fundraiser to be 15% of the proceeds. If the fundraiser will cost $500, how much money do they need to raise at the fundraiser?

46. COFFEE A barista is mixing a house blend of coffee that is 25% light roast. If there are 8 pounds of the light roast available, how much of the blend can the barista make?

47. CHEMISTRY A chemistry teacher needs to mix an acid solution for an experiment. How much hydrochloric acid needs to be mixed with 1500 milliliters of water to make a solution that is 12% acid?

48. LEMONADE Laronda wants to make fresh lemonade. The recipe she finds online recommends that the fresh lemon juice should be 20% of the total volume. She has 18 ounces of fresh lemon juice. How much water should she mix with the lemon juice?

Mixed Exercises

Solve each proportion. If necessary, round to the nearest hundredth.

49. $\dfrac{9}{g} = \dfrac{15}{10}$

50. $\dfrac{3}{a} = \dfrac{1}{6}$

51. $\dfrac{6}{z} = \dfrac{3}{5}$

52. $\dfrac{5}{f} = \dfrac{35}{21}$

53. $\dfrac{12}{7} = \dfrac{36}{m}$

54. $\dfrac{6}{23} = \dfrac{y}{69}$

55. $\dfrac{42}{56} = \dfrac{6}{f}$

56. $\dfrac{7}{b} = \dfrac{1}{9}$

57. $\dfrac{10}{14} = \dfrac{30}{m}$

58. $\dfrac{3}{4} = \dfrac{n}{20}$

59. $\dfrac{6}{4} = \dfrac{x}{18}$

60. $\dfrac{33}{b} = \dfrac{15}{45}$

61. $\dfrac{m-2}{4} = \dfrac{5}{20}$

62. $\dfrac{9}{5} = \dfrac{3}{x+7}$

63. $\dfrac{5}{b} = \dfrac{3}{b-6}$

64. $\dfrac{2p+3}{3} = \dfrac{4p-7}{2}$

65. $\dfrac{3y+4}{5} = \dfrac{y-1}{4}$

66. $\dfrac{2}{w} = \dfrac{7}{w+5}$

67. $\dfrac{7n-2}{6} = \dfrac{3n-2}{4}$

68. $\dfrac{-a-8}{10} = \dfrac{-a+3}{2}$

69. $\dfrac{c+2}{c-2} = \dfrac{4}{8}$

70. USE A SOURCE Find the heights of Willis Tower and the John Hancock Center in Chicago including the tip. Suppose you build a scale model of each building. If you make the model of the Willis Tower 3 meters tall, what would be the approximate height of the John Hancock Center model? Round to the nearest hundredth.

71. USE TOOLS A map of Waco, Texas and neighboring towns is shown.

a. Use a metric ruler to measure the distances between Robinson and Neale on the map.

b. If the scale on the map is 1 cm = 3 mi, find the actual distance between Robinson and Neale.

c. How many square miles are shown on this map?

72. BUDGET Shawnda spent $259.20 on a scooter. This was 80% of her budget for a new scooter. How much was the total budget?

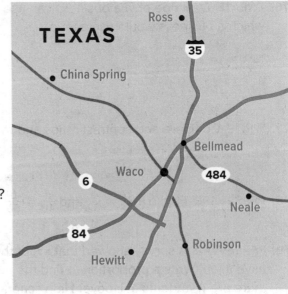

73. USE TOOLS On average, 8 potatoes cost $1.50 at a farmer's market. Fernando needs to buy 22 potatoes.

   a. Estimate the cost of 22 potatoes. Explain.

   b. How much will it cost Fernando to buy 22 potatoes? How does this compare to your estimate?

   c. How many potatoes could Fernando buy with $7?

   d. What is the unit cost per potato?

74. USE A MODEL Kina and Aiesha started walking from the same location at the same time. Kina walked 6 miles. Aiesha walked 8 miles and walked 1 mile per hour faster than Kina. They each walked for the same amount of time.

   a. Describe how a proportion could be used to find the rate that each person walked.

   b. Solve the proportion. Explain the meaning of the solution.

75. PERSEVERE If $\frac{a+1}{b-1} = \frac{5}{1}$ and $\frac{a-1}{b+1} = \frac{1}{1}$, find the value of $\frac{b}{a}$. (Hint: Choose values of a and b for which the proportions are true and evaluate $\frac{b}{a}$.).

76. CREATE Describe how a business can use ratios. Include a real-world situation in which a business would use a ratio.

77. WRITE Compare and contrast ratios and rates.

78. PERSEVERE Find b if $\frac{17}{34} = \frac{a}{32}$ and $\frac{a}{40} = \frac{b}{60}$.

79. PERSEVERE A survey showed that $x\%$ of the students at Hoover High school have a job. Write a proportion to find the number of students that have a job, z, if there are y students at Hoover High school.

Using Formulas

Explore Centripetal Force

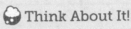 **Online Activity** Use a video to complete the Explore.

> **⊘ INQUIRY** Why might you want to solve a formula for a specified value? ✕

Learn Solving Equations for Given Variables

A **formula** is an equation that expresses a relationship between certain quantities. A formula or equation that involves more than one variable is called a **literal equation**.

Example 1 Solve for a Specific Variable

Solve 5a − 2b = 15 for a.

$$5a - 2b = 15$$ Original equation

$$5a - 2b \rule{1cm}{0.15mm} = 15 \rule{1cm}{0.15mm}$$ Add 2b to each side.

$$5a = 15 + 2b$$ Simplify.

$$\frac{\rule{1cm}{0.15mm}}{\rule{0.5cm}{0.15mm}} = \frac{15 + 2b}{5}$$ Divide each side by 5.

$$a = \frac{15}{5} + \frac{2b}{5}$$ Simplify.

$$a = \rule{1cm}{0.15mm} + \rule{1cm}{0.15mm}$$ Simplify.

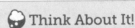 **Go Online** You can complete an Extra Example online.

Today's Goals
- Solve equations for specific variables.
- Convert units of measure.

Today's Vocabulary
formula

literal equation

dimensional analysis

> **☁ Think About It!**
> How do you solve an equation for a specific variable?

> **☁ Think About It!**
> Describe how you would solve for *b* instead of *a*. How would the answer change?

Example 2 Solve for a Specific Variable When the Variable Is on Each Side

Solve $4p - 7r = pq + 16$ for p.

$4p - 7r = pq + 16$	Original equation
$4p - 7r \underline{\qquad} = pq + 16 \underline{\qquad}$	Add $7r$ to each side.
$\underline{\qquad} = pq + 16 + 7r$	Simplify.
$4p \underline{\qquad} = pq + 16 + 7r \underline{\qquad}$	Subtract pq from each side.
$\underline{\qquad} = \underline{\qquad}$	Simplify.
$p(4 - q) = 16 + 7r$	_____
$\dfrac{p(4 - q)}{4 - q} = \dfrac{16 + 7r}{4 - q}$	Divide each side by $4 - q$.
$p = \dfrac{16 + 7r}{4 - q}$	Simplify.

Check

Solve $2v = \dfrac{w + v}{t}$ for v.

$v = \underline{\qquad}$

🌐 Example 3 Solve Literal Equations for a Given Variable

GEOMETRY **The area of a trapezoid is $A = \dfrac{h(b_1 + b_2)}{2}$. A represents the area, h represents the height, b_1 represents the length of one base, and b_2 represents the length of the other base.**

Part A

Solve the formula for h.

$A = \dfrac{h(b_1 + b_2)}{2}$	Area Formula
$\underline{\quad} A = \underline{\quad} \dfrac{h(b_1 + b_2)}{2}$	Multiply each side by 2.
$2A = h(b_1 + b_2)$	Simplify.
$\dfrac{2A}{b_1 + b_2} = \dfrac{h(b_1 + b_2)}{b_1 + b_2}$	Divide each side by $b_1 + b_2$.
$\underline{\qquad} = h$	Simplify.

Part B

Find the height of a trapezoid with an area of 70 square feet and bases that are 22 feet and 18 feet.

$h = \dfrac{2A}{(b_1 + b_2)}$ Formula for height

$h = \dfrac{2(70)}{(22 + 18)}$ $A = 70,\ b_1 = 22,\ b_2 = 18$

$h = \dfrac{140}{40}$ Simplify.

$h = \underline{\quad}$ Divide.

The height of the trapezoid is ____ feet.

Check

BUSINESS A Mason jar company wants to increase the volume of its cylindrical jars by 6 cubic inches. The company's designer wants a formula that states the height h of a jar given its volume V and radius r. The volume of a cylindrical jar is modeled by the equation $V = \pi r^2 h$.

Part A

What formula should be used to find the height h? ____

A. $h = \pi \cdot r^2$

B. $h = \sqrt{\dfrac{V}{\pi r}}$

C. $h = \dfrac{V}{\pi r^2}$

D. $h = \dfrac{V}{r^2}$

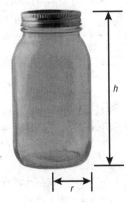

Part B

If the radius of the jar is 1.5 inches and the original volume is 30 cubic inches, then what height should the company make its new jar to increase the volume by 6 cubic inches? _____

Go Online You can complete an Extra Example online.

Watch Out!

Dividing by a Quantity
Do not forget to divide each side by the entire quantity $b_1 + b_2$.

💭 **Think About It!**
How would the height of the trapezoid change if the area were doubled and all other measures remained the same?

Study Tip

Solving for a Specific Variable
When an equation has more than one variable, it can be helpful to highlight the variable for which you are solving on a piece of paper.

🌐 Example 4 Use Literal Equations

PARTY The total amount of money Kishi spends on pizza for a party is $T = 13.49c + 15.49p$. T represents the total amount of money she spends, c represents the number of cheese pizzas she buys, and p represents the number of pepperoni pizzas she buys.

Pepperoni
$15.49

Cheese
$13.49

Part A

If Kishi has \$85 to spend on pizza, describe the constraints on $T = 13.49c + 15.49p$.

- The maximum number of cheese pizzas Kishi can buy is 6, because she can buy 6 cheese pizzas and no pepperoni pizzas without exceeding \$85.

- The maximum number of pepperoni pizzas Kishi can buy is ____, because she can buy ____ pepperoni and ____ cheese pizzas without going over her budget.

- The minimum number of each type of pizza she can buy is ____, because you cannot buy a negative number of pizzas.

Part B

Solve $T = 13.49c + 15.49p$ for c.

$$T = 13.49c + 15.49p \qquad \text{Original equation}$$

$$T \underline{\hspace{1cm}} = 13.49c + 15.49p \underline{\hspace{1cm}} \qquad \text{Subtract } 15.49p.$$

$$T - 15.49p = 13.49c \qquad \text{Simplify.}$$

$$\frac{T - 15.49p}{13.49} = \frac{13.49c}{13.49} \qquad \text{Divide by 13.49.}$$

$$\frac{T - 15.49p}{13.49} = c \qquad \text{Simplify.}$$

💭 **Think About It!**

Why did you round your answer in Part C to 2 instead of 3?

Part C

If Kishi has \$85 to spend on pizza and she needs to buy 3 pepperoni pizzas, find the maximum number of cheese pizzas she can buy.

$$c = \frac{T - 15.49p}{13.49} \qquad \text{Original equation solved for } c$$

$$c = \frac{85 - 15.49(3)}{13.49} \qquad T = 85, p = 3$$

$$c = \frac{38.53}{13.49} \qquad \text{Simplify.}$$

$$c \approx 2.86 \qquad \text{Divide.}$$

Kishi can buy a maximum of ____ cheese pizzas.

🔎 **Go Online** You can complete an Extra Example online.

Check

VIDEO GAMES Ella makes a video game that becomes very popular. She creates a formula, $P = 40c - 300$, to model her profit P given the number of copies sold c, taking into account the $300 fee that she has paid a retailer to sell her game. Which equation would model how many copies c must be sold to yield a specific amount of profit? _____

A. $c = \dfrac{P + 40}{300}$

B. $c = \dfrac{P}{40} + 300$

C. $c = \dfrac{P}{40} - 300$

D. $c = \dfrac{P + 300}{40}$

Think About It!

Why might you want to convert units?

Explore Using Dimensional Analysis

Online Activity Use a real-world situation to complete the Explore.

> ×
>
> **INQUIRY** Why might you want to convert the units for a given quantity or measurement?

Learn Dimensional Analysis

When using formulas, you may want to use dimensional analysis. **Dimensional analysis** or **unit analysis** is the process of performing operations with units.

As you plan your solution method, think about

- what units were given,
- what units you need for the solution, and
- the step(s) you need to take to convert your units from what you are given to what you will need for the solution.

Example 5 Multiply by a Conversion Factor

POOLS **Mark is purchasing an above-ground swimming pool. The salesperson says that the pool will hold 97,285 liters of water. If 1 gallon = 3.785 liters, determine approximately how many gallons of water Mark's pool will hold.**

Two ratios can be used to compare liters and gallons, $\dfrac{1 \text{ gallon}}{3.785 \text{ liters}}$ and $\dfrac{3.785 \text{ liters}}{1 \text{ gallon}}$. To convert from liters to gallons, multiply by $\dfrac{1 \text{ gallon}}{3.785 \text{ liters}}$.

Number of liters the pool will hold × gallons to liters

$$\underline{\hspace{1cm}} \text{ liters} \times \dfrac{1 \text{ gallon}}{3.785 \text{ liters}}$$

$$97,285 \text{ liters} \times \dfrac{1 \text{ gallon}}{3.785 \text{ liters}} \approx \underline{\hspace{1cm}} \text{ gallons}$$

Mark's pool will hold approximately _____ gallons of water.

Study Tip

Precision
Notice that the question asks for an estimate, not an exact answer.

Copyright © McGraw-Hill Education

Check

COOKING For the chefs to prepare the dishes for the next hour, Adelina needs to provide them with 96 cloves of garlic. However, she can only purchase bags of whole heads of garlic. If each bag of garlic contains 3 heads and each head has about 8 cloves, then how many bags of garlic should she purchase?

Part A

What assumption must Adelina make when calculating how many bags of garlic to buy? _____

A. Each head of garlic has exactly 8 cloves.

B. The chefs need 96 heads of garlic.

C. The chefs need 96 cloves of garlic.

D. Each bag has 3 heads of garlic.

Part B

How many bags of garlic should Adelina purchase? _____

A. 4   B. 12

C. 32   D. 96

🌐 **Example 6** Use Dimensional Analysis to Convert Units

RECIPE
BREAD PUDDING
1.5 tbsp. butter
12 oz. bread
20 fl. oz. milk
3 eggs
1 tbsp. vanilla
2 cups sugar

COOKING A recipe calls for 20 fluid ounces of milk. If Nita buys a half gallon of milk, how many batches of that recipe can she make?
(Hint: 8 fluid ounces = 1 cup)

First, convert gallons to fluid ounces.

total amount of milk	× gallons to quarts	× quarts to pints	× pint to cups	× cups to fluid ounces

$$0.5 \text{ gallon} \times \frac{4 \text{ quarts}}{1 \text{ gallon}} \times \frac{2 \text{ pints}}{1 \text{ quart}} \times \frac{2 \text{ cups}}{1 \text{ pint}} \times \frac{8 \text{ fl. oz.}}{1 \text{ cup}} = \underline{\quad} \text{ fl. oz.}$$

Use the following conversion factors to change gallons to ounces.

1 gallon = 4 quarts   1 quart = 2 pints
1 pint = 2 cups   1 cup = 8 fluid ounces

Nita has 64 ounces of milk. Each batch calls for 20 fluid ounces, so to find the number of batches she can make, divide by 20 fluid ounces.

$$64 \text{ fl. oz.} \times \frac{1 \text{ batch}}{20 \text{ fl. oz.}} = \underline{\qquad\qquad}$$

Nita has enough milk to make _____ batches with some milk left over.

🔎 **Go Online** You can complete an Extra Example online.

Avoid a Common Error

Remember that a unit will only cancel when you divide it by itself. What error does this solution make?

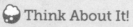

$$97{,}285 \text{ liters} \times \frac{3.785 \text{ liters}}{1 \text{ gallon}}$$

$$\approx 368{,}224 \text{ gallons}$$

🧁 Think About It!

When you are converting, how do you know which unit goes in the numerator and which unit goes in the denominator?

Check

AGRICULTURE On average, a dairy cow produces 832 ounces of milk a day. About how many gallons of milk does a dairy cow produce each year? (Hint: 1 cup = 8 ounces, 1 quart = 4 cups, and 1 gallon = 4 quarts) _____

A. 6.5 gallons per year

B. 2372.5 gallons per year

C. 4357.2 gallons per year

D. 303,680 gallons per year

🌐 Example 7 Use Dimensional Analysis to Convert Rates

SPEED In a novel, the main character, Aiko, can run long distances at 16.5 *kanejaku* per second. Carla knows that the Olympic record for running a marathon distance of 26.2 miles is about 126.5 minutes. She wonders if Aiko could beat that record. If 1 *kanejaku* = $\frac{10}{33}$ meters, find how far Aiko could run, in miles, in that amount of time. (Hint: 1 mile ≈ 1609.344 meters)

Use the formula $d = rt$ that relates distance d, rate r, and time t to find the distance Aiko could run in 126.5 minutes.

$$d = rt \qquad \text{Distance equation}$$

$$d = \left(\frac{16.5 \text{ kanejaku}}{1 \text{ second}}\right) \cdot t \qquad \text{Substitute Aiko's rate.}$$

In order to compare Aiko to the Olympic runner, convert Aiko's rate in *kanejaku* to miles per minute.

Step 1 Convert distance.

You want distance in miles, but Aiko's distance is in *kanejaku*. Use the given conversion rates that relate to distance to convert Aiko's rate in *kanejaku* per second to _____.

$$\frac{16.5 \text{ kanejaku}}{1 \text{ second}} \times \frac{\frac{10}{33} \text{ meters}}{1 \text{ kanejaku}} \times \frac{1 \text{ mile}}{1609.344 \text{ meters}} = \frac{5 \text{ miles}}{1609.344 \text{ seconds}}$$

Step 2 Convert time.

You want time in _____, but Aiko's time is in _____. Use the resulting rate from step 2 to convert seconds to minutes.

$$\frac{5 \text{ miles}}{1609.344 \text{ seconds}} \times \frac{60 \text{ seconds}}{1 \text{ minute}} = \frac{300 \text{ miles}}{1609.344 \text{ minutes}}$$

(continued on the next page)

🔆 **Go Online** You can complete an Extra Example online.

Problem-Solving Tip

Make a Plan Before you solve a problem, think about what the question is asking and what information will apply to the solution.

Watch Out!

Canceling Units Do not forget to cancel your units as you multiply so that you can see what units are left. The units that are left are the units of your final answer.

Step 3 Substitute.

Substitute Aiko's rate in miles per minute and the given time into the formula and simplify.

$$d = \frac{300 \text{ miles}}{1609.344 \text{ minutes}} = \underline{\qquad} \text{ minutes} \approx 23.6 \text{ miles}$$

Aiko would run approximately 23.6 miles in the time it took the Olympic runner to complete 26.2 miles. So, Aiko would not beat the Olympic marathon record time.

Go Online
An alternate method is available for this example.

Pause and Reflect

Did you struggle with anything in this lesson? If so, how did you deal with it?

Record your observations here.

Go Online You can complete an Extra Example online.

Practice

Go Online You can complete your homework online.

Examples 1 and 2

Solve each equation or formula for the variable indicated.

1. $x - 2y = 1$, for y

2. $d + 3n = 1$, for n

3. $7f + g = 5$, for f

4. $3c - 8d = 12$, for c

5. $7t = x$, for t

6. $r = wp$, for p

7. $q - r = r$, for r

8. $4m - t = m$, for m

9. $7a - b = 15a$, for a

10. $-5c + d = 2c$, for c

Solve each equation or formula for the variable indicated.

11. $u = vw + z$, for v

12. $x = b - cd$, for c

13. $fg - 9h = 10j$, for g

14. $10m - p = -n$, for m

15. $r = \frac{2}{3}t + v$, for t

16. $\frac{5}{9}v + w = z$, for v

17. $\frac{10ac - x}{11} = -3$, for a

18. $\frac{df + 10}{6} = g$, for f

Example 3

19. RECTANGLES The formula $P = 2\ell + 2w$ represents the perimeter of a rectangle. In this formula, ℓ is the length of the rectangle and w is the width.

    a. Solve the formula for ℓ.

    b. Find the length when the width is 4 meters and the perimeter is 36 meters.

20. BASEBALL The formula $a = \frac{h}{b}$ can be used to find the batting average a of a batter who has h hits in b times at bat.

    a. Solve the formula for b.

    b. If a batter has a batting average of 0.325 and has 39 hits, how many times has the player been at bat?

21. SHOPPING Thomas went to the store to buy videogames for $13.50 each and controllers. The total amount Thomas spent can be represented by $c = 13.50g + p$, where c is the total cost, g is the number of games he bought, and p is the cost of the controllers. The controllers cost $55 and Thomas spent $136 total.

    a. Solve the equation for g.

    b. Find how many games Thomas bought.

22. GEOMETRY The volume of a box V is given by the formula $V = \ell wh$, where ℓ is the length, w is the width, and h is the height.

    a. Solve the formula for h.

    b. What is the height of a box with a volume of 50 cubic meters, length of 10 meters, and width of 2 meters?

Example 4

23. COFFEE SHOP Consuelo is buying flavored coffee and plain coffee. The total amount of money she spends on coffee is $T = 5.50p + 7f$, where p represents the cost of a package of plain coffee and f represents the cost of a package of flavored coffee.

    a. If Consuelo has \$40 to spend on coffee, describe the constraints on the formula.

    b. Solve for f.

    c. If Consuelo needs to buy 3 packages of plain coffee, what is the maximum number of packages of flavored coffee she can buy?

24. SHOPPING Kimberly is ordering bath towels and washcloths for the inn where she works. The total cost of the order is $T = 6b + 2w$, where T is the total cost, b is the number of bath towels, and w is the number of washcloths.

    a. Her budget is \$85. Describe the constraints.

    b. Solve for b.

    c. She needs to order at least 20 washcloths. How many bath towels can she order and stay under budget?

Examples 5–7

25. ENVIRONMENT The United States released 5.877 billion metric tons of carbon dioxide into the environment through the burning of fossil fuels in a recent year. If 1 trillion pounds = 0.4536 billion metric tons, how many trillion pounds of carbon dioxide did the United States release in that year?

26. EUROS Trent purchases 44 euros worth of souvenirs while on vacation in France. If \$1 U.S. = 0.678 euros, find the cost of the souvenirs in United States dollars.

27. LENGTH A pencil is 13.5 centimeters long. If 1 centimeter = 0.39 inch, what is the length of the pencil in feet, to the nearest hundredth?

28. TRACK If a track is 400 meters around, how many laps around the track would it take to run 3.1 miles? Round to the nearest tenth. (*Hint:* 1 foot = 0.3048 meter)

29. BIKING Imelda rode her bicycle 39 kilometers. If 1 meter = 1.094 yards, find the distance Imelda rode her bicycle to the nearest mile. (*Hint:* 1 mi = 1760 yd)

30. MANUFACTURING Aluminum, Inc. produces cans at a rate of 0.04 per hundredth of a second. How many cans can be produced in a 7-hour day?

31. WATER USAGE Each minute, 8.8 quarts of water flow from a shower. If the average person spends 8.2 minutes in the shower, how many gallons of water will the average person have used after taking five showers?

32. TRAVEL The swim team is going to finals. If the meet is in 85 days, determine how many seconds there are until the meet.

33. PRECISION The chemistry teacher set out a 5-pound jar of salt at the beginning of the day. If each student needs 27.6 grams of salt for an experiment, how many students can perform the experiment before the jar is empty? (*Hint:* 1 lb = 454 g)

Mixed Exercises

Solve each equation for the variable indicated.

34. $rt - 2n = y$, for t

35. $bc + 3g = 2k$, for c

36. $kn + 4f = 9v$, for n

37. $8c + 6j = 5p$, for c

38. $\frac{x - c}{2} = d$, for x

39. $\frac{x - c}{2} = d$, for c

40. $-14n + q = rt - 4n$, for n

41. $18t + 11v = w - 13t$, for t

42. $ax + z = aw - y$, for a

43. $10c - f = -13 + cd$, for c

44. STRUCTURE Jethro used dimensional analysis to convert from one rate of speed to another. Two of the conversion factors he used are $\frac{5280 \text{ ft}}{1 \text{ mi}}$ and $\frac{1 \text{ hr}}{60 \text{ min}}$. What could be the units of the initial rate of speed and final rate of speed? Justify your answer.

45. REASONING The formula $A = P(1 + r)$ represents the amount of money A in an account after 1 year, where P is the amount initially deposited and r is the interest rate. Note that the interest rate is written as a decimal. Dennis deposits $2150 into a savings account. After 1 year, he has $2182.25 in his account. Solve the equation for r, and determine the interest rate. Show your work.

46. REASONING The regular octagon shown is divided into 8 congruent triangles. Each triangle has an area of 21.7 square centimeters. The perimeter of the octagon is 48 centimeters.

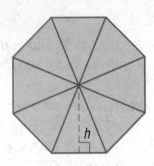

  a. What is the length of each side of the octagon?

  b. Solve the formula for the area of a triangle for h.

  c. What is the height of each triangle? Round to the nearest tenth.

47. Consider the equation $\frac{ry + z}{m} - t = x$.

  a. Solve the equation for y.

  b. Would there be any restrictions on the value of each variable? If so, explain the restrictions.

🧁 **Higher Order Thinking Skills**

48. CREATE Think about the area formula of some geometric figures that involve a fraction.

  a. Write a formula for A, the area of a geometric figure that includes a fraction.

  b. Solve the formula for a variable other than A.

49. FIND THE ERROR The formula represents the relationship between temperatures in degrees Fahrenheit F and degrees Celsius C. Sasha solves the formula for C. Her solution is shown at the right. Is Sasha's solution correct? Explain your reasoning.

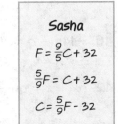

Sasha

$F = \frac{9}{5}C + 32$

$\frac{5}{9}F = C + 32$

$C = \frac{5}{9}F - 32$

50. PERSEVERE The formula $A = P(1 + r)^t$ represents the amount of money A in an account after t years, where P is the amount initially deposited and r is the interest rate. Patricia currently has $1839.79 in an account that has an interest rate of 2.5%. She opened the account 8 years ago and has made no additional deposits since then.

  a. Solve the formula for P and find the amount of Patricia's initial deposit.

  b. Nia says that for the formula in **part a**, A is always greater than P when r is positive and t is a positive integer. Do you agree? Why or why not?

Essential Question

How can writing and solving equations help you solve problems in the real world?

Module Summary

Lessons 2-1 and 2-2

One-Step Equations

- To write an equation, first identify each unknown and assign a variable to it. Then identify the givens and their relationships. Finally, write the sentence as an equation.

- Solving an equation is the process of finding all values of the variable that make the equation true.

- If a number is added to or subtracted from each side of a true equation, the resulting equivalent equation is also true.

- If an equation is true and each side is multiplied or divided by the same nonzero number, the resulting equation is equivalent.

Lessons 2-3 and 2-4

Multi-Step Equations

- To solve a multi-step equation, you can undo each operation using properties of equality. Working backward in the order of operations makes this process simpler. Each step in this process results in equivalent equations.

- To solve an equation with the variable on each side, write an equivalent expression with all of the variable terms on one side and the constants on the other side.

- When a grouping symbol appears in an equation, it must first be removed before continuing to solve the equation.

Lessons 2-5 and 2-6

Absolute Value Equations and Proportions

- When solving equations that involve absolute values, there are two cases to consider.

  Case 1: The expression inside the absolute value symbol is positive or zero.

  Case 2: The expression inside the absolute value symbol is negative.

- A proportion is a statement that two ratios are equivalent.

Lesson 2-7

Formulas

- A formula is an equation that expresses a relationship between certain quantities.

- A formula or equation that involves several variables is called a literal equation. To solve a literal equation, solve for a specific variable.

- Dimensional analysis is the process of performing operations with units.

Study Organizer

 Foldables

Use your Foldable to review the module. Working with a partner can be helpful. Ask for clarification of concepts as needed.

Test Practice

1. MULTIPLE CHOICE Which equation represents this sentence? (Lesson 2-1)

The sum of 5 times a number m and 12 is equal to 27.

(A) $5m + 12 = 27$

(B) $5(m + 12) = 27$

(C) $5 + 12m = 27$

(D) $5(12m) = 27$

2. OPEN RESPONSE A concert venue surveyed 680 concert attendees about the concession stand. Of those that visited the concession stand, 527 said the concession stand prices are excessive, and the remaining 44 did not. Write an equation to find the number of attendees *a* who did not visit the concession stand. (Lesson 2-1)

3. MULTIPLE CHOICE Write a verbal sentence for the algebraic equation $5x^2 + 2 = 22$. (Lesson 2-1)

(A) 5 times *x* squared less 2 is 22.

(B) Five plus *x* squared plus 2 is 22.

(C) The product of 5 times *x* squared and 2 is 22.

(D) Five times *x* squared plus 2 is 22.

4. MULTIPLE CHOICE Solve $n - 8 = 5$. (Lesson 2-2)

(A) $n = -13$

(B) $n = -3$

(C) $n = 3$

(D) $n = 13$

5. MULTIPLE CHOICE Solve $\frac{1}{5}t = 10$ for *t*. (Lesson 2-2)

(A) 2

(B) 5

(C) 15

(D) 50

6. OPEN RESPONSE Solve $z + 12 = -3$ for *z*. Explain. (Lesson 2-2)

7. MULTIPLE CHOICE If $3x - 6 = 42$, what is the value of *x*? (Lesson 2-3)

(A) 8

(B) 12

(C) 16

(D) 20

8. OPEN RESPONSE Solve $8 = 11 - 3v$. (Lesson 2-3)

$v = \boxed{}$

9. MULTI-SELECT Jaime bought a notebook and a box of pencils for $5.00. The notebook cost $3.00 and there are 10 pencils in a box. The equation $3 + 10p = 5$ can be used to find the cost of one pencil. Select all equations that are equivalent. (Lesson 2-3)

(A) $10p = 8$

(B) $10p = 2$

(C) $p = 0.80$

(D) $p = 0.20$

(E) $10p = 5$

10. MULTIPLE CHOICE Solve the equation
$3x + 1 = 4x - 8$ for x. (Lesson 2-4)

Ⓐ $x = -9$

Ⓑ $x = -1$

Ⓒ $x = 1$

Ⓓ $x = 9$

11. MULTIPLE CHOICE Solve the equation
$-2(x + 4) + 3x = x - 8$. (Lesson 2-4)

Ⓐ $x = 2$

Ⓑ $x = 4$

Ⓒ all real numbers

Ⓓ no solution

12. MULTIPLE CHOICE Solve the equation
$5(2y + 1) = 4y + 10$. (Lesson 2-4)

Ⓐ $y = \frac{5}{6}$

Ⓑ $y = \frac{3}{2}$

Ⓒ $y = \frac{15}{14}$

Ⓓ $y = \frac{5}{2}$

13. OPEN RESPONSE A park has a ginkgo tree, a dogwood tree, and 2 blue spruce trees. The blue spruce trees are 8 years old. The ginkgo tree is 2 years less than three times the age of the dogwood tree. The ginkgo tree is also half the sum of the ages of the dogwood tree and both of the blue spruce trees. Write and solve an equation to find the ages of the ginkgo and dogwood trees.
(Lesson 2-4)

14. MULTI-SELECT Select all the values of x that are solutions of $|3x - 6| = 12$. (Lesson 2-5)

Ⓐ -18

Ⓑ -6

Ⓒ -2

Ⓓ 2

Ⓔ 6

Ⓕ 18

15. MULTIPLE CHOICE A thermometer is accurate to $\pm 2°F$. Which absolute value equation can be used to find the greatest and least possible temperatures if the thermometer reading is $17°F$? (Lesson 2-5)

Ⓐ $|t - 17| = 2$

Ⓑ $|t + 17| = 2$

Ⓒ $|t - 2| = 17$

Ⓓ $|t + 2| = 17$

16. OPEN RESPONSE Solve the absolute value equation $-5|x + 1| + 2 = 12$. If there is no solution, state no solution. (Lesson 2-5)

17. OPEN RESPONSE Solve $\frac{b}{12} = \frac{10}{15}$. (Lesson 2-6)

18. MULTIPLE CHOICE Solve $\dfrac{3}{x+4} = \dfrac{2}{x-4}$.

(Lesson 2-6)

(A) $x = -4$

(B) $x = 4$

(C) $x = 8$

(D) $x = 20$

19. MULTIPLE CHOICE A biologist estimated that 5% of the seagulls in a flock have been banded. There were 22 seagulls that have been banded. Which equation and solution represent the approximate number of seagulls, g, that were in the flock? (Lesson 2-6)

(A) $0.005g = 22; g = 4400$

(B) $0.05g = 22; g = 440$

(C) $22g = 500; g = 22$

(D) $\dfrac{5}{g} = 22; g = 227$

20. MULTIPLE CHOICE On a map of Texas, the distance between Dallas and Houston is 4.8 inches. If 1 inch = 50 miles, what is the distance, in miles, between the two cities? (Lesson 2-6)

(A) 96 miles

(B) 104 miles

(C) 240 miles

(D) 250 miles

21. MULTIPLE CHOICE If 1 foot ≈ 0.305 meter, approximately how many feet are in 8 meters? (Lesson 2-6)

(A) 2.44

(B) 2.62

(C) 24.4

(D) 26.2

22. OPEN RESPONSE Solve the formula for the circumference of a circle, $C = 2\pi r$, for r.

(Lesson 2-7)

23. OPEN RESPONSE The volume of a right pyramid is given by the formula $V = \dfrac{1}{3}Bh$, where B is the area of the base and h is the height. (Lesson 2-7)

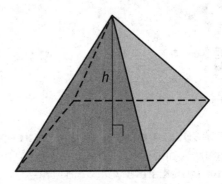

Part A Solve the formula for h.

Part B Find the height, in inches, of a right pyramid with a volume of 900 cubic inches and a base area of 225 square inches.

Relations and Functions

e Essential Question

Why are representations of relations and functions useful?

What Will You Learn?

Place a check mark (✓) in each row that corresponds with how much you already know about each topic **before** starting this module.

KEY

👎 — I don't know.   👈 — I've heard of it.   👍 — I know it!

	Before			After		
	👎	👈	👍	👎	👈	👍
represent relations using ordered pairs, tables, graphs, and mappings						
analyze graphs of relations						
choose and interpret the scale on a coordinate graph						
determine whether relations are functions						
use function notation						
determine whether a graph is discrete, continuous, or neither						
determine whether a function is linear or nonlinear						
write linear functions in standard form						
find x- and y-intercepts of graphs						
interpret intercepts of graphs of functions						
determine whether a graph has line symmetry						
identify where a graph is increasing and where it is decreasing						
find extrema of a function						
describe the end behavior of a function						
sketch graphs of functions						
solve equations by graphing						

📙 **Foldables** Make this Foldable to help you organize your notes about expressions. Begin with one sheet of 11″ × 17″ paper.

1. **Fold** the short sides to meet in the middle.

2. **Fold** the booklet in thirds lengthwise.

3. **Open and cut** the booklet in thirds lengthwise.

4. **Label** the tabs as shown.

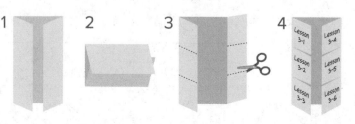

What Vocabulary Will You Learn?

Check the box next to each vocabulary term that you may already know.

- ☐ continuous function
- ☐ decreasing
- ☐ dependent variable
- ☐ discrete function
- ☐ domain
- ☐ end behavior
- ☐ extrema
- ☐ function
- ☐ function notation

- ☐ increasing
- ☐ independent variable
- ☐ line symmetry
- ☐ linear equation
- ☐ linear function
- ☐ mapping
- ☐ negative
- ☐ nonlinear function
- ☐ positive

- ☐ range
- ☐ relation
- ☐ relative maximum
- ☐ relative minimum
- ☐ root
- ☐ scale
- ☐ *x*-intercept
- ☐ *y*-intercept
- ☐ zero

Are You Ready?

Complete the Quick Review to see if you are ready to start this module.
Then complete the Quick Check.

Quick Review

Example 1

Graph and label the point *A*(3, 5) on the coordinate plane.

Start at the origin. Since the *x*-coordinate is positive, move 3 units to the right. Then move 5 units up since the *y*-coordinate is positive. Draw a dot and label it *A*.

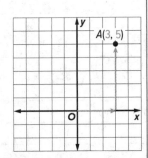

Example 2

Evaluate $5x + 13$ for $x = 4$.

Substitute the known value for *x*. Then follow the order of operations.

$5x + 13$	Original expression
$= 5(4) + 13$	Substitute 4 for *x*.
$= 20 + 13$	Multiply.
$= 33$	Add.

Quick Check

Graph and label each point on the coordinate plane.

1. *B*(4, 2)
2. *C*(0, 3)
3. *G*(3, 1)
4. *H*(2, 0)

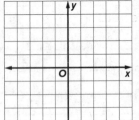

Evaluate each expression for the given value.

5. $3x - 1$ for $x = 7$
6. $\frac{1}{2}x$ for $x = 12$
7. $5x + 2$ for $x = -1$
8. $2x - 9$ for $x = -3$

How Did You Do?

Which exercises did you answer correctly in the Quick Check? Shade those exercise numbers below.

       ⑧

Representing Relations

Learn Relations

You can use math to represent the relationship between two sets of numbers. For example, suppose you recorded the number of minutes you spent driving for your driver's training course each day for a week. You may use the set {1, 2, 3, 4, 5, 6, 7} to represent the days and the set {30, 45, 20, 40, 45, 90, 60} to represent the minutes. The relationship between the sets pairs each day with the time driven that day. This pairing of the numbers is called a **relation**. The set of days is the **domain** of the relation and the set of times is the **range**.

A relation is a set of ordered pairs. The set of the first numbers of the ordered pairs in a relation is called the domain. The set of second numbers of the ordered pairs in a relation is called the range.

A relation can be represented in multiple ways. A **mapping** illustrates the relationship between the domain and range by showing how each element of the domain is paired with an element in the range. An equation shows the relationship between the domain and range of a relation where substituting each value in the domain for x results in the corresponding y-value of the range. The x-values in the domain and the resulting y-values can be written as ordered pairs and are called solutions of the equation because they make the equation true. Below is a mapping, table, graph, and equation of the relation {(3, 5), (−4, −2), (0, 2)}.

Today's Goals
• Represent relations.
• Interpret graphs of relations.
• Choose and interpret appropriate scales for the axes and origins of graphs.

Today's Vocabulary
relation
domain
range
mapping
independent variable
dependent variable
scale

Think About It!

Compare the table and mapping of the relation. What conclusions can you draw about the x- and y-coordinates and the domain and range?

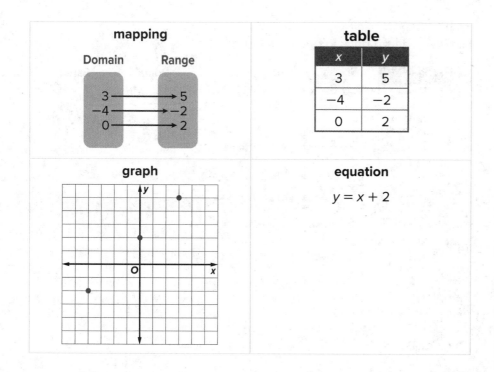

mapping

Domain   Range

3 → 5
−4 → −2
0 → 2

table

x	y
3	5
−4	−2
0	2

graph

equation

$$y = x + 2$$

Example 1 Representations of a Relation

Use {(2, 0), (0, 4), (3, −5), (−3, −5)}.

Part A Express the relation as a table, a graph, and a mapping.

table

graph

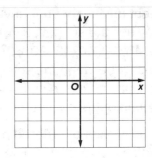

mapping

Part B Determine the domain and the range of the relation.

The domain of the relation is {−3, 0, 2, 3}.

The range of the relation is {−5, 0, 4}.

Check

List the ordered pairs in each relation.

x	y
−1	−5
4	7
−2	3

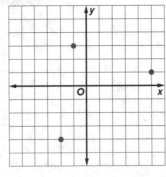

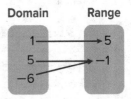

Table: _____

Graph: _____

Mapping: _____

🐺 **Go Online** You can complete an Extra Example online.

Learn Analyzing Graphs of Relations

Graphing the total time driven during your driving course can help you visualize the progress.

A relation can be graphed without a scale on either axis to show the relationship between the independent and dependent variables. These graphs can be interpreted by analyzing their shapes.

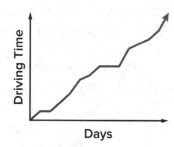

The values in the domain correspond to the independent variable in a relation. The **independent variable**, usually x, has a value that is subject to choice. In the graph above, the independent variable is the days.

The values in the range correspond to the dependent variable of the relation. The **dependent variable** is the variable in a relation, usually y, with values that depend on x. In the graph above, the dependent variable is the driving time.

Example 2 Analyze Graphs

TEXTING **The graph represents the number of text messages sent by Nora throughout the day.**

Part A Identify the independent and dependent variables of the relation. independent variable: _____ dependent variable: _____

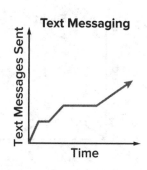

Think About It!

What can you conclude from the graph about the rates at which Nora sent text messages throughout the day?

Part B Describe what happens in the graph.

As you move from left to right along the graph, time _____ and the number of text messages sent _____ until the graph becomes a _____.

The horizontal line means that time is _____, but the number of text messages sent _____. During this time, Nora _____ sending text messages.

Then she continued to send text messages until she _____ again for a period of time.

Finally, Nora began _____ text messages again.

Check

Identify the independent and dependent variables of each relation.

a. The average price of a ticket to an amusement park has steadily increased over time.

The average price of a ticket is the _____ variable.
Time is the _____ variable.

b. The air pressure inside a soccer ball decreases with time.

Time is the _____ variable.
Air pressure is the _____ variable.

 Go Online You can complete an Extra Example online.

Check

AIRPLANES After an airplane takes off, its altitude increases rapidly. Once it reaches the desired altitude, it continues to fly at that level for a short period of time. Then the plane's altitude fluctuates slightly due to turbulence. Finally, the altitude decreases quickly as the plane lands. Which graph best represents this situation? _____

A.
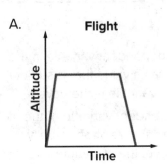

B.
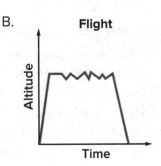

C.
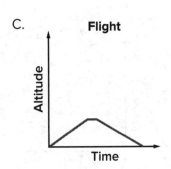

D.
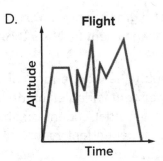

TRAFFIC Kane notices that traffic congestion on his street on weekday mornings depends on the time of day. He draws a graph to represent the relationship between the time of day and the volume of traffic congestion on his street on a weekday morning. Analyze the orange segment. Select the statement that describes what is happening during this time. _____

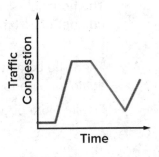

A. Traffic congestion is steadily increasing at a very fast rate.

B. Traffic congestion is constant and very light.

C. Traffic congestion is constant and very heavy.

D. Traffic congestion is decreasing at a steady rate.

Go Online You can complete an Extra Example online.

Explore Choosing Scales

Online Activity Use a real-world situation to complete the Explore.

> **INQUIRY** How can you tell if an appropriate scale is being used to represent a relationship?

Learn The Coordinate System

When graphing on the coordinate system, the **scale** of a graph refers to the distance, or interval, between tick marks on the x- and y-axes. For example, if one tick mark represents 5 units, then the scale of the graph is 5. Each axis may have a different scale.

A scale of 1 tick mark = 1 unit is frequently used in mathematics. However, using a different scale may make it easier to graph a given set of ordered pairs.

Example 3 Use Appropriate Scales

Graph (5, 80), (−36, 48), (25, −91), (38, 95), (−10, −50), and (1, 22).

Step 1 The x-coordinates are between _____ and _____.
The y-coordinates are between _____ and _____.

Step 2 The x-axis should include values from about _____ to _____.
The y-axis should include values from about _____ to _____.

A scale of _____ on the x-axis is appropriate.
A scale of _____ on the y-axis is appropriate.

Step 3 Graph.

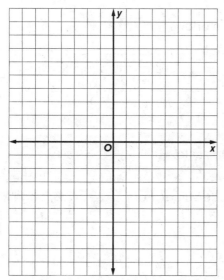

Go Online You can complete an Extra Example online.

Talk About It!
Describe a real-world situation in which it might be easier to count by a value other than 1.

Watch Out!
Large Scales Be careful not to choose scales that are too large. If a scale is too large, it can be difficult to accurately graph points.

Think About It!
Gwen used a scale of 4 on the x-axis. Is her scale appropriate? Explain your reasoning.

Study Tip

Notation Scales of graphs are sometimes written as [−40, 40] scl: 5 or −40 to 40; scale: 5, where −40 and 40 are the minimum and maximum values and 5 is the scale.

Check

When graphing (16, 32), (−10, 11), (4, −27), and (−7, −5), select the most appropriate scale for

a. the *x*-axis _____

A. −20 to 20; scale: 1

B. −12 to 18; scale: 2

C. −12 to 18; scale: 6

D. −20 to 20; scale: 10

b. the *y*-axis _____

A. −30 to 35; scale: 1

B. −30 to 30; scale: 5

C. −30 to 35; scale: 5

D. −30 to 40; scale: 10

🌐 **Example 4** Choose an Appropriate Origin

WEATHER **The table shows the total snowfall in January for Boston.**

Year	Total Snowfall (inches)
2005	43.30
2006	8.10
2007	1.00
2008	8.30
2009	23.70
2010	13.20
2011	38.30
2012	6.80
2013	5.00
2014	21.80
2015	34.30

Study Tip

Appropriate Origins When choosing an appropriate origin, you are still using the point (0, 0) as the origin, but are changing what it represents.

Part A Choose an appropriate origin.

Let the *x*-axis represent the _____ and the *y*-axis represent _____ . Then the origin (0, 0) represents _____ and _____.

Part B Choose an appropriate scale.

The total snowfall is between _____ and _____ inches, so the *y*-axis should include values from _____ to _____ and have a scale of _____ inches.

🔁 **Go Online** You can complete an Extra Example online.

Part C Graph the data points on the coordinate plane.

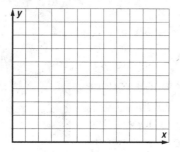

🌐 **Example 5** Interpret Scales and Origins

SOCIAL MEDIA **The average number of posts per day on a social media site each year is given in the table. Interpret the meaning of the axes, scale, and origin of the corresponding graph of the data.**

Year	Average Posts Per Day (millions)
2008	15
2009	20
2010	35
2011	50
2012	100
2013	200
2014	340
2015	500
2016	560
2017	625

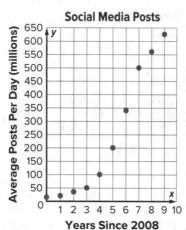

x- and y-axes

The x-axis represents the _____ , because it is not appropriate to start at the year 0.

The y-axis represents _____ on the social media site.

Scale

The x-axis has a scale of 1 mark = _____.
The y-axis has a scale of 1 mark = _____ .

Origin

The origin (0, 0) represents the year _____ and ____ posts per day.

Use a Source

Find data about the number of participants in a sport or other activity of interest to you in recent years. If a graph is provided, interpret the scales of the axes. If no graph is provided, create one with appropriate scales.

Check

MONEY The United States Mint is responsible for producing and distributing circulating coins that are used by people every day to buy and sell goods. Facilities in Denver and Philadelphia produce billions of coins each year. The table and graph show how many circulating coins were produced each year at these facilities from 2001 to 2011.

Year	Circulating Coins Produced (billions)
2001	19.4
2002	14.4
2003	12
2004	13.2
2005	15.3
2006	15.5
2007	14.4
2008	10.1
2009	3.5
2010	6.4
2011	8.2

Circulating Coins Production

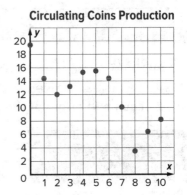

Part A

Interpret the meaning of the x- and y-axes in the context of the situation.

The x-axis represents _____ and has a scale of 1 mark = _____.

The y-axis represents _____ and has a scale of 1 mark = _____.

Part B

Interpret the meaning of the origin in the context of the situation.

The origin (0, 0) represents the year _____ and ____ coins produced.

⚫ **Go Online** You can complete an Extra Example online.

Practice

Go Online You can complete your homework online.

Example 1

Express each relation as a table, a graph, and a mapping. Then determine the domain and range.

1. {(−1, −1), (1, 1), (2, 1), (3, 2)}

2. {(0, 4), (− 4, − 4), (−2, 3), (4, 0)}

3. {(3, −2), (1, 0), (−2, 4), (3, 1)}

Example 2

4. PAYCHECK The graph represents the amount of Seth's paycheck for different numbers of hours he works.

  a. Identify the independent and dependent variables of the relation.

  b. Describe what happens in the graph.

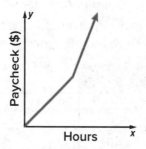

5. DEMAND The graph represents the price of an item and the number of items purchased.

  a. Identify the independent and dependent variables of the relation.

  b. Describe what happens in the graph.

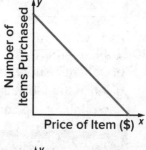

6. AIRPLANES The graph represents the number of hours of a flight and the distance an airplane is from the ground.

  a. Identify the independent and dependent variables of the relation.

  b. Describe what happens in the graph.

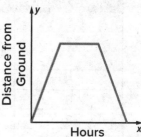

Examples 3 and 4

7. HEALTH The American Heart Association recommends that your target heart rate during exercise should be between 50% and 75% of your maximum heart rate. Use the data in the table below to graph the approximate maximum heart rates for people of given ages.
Source: American Heart Association

Age (years)	20	25	30	35	40
Maximum Heart Rate (beats per minute)	200	195	190	185	180

8. USE TOOLS The following ordered pairs give the length in feet and the weight in pounds of five snakes at the reptile house at a zoo. Graph the data. {(5.5, 4.5), (3, 0.5), (3, 2), (8, 4.5), (2, 0.5)}

Example 5

9. ELEVATOR The height of an elevator above the ground is given in the table. Interpret the meaning of the axes, scale, and origin of the corresponding graph of the data.

Time (s)	Height (ft)
0	0
1	20
2	40
3	60
4	80

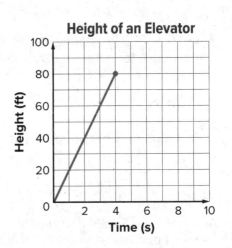

Height of an Elevator

10. SUPERMARKET The number of items that eight customers bought at a supermarket and the total cost of the items is given in the table. Interpret the meaning of the axes, scale, and origin of the corresponding graph of the data.

Number of Items	Total Cost ($)
2	2
3	6
5	8
5	10
6	16
8	12
8	18
9	14

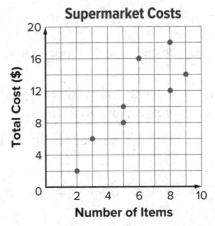

Supermarket Costs

Mixed Exercises

For Exercises 11–14, express the relation in each table, graph, or mapping as a set of ordered pairs.

11.

x	y
1	7
3	45
5	11
13	15

12.

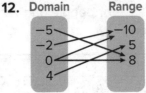

Domain   Range
−5   −10
−2   5
0   8
4

13.

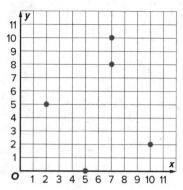

14.

x	y
4	2
7	−7
15	9
12	0

15. NATURE Maple syrup is made by collecting sap from sugar maple trees and boiling it down to remove excess water. The graph shows the number of gallons of tree sap required to make different quantities of maple syrup. Express the relation as a set of ordered pairs. Interpret the meaning of the axes, scale, and origin of the corresponding graph of the data.
Source: Vermont Maple Sugar Makers' Association

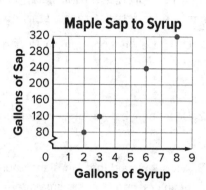

16. TRAVEL Omari drives a car that gets 18 miles per gallon of gasoline. The car's gasoline tank holds 15 gallons. The distance Omari drives before refueling is a function of the number of gallons of gasoline in the tank. Identify a reasonable domain for this situation.

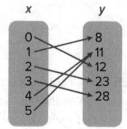

17. DATA COLLECTION Rafaella collected data to determine the number of books her schoolmates were bringing home each evening. Her data is shown in the mapping. She let x be the number of textbooks brought home after school and y be the number of students with x textbooks.

a. Express the relation as a set of ordered pairs.

b. What is the domain of the relation?

c. What is the range of the relation?

18. COOKIES Identify the graph that best represents the relationship between the number of cookies and the equivalent number of dozens.

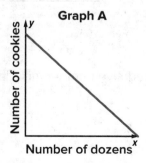

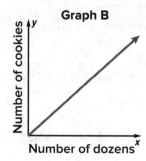

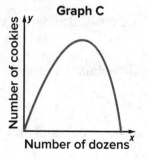

19. MOWING Cordell is mowing his front lawn. His mailbox is on the edge of the lawn. Draw a reasonable graph that shows the distance Cordell is from the mailbox as he mows. Let the horizontal axis show the time and the vertical axis show the distance from the mailbox.

20. STRUCTURE Express the relation in the mapping as a set of ordered pairs.

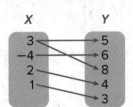

Tim and Lauren use their cars to deliver pizzas. The graph represents their distance from the pizzeria starting at 6 P.M. Use the graph for Exercises 21–24.

21. Describe what happens in Tim's graph.

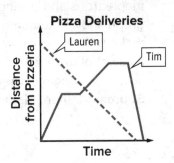

22. Describe what happens in Lauren's graph.

23. A student said that Tim's and Lauren's graphs intersect, so their cars must have crashed at some time after 6 p.m. Do you agree or disagree? Explain.

24. After 6 p.m., which delivery person was the first to return to the pizzeria? How do you know?

25. ANALYZE Cameron said that for any relation, the number of elements in the domain must be greater than or equal to the number of elements in the range. Do you agree? If so, explain why. If not, give a counterexample.

26. CREATE Think of a situation that could be modeled by this graph. Then label the axes of the graph and write several sentences describing the situation.

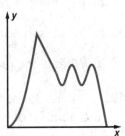

27. CREATE Use the set {−1, 0, 1, 2} as a domain and the set {−3, −1, 4, 5} as a range.
   a. Create a relation. Express the relation as a set of ordered pairs.

   b. Express the relation you created in **part a** as a table, a graph, and a mapping.

28. ANALYZE Describe a real-life situation where it is reasonable to have a negative number included in the domain or range.

29. WRITE Compare and contrast dependent and independent variables.

Functions

Explore Vertical Line Test

 Online Activity Use graphing technology to complete the Explore.

> ⊘ **INQUIRY** How can you tell whether a relation is a function?

Copyright © McGraw-Hill Education

Learn Functions

A **function** is a relationship between input and output. In a function, there is exactly one output for each input. The relation shown is a function because each element of the domain is paired with *exactly* one element in the range.

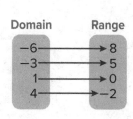

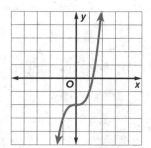

You can use the vertical line test to see if a graph represents a function.

Key Concept • Vertical Line Test	
function	**not a function**
A relation is a function if it passes the vertical line test, meaning a vertical line intersects the graph no more than once.	A relation is not a function if it fails the vertical line test, meaning that a vertical line intersects the graph more than once.
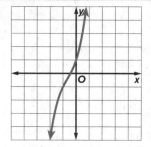	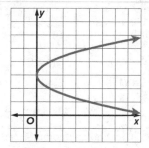

Today's Goals
- Determine whether relations are functions.
- Evaluate functions in function notation for given values.

Today's Vocabulary
function

function notation

💬 **Talk About It!**

Describe a way, other than using the vertical line test, that you could use to determine that a relation is not a function.

Example 1 Identify Functions

Determine whether each relation is a function. Explain.

a.

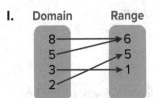

For each element of the domain, there is _____ element of the range. So this mapping _____ a function.

b.

Domain	2	6	8	2
Range	3	5	5	−3

The element _____ in the domain is paired with both 3 and −3 in the range. This relation _____ a function.

c. (2, 5), (4, 7), (8, 11), (4, 13)

The element _____ in the domain is paired with both 7 and 13 in the range. This relation _____ a function.

Check

Which of these relations are functions? _____

I.

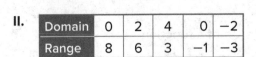

II.

Domain	0	2	4	0	−2
Range	8	6	3	−1	−3

III. {(−4, 5), (−2, 1), (1, −5), (−2, −7), (5, −13)}

A. I only

B. III only

C. I and II

D. I, II, and III

 Go Online You can complete an Extra Example online.

Think About It!

Suppose the last element in the domain in part **b** was 4 as shown in the table below. Would this relation be a function? Justify your argument.

Domain	2	6	8	4
Range	3	5	5	−3

⊕ Example 2 Analyze Data

LONG JUMP **Five schools are competing in the long jump portion of a track meet. The distances of the players with the best jump on each team are as follows: Team 1, 20.6 feet; Team 2, 21.5 feet; Team 3, 20.9 feet; Team 4, 19.4 feet; Team 5, 20.2 feet.**

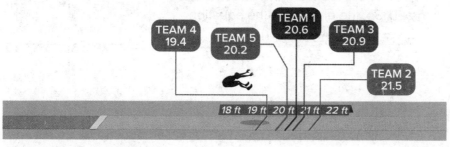

Part A Make a table.

Team Number	1	2	3	4	5
Best Jump (ft)	20.6				

Part B Determine the domain and range of the relation.

Domain:

Range:

Part C Determine whether the relation is a function.

For each element of the domain, there is _____ element of the range. So, this relation _____ a function.

Check

HEIGHT **Bailey recorded the heights of five of her friends. The heights, in inches, are as follows: Hunter, 62; Ling, 66; Ela, 65; Omar, 66; Alma, 67.**

Part A Find the domain and range of the relation. _____

A. D: {Hunter, 62, Ling, 66, Ela}; R: {65, Omar, 66, Alma, 67}

B. D: {65, Omar, 66, Alma, 67}; R: {Hunter, 62, Ling, 66, Ela}

C. D: {Hunter, Ling, Ela, Omar, Alma}; R: {62, 65, 66, 67}

D. D: {62, 65, 66, 67}; R: {Hunter, Ling, Ela, Omar, Alma}

Part B Is the relation a function? Explain your reasoning. _____

A. Yes; for each element of the domain, there is only one element of the range.

B. No; it fails the vertical line test.

C. No; an element in the domain is paired with more than one element in the range.

D. No; the domain and range are all real numbers.

⊙ **Go Online** You can complete an Extra Example online.

💭 **Think About It!**

Explain why the domain and range of the function are not all real numbers.

Study Tip

Domain and Range
Remember the domain is related to the independent variable, and the range is related to the dependent variable.

Example 3 Equations as Functions

Determine whether $6x + 2y = 14$ is a function. Explain.

For any value x, the vertical line passes through no more than one point on the graph. So, the graph and the equation _____ a function.

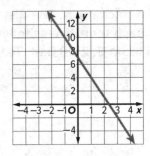

Check

Which relations described in the chart represent functions? Select Yes or No in each row.

Is this a function?	Yes	No
The relation passes the vertical line test.		
An element of the domain is paired with only two elements of the range.		
A vertical line passes through more than one point on the graph.		
The domain represents each student in class, while the range represents the age of each student.		
The domain represents the number of a day in May, while the range represents the high temperature for the day.		
The relation is of the form $y = x^2 + b$.		

Learn Function Values

Equations that are functions can be written in a form called **function notation**. Function notation is a way of writing an equation so that $y = f(x)$.

In a function, x represents the elements of the domain, and $f(x)$ represents the elements of the range. The graph of $f(x)$ is the graph of the equation $y = f(x)$.

 Go Online You can complete an Extra Example online.

Example 4 Find Function Values

For $f(x) = -2x + 9$, find each value.

$f(4) + 3$

$f(4) + 3 = [-2(\_\_\_\_) + 9] + 3$ $x = 4$

$\quad\quad = (\_\_\_\_ + 9) + 3$ Multiply.

$\quad\quad = \_\_\_\_ + 3$ Add.

$\quad\quad = \_\_\_\_$ Add.

$f(5) - f(1)$

$f(5) - f(1) = [-2(\_\_\_\_) + 9] - [-2(\_\_\_\_) + 9]$ Substitute for x.

$\quad\quad = (\_\_\_\_ + 9) - (\_\_\_\_ + 9)$ Multiply.

$\quad\quad = \_\_\_\_ - \_\_\_\_$ Add.

$\quad\quad = \_\_\_\_$ Subtract.

Think About It!
Is $f(3) - f(-2)$ the same as $f(3) + f(2)$? Justify your argument.

Check

For $f(x) = -3x + 2$, find each value.

a. $f(-4) = \_\_\_\_$ **b.** $f(5) + 8 = \_\_\_\_$ **c.** $f(3) - f(6) = \_\_\_\_$

Example 5 Evaluate Functions

For $h(x) = -6x^2 + 18x + 36$, find each value.

a. $h(2)$

$h(2) = -6(\_\_\_\_)^2 + 18(\_\_\_\_) + 36$ $x = 2$

$\quad\quad = \_\_\_\_ + \_\_\_\_ + 36$ Multiply.

$\quad\quad = \_\_\_\_$ Add.

b. $h(4) - h(1)$

$h(4) - h(1) = [-6(\_\_\_\_)^2 + 18(\_\_\_\_) + 36] - [-6(\_\_\_\_)^2 + 18(\_\_\_\_) + 36]$

$\quad\quad = (\_\_\_\_ + \_\_\_\_ + 36) - (\_\_\_\_ + \_\_\_\_ + 36)$ Multiply.

$\quad\quad = \_\_\_\_ - \_\_\_\_$ Add.

$\quad\quad = \_\_\_\_$ Subtract.

c. $h(5) - 7$

$h(5) - 7 = [-6(\_\_\_\_)^2 + 18(\_\_\_\_) + 36] - 7$ $x = 5$

$\quad\quad = (\_\_\_\_ + \_\_\_\_ + 36) - 7$ Multiply.

$\quad\quad = \_\_\_\_ - 7$ Add.

$\quad\quad = \_\_\_\_$ Subtract.

Think About It!
Find $h(3)$. Then find $h(-3)$. What was similar and what was different about finding these values?

Go Online You can complete an Extra Example online.

Check

For $g(x) = 2x^2 - 8x - 10$, find each value.

a. $g(2) = $ _____

b. $g(5) + 12 = $ _____

c. $g(8) - 18 = $ _____

d. $g(-3) - g(4) = $ _____

🌐 Example 6 Interpret Function Values

EMPLOYMENT **Mason works at the movie theater after school. The function $g(x) = 9.25x$ represents the amount of money he makes before taxes for each hour that he works. Evaluate and interpret the function for each value.**

a. $g(8)$

$g(x) = 9.25(\underline{\hspace{0.5cm}}) = \underline{\hspace{0.5cm}}$

This means that if Mason works for 8 hours, he will make _____ before taxes.

b. $g(14)$

$g(x) = 9.25(\underline{\hspace{0.5cm}}) = \underline{\hspace{1cm}}$

This means that if Mason works for 14 hours, he will make _____ before taxes.

c. $g(27.5)$

$g(x) = 9.25(\underline{\hspace{0.5cm}}) = \underline{\hspace{1.5cm}}$

This means that if Mason works for 27.5 hours, he will make _____ before taxes.

Check

POOLS A swimming pool that holds 10,000 gallons of water is being filled at a rate of 600 gallons per hour. The function $h(x) = 600x$ represents the amount of water in the pool after x hours.

Part A Find $h(3.25)$.

$h(3.25) = $ _____

Part B Describe the meaning of $h(3.25)$ in the context of the situation.

After _____ the total amount of water in the pool will be _____.

 Go Online You can complete an Extra Example online.

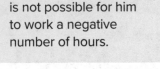 **Think About It!**

Find $g(-4)$. What does it mean in the context of the situation?

Watch Out!

Appropriate Domain Remember, it is possible for Mason to work for zero hours and make no money, but it is not possible for him to work a negative number of hours.

Practice
Examples 1 and 3

⟡ Go Online You can complete your homework online.

Determine whether each relation is a function. Explain.

1.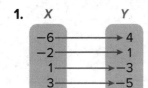

2.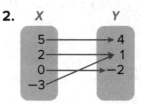

3.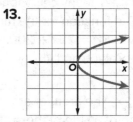

4.
x	y
4	−5
−1	−10
0	−9
1	−7
9	1

5.
x	y
2	7
5	−3
3	5
−4	−2
5	2

6.
x	y
3	7
−1	1
1	0
3	5
7	3

7. {(2, 5), (4, −2), (3, 3), (5, 4), (−2, 5)}

8. {(6, −1), (−4, 2), (5, 2), (4, 6), (6, 5)}

9. $y = 2x − 5$

10. $y = 11$

11.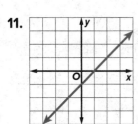

12.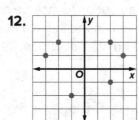

13.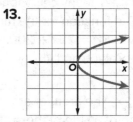

Example 2

14. TURNPIKE The total cost for cars entering President George Bush Turnpike at Beltline Road is related to the number of cars entering the turnpike at Beltline Road. The costs of different numbers of cars entering the turnpike at Beltline Road are as follows: 1 car, $0.75; 2 cars, $1.50; 3 cars, $2.25; 4 cars, $3.00; 5 cars, $3.75.

   a. Make a table.

   b. Determine the domain and range of the relation.

   c. Determine whether the relation is a function.

15. HOME VALUE The average value of a house changes over time. The average values of a house in January in Denver, Colorado, for different years are as follows: 2014, $254,000; 2015, $293,000; 2016, $338,000; 2017, $372,000.

   a. Make a table.

   b. Determine the domain and range of the relation.

   c. Determine whether the relation is a function.

If $f(x) = 3x + 2$ and $g(x) = x^2 - x$, find each value.

16. $f(4)$ _____

17. $f(8)$ _____

18. $f(-2)$ _____

19. $g(2)$ _____

20. $g(-3)$ _____

21. $g(-6)$ _____

22. $f(2) + 1$ _____

23. $f(1) - 1$ _____

24. $g(2) - 2$ _____

25. $g(-1) + 4$ _____

26. $f(x) + 1$ _____

27. $g(3b)$ _____

Example 6

28. CELL PHONES Many cell phone plans have an option to include more than one phone. The function for the monthly cost of cell phone service from a wireless company is $f(x) = 25x + 200$, where x is the number of phones on the plan. Find and interpret $f(3)$ and $f(5)$.

29. GEOMETRY The area for any square is given by the function $f(x) = x^2$, where x is the length of a side of the square. Find and interpret $f(3.5)$.

30. TRANSPORTATION The cost of riding in a cab is $3.00 plus $0.75 per mile. The equation that represents this relation is $y = 0.75x + 3$, where x is the number of miles traveled and y is the cost of the trip. Find and interpret $f(17)$.

31. GYM MEMBERSHIP The cost of a gym membership is $75 plus $30 per month. The equation that represents this relation is $y = 30x + 75$, where x is the number of months and y is the cost. Find and interpret $f(12)$.

Mixed Exercises

32. USE A MODEL Mario collected data about some of the players on a women's basketball team. The data are shown in the table, mapping, and graph. Is each relation a function? Why or why not?

Stats from Last Game

Team History	
Years on Team	Games Played
1	24
2	45
3	82
3	88
5	120

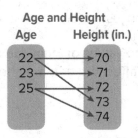

Age and Height

If $f(x) = -2x - 3$ and $g(x) = x^2 + 5x$, find each value.

33. $f(-1)$ _____

34. $f(6)$ _____

35. $g(2)$ _____

36. $g(-3)$ _____

37. $g(-2) + 2$ _____

38. $f(0) - 7$ _____

39. $f(4y)$ _____

40. $g(-6m)$ _____

41. $f(c) - 5$ _____

42. $f(r) + 2$ _____

43. $5[f(d)]$ _____

44. $3[g(n)]$ _____

45. FINANCIAL LITERACY Aisha has $40 to spend for her ornithology club. She spends some of it buying birdseed and saves the rest. Her savings is given by $f(x) = 40 - 1.25x$, where x is the number of pounds of birdseed she buys at $1.25 per pound.

　　a. Graph the equation.

　　b. Is the relation a function? Explain.

　　c. Find $f(3)$, $f(18)$, and $f(36)$. What do these values represent?

　　d. How many pounds of birdseed can Aisha buy if she wants to save $30?

46. USE A MODEL A recipe for homemade pasta dough says that the number of eggs you need is always one more than the number of servings you are making.

　　a. Make a graph that shows the relationship between the number of servings and the number of eggs.

　　b. Is the relation a function? Explain.

　　c. What is the domain of the function? Describe its meaning in the context of the situation.

47. **REASONING** The height h of a balloon, in feet, t seconds after it is released is given by the function $h(t) = 2t + 6$.

    a. What is the value of $h(20)$, and what does it mean in the context of the situation?

    b. Explain how to use the function to find the height of the balloon 2 minutes after it is released.

    c. What is the height of the balloon just before it is released? How do you know?

    d. Are there any restrictions on the values of t that can be used as inputs for the function? If so, how would this affect the graph of the function? Explain.

🍥 Higher-Order Thinking Skills

48. **CONSTRUCT ARGUMENTS** The following set of ordered pairs represents a function, but one of the values is missing: {(−4, −1), (−3, −1), (3, 2), (5, 2), (?, 2)}. What conclusions can you make about the missing value? Justify your arguments.

49. **WRITE** How can you determine whether a relation represents a function?

50. **ANALYZE** Feng says that the set of ordered pairs {(0, 1), (3, 2), (3, −5), (5, 4)} represents a function. Determine whether his statement is *true* or *false*. Justify your argument.

51. **PERSEVERE** Consider $f(x) = -4.3x - 2$. Write $f(g + 3.5)$ and simplify by combining like terms.

52. **REASONING** For the function $y = 15x - 4$, assume the domain is only values of x from 0 to 5. What is the range of the function?

53. **PERSEVERE** If $f(3b - 1) = 9b - 1$, find one possible expression for $f(x)$.

Today's Goals
- Determine whether functions are continuous, discrete, or neither.
- Determine whether functions are linear or nonlinear.

Today's Vocabulary
discrete function

continuous function

linear function

linear equation

nonlinear function

Explore Representing Discrete and Continuous Functions

Online Activity Use a real-world situation to complete the Explore.

> **INQUIRY** How can you use the graph of a function to determine whether it is discrete? ✕

Learn Discrete and Continuous Functions

Discrete Function	Continuous Function	Neither Discrete nor Continuous
Points are not connected.	Points are connected to form a line or curve.	Some points are connected by a line or curve, but it is not continuous everywhere.
Domain: set of individual values	Domain: all real numbers	Domain: varies
Range: set of individual values	Range: one interval of real numbers	Range: varies

Think About It!

Can a function be both discrete and continuous? Justify your answer.

🌀 Think About It!

If Bargain Book Barn did not use a sliding scale, that is, 1 book costs $1.50 and 2 books cost $3.00, then would the function still be discrete? Explain your reasoning.

🌐 Example 1 Determine Continuity

BOOKS **The Bargain Book Barn sells young adult novels on a sliding scale. That is, the more books you buy, the cheaper they are. Let** *f(x)* **model the store's prices for given quantities. Is** *f(x)* ***discrete*** **or** ***continuous*****? Explain your reasoning.**

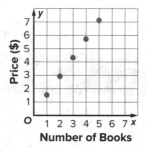

The Bargain Book Barn

Number of Books	Price ($)
1	1.50
2	2.90
3	4.30
4	5.70
5	7.10

Use the table and context of the situation. The quantity and price correspond to ordered pairs like (1, 1.50) which can be graphed. Because books are not sold in fractional quantities, the number of books and their corresponding prices cannot be between the points given. So the points should not be connected and the function is _____. The domain and range are _____.

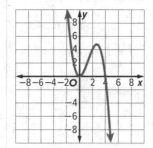

Number of Books

Example 2 Determine Continuity by Using Graphs

Determine whether *f(x)* **and** *g(x)* **are** ***continuous, discrete,*** **or** ***neither.*** **Explain your reasoning.**

a. *f(x)*

b. *g(x)*

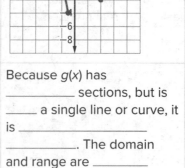

Because *f(x)* is graphed with a _____, it is a _____ function. The domain and range are both _____.

Because *g(x)* has _____ sections, but is _____ a single line or curve, it is _____. The domain and range are _____ _____.

🌐 **Go Online** You can complete an Extra Example online.

Check

Determine whether $f(x)$ is *discrete*, *continuous*, or *neither*.

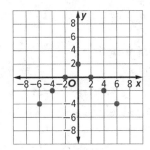

The function is _____.

🌐 Example 3 Apply Discrete and Continuous Functions

DETECTIVE As a private investigator, Tia charges $25 per hour for any amount of time up to eight hours and then a flat rate of $250 per day. Use the graph to determine if the function that models this situation is *discrete*, *continuous*, or *neither*.

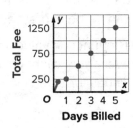

Discrete?

No; the function is not made up entirely of _____.

Continuous?

No; the function cannot be drawn with a _____ line or _____.

Neither?

Yes; the function is neither _____ nor _____.

Watch Out!

Continuous Intervals
Recall that a function can have a continuous interval, but the function itself is not continuous unless it is continuous over its entire domain.

💭 **Think About It!**

How could Tia alter her pricing model to make it a discrete function? a continuous function?

Check

PLANTS Circadian rhythms are cycles of behavior that occur over twenty-four hours, based on a day-night cycle. One aspect of a plant's circadian rhythm is the percentage of its flowers that are open or closed.

If $f(x)$ is represented by the curve, then is $f(x)$ *discrete*, *continuous*, or *neither*?

The function is _____.

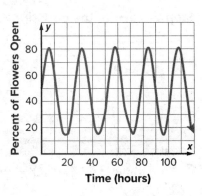

🧭 **Go Online** You can complete an Extra Example online.

Copyright © McGraw-Hill Education

Learn Linear and Nonlinear Functions

A **linear function** is a function that has a graph that is a line. If the domain of the function is all real numbers, then the function is continuous.

A **linear equation** can be used to describe a linear function.

Linear equations are often written in standard form.

Key Concept • Standard Form of a Linear Equation	
Words	The standard form of a linear equation is $Ax + By = C$, where $A \geq 0$, A and B are not both zero, and A, B, and C are integers with a greatest common factor of 1.
Examples	In $2x + 5y = 7$, $A = 2$, $B = 5$, and $C = 7$.
	In $x = -3$, $A = 1$, $B = 0$, and $C = -3$.

A **nonlinear function** has a graph with a set of points that cannot all lie on the same line. An equation that represents a nonlinear function cannot be expressed in the form $Ax + By = C$.

The function values of a linear function change at a constant rate for every equivalent change in the x-values. The change in the function values for a nonlinear function will vary for equivalent changes in x.

Example 4 Linear and Nonlinear Functions

Determine whether $y = 4x^2 - (2x)^2 + 3x - 5$ is an equation for a *linear* or *nonlinear* function.

Step 1 Simplify the equation.

$y = 4x^2 - (2x)^2 + 3x - 5$	Original equation
$= 4x^2 - \underline{\hphantom{xx}} + 3x - 5$	Simplify exponents.
$= \underline{\hphantom{xx}} + 3x - 5$	Subtract.
$= \underline{\hphantom{xxx}}$	Simplify.

Step 2 Rewrite the equation.

$y = 3x - 5$	Simplified equation
$y \underline{\hphantom{xx}} = 3x \underline{\hphantom{xx}} - 5$	Subtract $3x$ from each side.
$y - 3x = -5$	Simplify.
$3x - y = 5$	Multiply each side by -1.

Because $3x - y = 5$ is in the form $Ax + By = C$, where $A = \underline{\hphantom{xx}}$, $B = \underline{\hphantom{xx}}$, and $C = \underline{\hphantom{xx}}$, $y = 4x^2 - (2x)^2 + 3x - 5$ is $\underline{\hphantom{xxx}}$.

Check

The function $3x^2 - \sqrt{9}x^2 - y = 17$ is $\underline{\hphantom{xxx}}$.

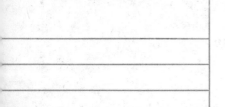 **Go Online** You can complete an Extra Example online.

Think About It!

Can a function be both linear and nonlinear? Explain your reasoning.

Think About It!

Can $y = -5x^2$ be written in standard form? If so, write the standard form of the equation.

Example 5 Identify Linear and Nonlinear Functions

Determine whether $y = 3x^3 - x^3 + 3x + 6$ **is an equation for a _linear_ or _nonlinear_ function.**

Step 1 Simplify the equation.

$y = 3x^3 - x^3 + 3x + 6$ Original equation

$= $ _____   Subtract.

Step 2 Rewrite the equation.

$y = 2x^3 + 3x + 6$ Simplified equation

y _____ $= 2x^3 + 3x + 6$ _____

Subtract $2x^3$ and $3x$ from each side.

$y - 2x^3 - 3x = $ _____   Simplify.

$2x^3 + 3x - y = -6$ Multiply each side by –1.

Because $2x^3 + 3x - y = -6$ is _____ in the form $Ax + By = C$, $2x^3 + 3x - y = -6$ is _____.

Check

The function $4x - (2y)^2 = 3$ is _____.

⊕ Example 6 Functions in Table Form

SOCCER Salina kicks a soccer ball. The height of the ball after each half second is recorded in the table. Is the function that models the height of the ball a _linear_ or _nonlinear_ function?

Time (s)	Height (ft)
0	2
0.5	28
1	46
1.5	56
2	58
2.5	52
3	38
3.5	16

First Half-Second Interval

During the first half-second interval, the

ball goes from a height of _____ feet to a height of _____ feet. That is an increase of _____ feet.

Second Half-Second Interval

During the second half-second interval, the ball goes from _____ feet to _____ feet. That is an increase of _____ feet.

Because the change in the height _____ over the two equivalent intervals, the height of the soccer ball must be modeled by a

_____.

▶ **Go Online** You can complete an Extra Example online.

🍇 **Think About It!**

Is the function represented by $2x + 9 = 5y - 3xy$ a linear or nonlinear function? Justify your argument.

🍇 **Think About It!**

What do you notice about the height of the soccer ball over time that might indicate that the function is not linear?

Check

Determine whether the values in each table are best modeled by a *linear* or *nonlinear* function.

x	y
−5	−15
−1	−3
0	0
2	6
3	9
7	21

x	y
0	0
1	−1
2	−8
3	−27
5	−125
7	−343

🌐 Example 7 Identify Linear Functions by Graphing

POOL Fernando uses a garden hose to fill his empty pool. The table shows the amount of water in the pool after every five minutes.

Time (min)	Water (gal)
5	60
10	120
15	180
20	240
25	300

Part A Determine linearity.

The amount of water in the pool increases by _____ gallons during the first 5 minutes. It also increases by _____ gallons during each following 5-minute intervals up to 25 minutes. Because the change in the amount of water is _____ for every equivalent interval of 5 minutes, the numbers in the table can be modeled by a _____ function.

Part B Graph.

The points on the graph can be connected by a _____.

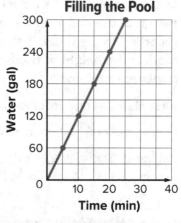

Filling the Pool

Check

MINIMUM WAGE The table shows the federal minimum wage rates during years in which the wage increased. Which statement best describes the function that models the wages over time? _____

Year	1990	1991	1996	1997	2007	2008	2009
Wage (dollars)	3.80	4.25	4.75	5.15	5.85	6.55	7.25

A. The function is linear because the increase is $0.70 between 1997 and 2007 and $0.70 between 2008 and 2009.

B. The function is nonlinear because the increase is $0.45 between 1990 and 1991 and $0.70 between 2007 and 2008.

C. The function is linear because it is constantly increasing.

D. The function is nonlinear because it is discrete.

🔎 **Go Online** You can complete an Extra Example online.

💬 Talk About It!

After 25 minutes, Fernando turns off the hose. He takes 5 minutes to remove any wrinkles from the pool liner. Then he returns to filling the pool at the same rate. Would the function that represents filling the pool from 0 to 40 minutes be a linear function? Explain your reasoning.

Practice

⬤ **Go Online** You can complete your homework online.

Examples 1 and 2

Determine whether the function is *discrete*, *continuous*, or *neither*. Explain.

1.

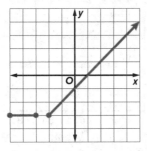

2.

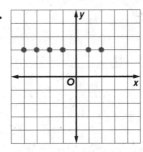

3.

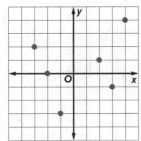

4.

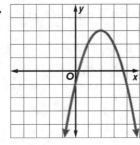

5.

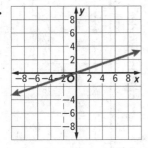

6.

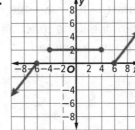

7.

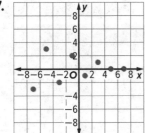

8.

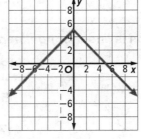

Example 3

9. **EMPLOYMENT** Kylie records the number of hours she works in her after-school job each week and then graphs the function that models the situation. Use the graph to determine if the function that models this situation is *discrete, continuous,* or *neither.*

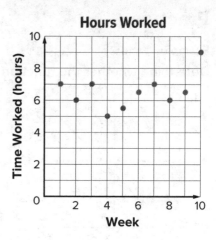

10. **BAKING** Kalynda keeps track of the number of cups of flour she has after baking batches of cookies and then graphs the function that models the situation. Use the graph to determine if the function that models this situation is *discrete, continuous,* or *neither.*

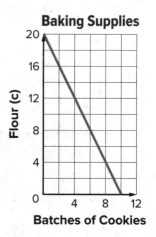

11. **HOMEWORK** Nayati records the number of hours of homework he has each day and then graphs the function that models the situation. Use the graph to determine if the function that models this situation is *discrete, continuous,* or *neither.*

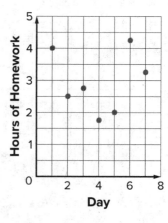

Examples 4, 5, and 6

Determine whether each function is *linear* or *nonlinear.*

12. $y - \frac{1}{x} = 11$

13. $x + (\sqrt{2}x)^2 + y - 2x^2 = 11$

14. $y = 2x + 5$

15. $y = 3x^2 - x + 5$

16. $2y + \frac{x}{3} = -6$

17. $9x - (9x)^2 = 19 - y$

18. $-9x + (\sqrt{4}x)^2 + y = -4 + 7x^2$

19. $y = -x^3 + 2x$

20.

x	y
1	100
2	125
3	150
4	175
5	200

21.

x	y
1	3
2	5
3	11
4	21
5	35

22.

x	y
−5	−10
−2	−4
0	0
2	4
3	6

23.

x	y
5	35
6	41
7	47
8	53
9	59

Example 7

24. ENDURANCE Tamika wants to improve her running endurance and thus tries to run for longer periods of time each day. The distance she runs, given the day, is provided in the table.

Day	1	2	3	4	5	6	7
Distance	2	2.25	2.75	3.5	4	4.5	5

a. Determine linearity.

b. Graph.

25. SCUBA DIVING Marco wants to know his elevation compared to sea level for periods of time each minute while scuba diving. His elevation is provided in the table.

Minute	1	2	3	4	5	6	7
Elevation	−3	−4.5	−6	−7.5	−9	−10.5	−12

a. Determine linearity.

b. Graph.

26. EXERCISE Rashona keeps track of the total number of minutes she exercises and then graphs the function that models the situation. Use the graph to determine if the function that models this situation is *discrete, continuous,* or *neither*.

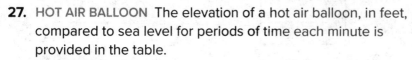

27. HOT AIR BALLOON The elevation of a hot air balloon, in feet, compared to sea level for periods of time each minute is provided in the table.

Minute	1	2	3	4	5	6	7	8
Elevation	10	15	12	18	25	45	40	42

   a. Determine linearity.

   b. Graph.

Determine whether the function is *discrete*, *continuous*, or *neither*. Then determine whether each function is *linear* or *nonlinear*.

28. $y - 14x = 3$

29. $(2x)^2 - 4y = 2$

30.

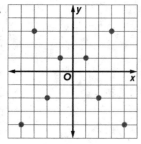

31.

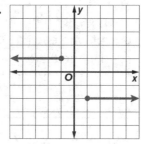

32.

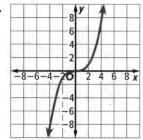

33.

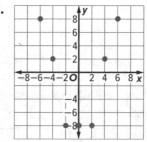

34. ANALYZE If $f(x)$ is a linear function, then is $f(x + a)$, where a is a real number, *sometimes, always,* or *never* a continuous function? Justify your argument.

35. WRITE Describe a real-world situation where the function that models the situation is neither discrete nor continuous.

36. ANALYZE A function that consists of a finite set of ordered pairs is *sometimes, always,* or *never* continuous? Justify your argument.

Intercepts of Graphs

Learn Intercepts of Graphs of Functions

The intercepts of graphs are points where the graph intersects an axis.

The **x-intercept** is the x-coordinate of a point where a graph crosses the x-axis.

The **y-intercept** is the y-coordinate of a point where a graph crosses the y-axis.

A function is **positive** when its graph lies *above* the x-axis.

A function is **negative** when its graph lies *below* the x-axis.

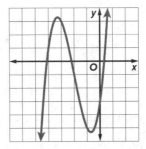

Today's Goals
- Identify the intercepts of functions and intervals where functions are positive and negative.
- Solve equations by graphing.

Today's Vocabulary
x-intercept
y-intercept
positive
negative
root
zero

Example 1 Intercepts of the Graph of a Linear Function

Use the graph to estimate the x- and y-intercepts of the function and describe where the function is positive and negative.

The x-intercept is the point where the graph crosses the x-axis, (____, ____).

The y-intercept is the point where the graph crosses the y-axis, (____, ____).

A function is positive when its graph lies above the x-axis, or when _____.

A function is negative when its graph lies below the x-axis, or when _____.

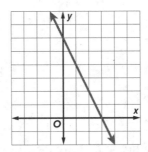

Study Tip

Intercepts Notice that *intercept* can be used to refer to either the point where the graph intersects the axis or the nonzero coordinate of the point where the graph intersects the axis.

🍎 **Think About It!**

Explain why this function is linear.

Check

Use the graph to estimate the x- and y-intercepts of the function and describe where the function is positive and negative. _____

A. x-intercept: (−2, 0); y-intercept: (0, −6); positive: $x > -2$; negative: $x < -2$

B. x-intercept: (0, −6); y-intercept: (−2, 0); positive: $x < -2$; negative: $x > -2$

C. x-intercept: (−2, 0); y-intercept: (0, −6); positive: $x < -2$; negative: $x > -2$

D. x-intercept: (0, −6); y-intercept: (−2, 0); positive: $x > -2$; negative: $x < -2$

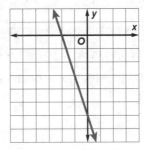
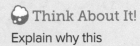

🐦 **Go Online** You can complete an Extra Example online.

Study Tip

x- and y-intercepts To help remember the difference between the x- and y-intercepts, remember that the x-intercept is where the graph intersects the x-axis, and the y-intercept is where the graph intersects the y-axis.

Example 2 Intercepts of the Graph of a Nonlinear Function

Use the graph to estimate the *x*- and *y*-intercepts of the function and describe where the function is positive and negative.

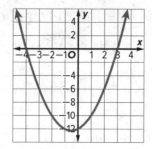

x-intercepts: _____ and _____.

y-intercept: _____.

positive: when _____ and when _____.

negative: *x* is between _____ and _____.

Check

Use the graph of the function to determine key features.

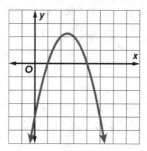

Part A Determine whether each ordered pair represents an *x-intercept*, a *y-intercept*, or *neither*.

(1, 0) _____

(0, 1) _____

Part B Describe where the function is positive and negative. _____

A. positive: *x* < 1 and *x* > 4; negative: *x* is between 1 and 4

B. positive: *x* > 1 and *x* < 4; negative: *x* is between 1 and 4

C. positive: *x* is between 1 and 4; negative: *x* > 1 and *x* < 4

D. positive: *x* is between 1 and 4; negative: *x* < 1 and *x* > 4

 Go Online You can complete an Extra Example online.

Example 3 Find Intercepts from a Graph

SPORTS The graph shows the height of a ball for each second *x* that it is airborne. Use the graph to estimate the *x*- and *y*-intercepts of the function, where the function is positive and negative, and interpret the meanings in the context of the situation.

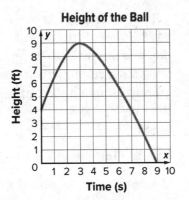

Height of the Ball

The *x*-intercept is ____. That means that the ball will hit the ground after ____ seconds.

The *y*-intercept is ____. This means that at time ____, the ball was at a height of ____ feet.

The function is positive when *x* is between ____ and ____, which means that the ball is in the air for ____ seconds.

No portion of the graph shows that the function is _____.

Think About It!
The function is only graphed from 0 to 9 seconds. What can you assume about the function when *x* > 9? Interpret this meaning. Does it make sense in the context of the situation?

Check

FITNESS The graph shows the number of people *y* at a gym *x* hours after the gym opens.

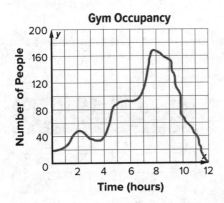

Gym Occupancy

Part A Use the graph to estimate the *x*- and *y*-intercepts.

x-intercept: (____, ____)

y-intercept: (____, ____)

Part B Which statements describe the meaning of the *x*- and *y*-intercepts in the context of the situation? Select all that apply. ____

A. There were 20 people at the gym when it opened.

B. The gym closed after 20 hours.

C. The gym closed after 12 hours.

D. There were 12 people at the gym when it opened.

E. There was no one at the gym when it opened.

Go Online You can complete an Extra Example online.

🌐 Example 4 Find Intercepts from a Table

LUNCH **Violet starts the semester with $150 in her student lunch account. Each day she spends $3.75 on lunch. The table shows the function relating the amount of money remaining in her lunch account to the number of days Violet has purchased lunch.**

Time (Days)	Balance ($)
x	y
0	150
2	142.50
5	131.25
10	112.50
15	93.75
30	37.50
40	0

Part A Find the intercepts.

The *x*-intercept is where *y* = ____, so the *x*-intercept is ____.

The *y*-intercept is where *x* = ____, so the *y*-intercept is ____.

Part B Describe what the intercepts mean in the context of the situation.

The *x*-intercept means that after buying lunch for ____ days, Violet will have $____ left in her lunch account, or it will take Violet ____ days to use all of the money in her lunch account. The *y*-intercept means that Violet's lunch account has $____ after buying lunch for ____ days, or the beginning balance of her lunch account is $____.

🌧 **Think About It!**

Explain why the *x*-coordinate of the *y*-intercept is always 0.

Check

MOVIES Ashley received a gift card to the movie theater for her birthday. The table shows the amount of money remaining on her gift card *y* after *x* trips to the movie theater.

Number of Trips	Balance ($)
x	y
0	90
1	81
2	72
3	63
5	45
7	27
10	0

Part A Find the *y*-intercept. (____, ____)

Part B Find the *x*-intercept and describe what it means in the context of the situation. ____

A. (10, 0); The initial balance on the gift card was $10.

B. (90, 0); The initial balance on the gift card was $90.

C. (10, 0); After 10 trips to the movies, there will be no money left on the gift card.

D. (90, 0); After 90 trips to the movies, there will be no money left on the gift card.

🧭 **Go Online** You can complete an Extra Example online.

Watch Out!

Intercepts The *y*-coordinate of the *x*-intercept will always be 0, not the *x*-coordinate. The *x*-coordinate of the *y*-intercept will always be 0, not the *y*-coordinate.

Learn Solving Equations by Graphing

The solution, or **root**, of an equation is any value that makes the equation true. A **zero** is an x-intercept of the graph of the function.

For example, the root of $3x = 6$ is 2. A linear equation, like $3x = 6$, has at most one root, while a nonlinear equation, like $x^2 + 4x - 5 = 0$, may have more than one.

Equation	Related Function
$3x = 6$	$f(x) = 3x - 6$ or $y = 3x - 6$
$x^2 + 4x - 5 = 0$	$f(x) = x^2 + 4x - 5$ or $y = x^2 + 4x - 5$

The graph of the related function can be used to find the solutions of an equation. The related function is formed by solving the equation for 0 and then replacing 0 with $f(x)$ or y.

Values of x for which $f(x) = 0$ are located at the x-intercepts of the graph of a function and are called the zeros of the function f. The roots of an equation are the same as the zeros of its related function. The solutions and roots of an equation are the same value as the zeros and x-intercepts of its related function. For the equation $3x = 6$:

- 2 is the solution of $3x = 6$.

- 2 is the root of $3x = 6$.

- 2 is the zero of $f(x) = 3x - 6$.

- 2 is the x-intercept of $f(x) = 3x - 6$.

Example 5 Solve a Linear Equation by Graphing

Solve $-2x + 7 = 1$ by graphing. Check your solution.

Find the related function.

$-2x + 7 = 1$	Original equation
$-2x + 7 \underline{\quad} = 1 \underline{\quad}$	Subtract 1 from each side.
$-2x + \underline{\quad} = \underline{\quad}$	Simplify.

Graph the left side of the equation. The related function is

$f(x) = $ _____, which can be graphed.

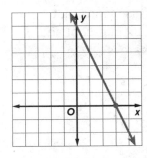

The graph intersects the x-axis at _____. This is the x-intercept, or zero, which is also the root of the equation. So, the solution of the equation is _____.

Check your solution by solving the equation algebraically.

 Go Online You can complete an Extra Example online.

 Think About It!
What is the difference between a *root* and a *zero*?

 Go Online
An alternate method is available for this example.

 Think About It!
Suppose you first solved the equation algebraically. How could you use your solution to graph the zero of the related function?

Example 6 Solve a Nonlinear Equation by Graphing

Solve $x^2 - 4x = -3$ by graphing. Check your solution.

Find the related function.

$$x^2 - 4x = -3 \qquad \text{Original equation}$$

$$x^2 - 4x \underline{\quad} = -3 \underline{\quad} \qquad \text{Add 3 to each side.}$$

$$x^2 - 4x + 3 = \underline{\quad} \qquad \text{Simplify.}$$

Graph the left side of the equation. The related function is
$f(x) =$ _____, which can be graphed.

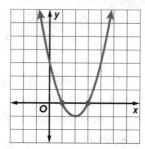

The graph intersects the x-axis at _____ and _____. These are the x-intercepts, or zeros, which are also the roots of the equation. So, the solutions of the equation are _____ and _____.

Example 7 Solve an Equation of a Horizontal Line by Graphing

Solve $4x + 3 = 4x - 5$ by graphing. Check your solution.

Find the related function.

$$4x + 3 = 4x - 5 \qquad \text{Original equation}$$

$$4x + 3 \underline{\quad} = 4x - 5 \underline{\quad} \qquad \text{Add 5 to each side.}$$

$$4x \underline{\quad} = 4x \qquad \text{Simplify.}$$

$$4x \underline{\quad} 8 = 4x \underline{\quad\quad} \qquad \text{Subtract } 4x \text{ from each side.}$$

$$8 = \underline{\quad} \qquad \text{Simplify.}$$

Graph the left side of the equation. The related function is
$f(x) =$ _____, which can be graphed.

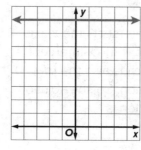

The graph does not intersect the x-axis. This means that there is _____ x-intercept and, therefore, there is _____ solution.

Go Online You can complete an Extra Example online.

Think About It!

Name two ways that you can tell that this is a nonlinear function.

_____

_____

_____

Go Online

You can watch a video to see how to use a graphing calculator with this example.

Talk About It!

Does solving the equation algebraically give a different solution? Explain your reasoning.

Check

Equations and the graphs of their related functions are shown. Write the related function and its zero(s) under the appropriate graph.

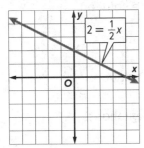

related function: _____

zeros: _____

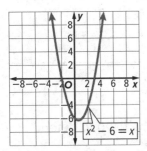

related function: _____

zeros: _____

related function: _____

zeros: _____

🌐 Apply Example 8 Estimate Solutions by Graphing

PARTY Haley is ordering invitations for her graduation party. She has \$40 to spend and each invitation costs \$0.96. The function $m = 40 - 0.96p$ represents the amount of money m Haley has left after ordering p party invitations. Find the zero of the function. Describe what this value means in the context of this situation.

1 What is the task?

Describe the task in your own words. Then list any questions that you may have. How can you find answers to your questions?

_____

_____

_____

2 How will you approach the task? What have you learned that you can use to help you complete the task?

_____

_____

_____

_____

🢒 **Go Online** You can complete an Extra Example online.

3 What is your solution?

Use your strategy to solve the problem.

Graph the function.

Graduation Party

Estimate the solution.

_____ invitations

Check the solution.

_____ invitations.

What does your solution mean
in the context of the situation?

_____

_____

_____

4 How can you know that your solution is reasonable?

Write About It! Write an argument that can be used to defend
your solution.

_____

_____

Check

DATA Blair's cell phone plan allows her to use
3 GB of data, and she uses approximately
0.14 GB of data each day. The function
$g = 3 - 0.14d$ represents the amount of
data g in GB she has left after d days.

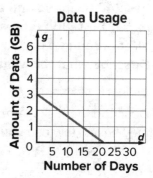

Part A Examine the graph of the function to
estimate its zero to the nearest day.

The graph appears to intersect the
x-axis at _____.

Part B Solve algebraically to check your answer. Round to the
nearest tenth.

$x =$ _____

Part C Describe what your answer to Part B means in this context.

After _____ days, Blair has _____ GB left.

Go Online You can complete an Extra Example online.

Practice

⊙ **Go Online** You can complete your homework online.

Examples 1 and 2

Use the graph to estimate the *x*- and *y*-intercepts of the function and describe where the function is positive and negative.

1.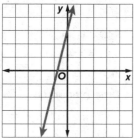

2.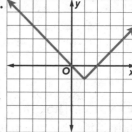

3.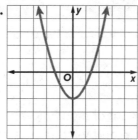

4.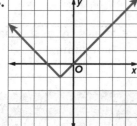

5.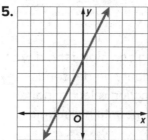

6.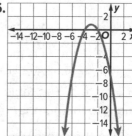

7.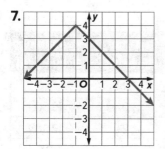

8.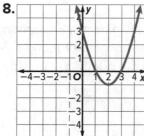

9.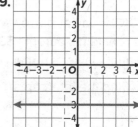

Example 3

10. FOOTBALL The graph shows the height of a football after being kicked. Use the graph to estimate the x- and y-intercepts of the function, where the function is positive and negative, and interpret the meanings in the context of the situation.

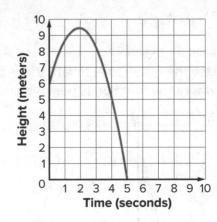

11. EARNINGS The graph shows the amount of money Ryan earns. Use the graph to estimate the x- and y-intercepts of the function, where the function is positive and negative, and interpret the meanings in the context of the situation.

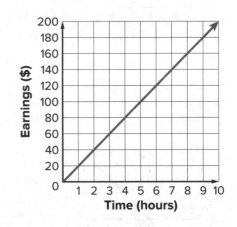

Example 4

12. CLIMBING Indira is mountain climbing and starts the day at 182.5 meters above sea level. Each hour she descends 36.5 meters. The table shows the function relating Indira's height to the number of hours she is mountain climbing.

a. Find the intercepts.

b. Describe what the intercepts mean in the context of the situation.

Time (hours)	Height (meters)
x	y
0	182.5
2	109.5
4	36.5
5	0

13. MONEY Javier borrowed $1950 from his parents. Each month he repaid his parents $325. The table shows the function relating Javier's remaining balance to the number of months.

a. Find the intercepts.

b. Describe what the intercepts mean in the context of the situation.

Time (months)	Remaining Balance ($)
x	y
0	1950
2	1300
5	325
6	0

Examples 5, 6, and 7

Solve each equation by graphing. Check your solution.

14. $2x - 3 = 3$

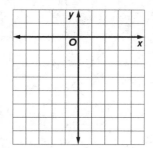

15. $-4x + 2 = -4x + 1$

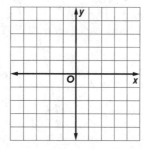

16. $4 = \frac{1}{2}x + 5$

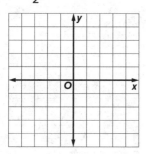

17. $x^2 = 6x - 8$

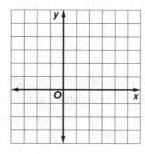

18. $3x - 5 = 3x - 3$

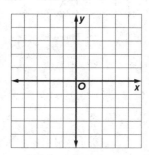

19. $x^2 + x = 6$

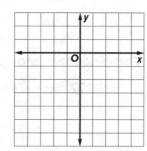

Example 8

20. INVITATIONS Moesha and Keyon are mailing invitations for their wedding. They have $50 to spend and each invitation costs $1.25 to mail. The function $m = 50 - 1.25w$ represents the amount of money m Moesha and Keyon have left after mailing w wedding invitations. Find the zero of the function. Describe what this value means in the context of this situation.

21. GIFT BAGS Juanita is tying ribbon on gift bags. She has 24 feet of ribbon and each gift bag uses 0.75 foot of ribbon. The function $r = 24 - 0.75g$ represents the amount of ribbon r Juanita has left after tying ribbon on g gift bags. Find the zero of the function. Describe what this value means in the context of this situation.

Mixed Exercises

Use the graph to estimate the *x*- and *y*-intercepts of the function and describe where the function is positive and negative.

22.

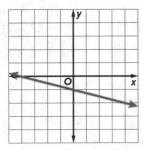

23.

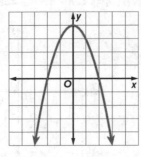

24.

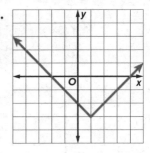

25. REASONING The graph shows the height of a bird compared to sea level over time. Use the graph to estimate the *x*- and *y*-intercepts of the function, where the function is positive and negative, and interpret the meanings in the context of the situation.

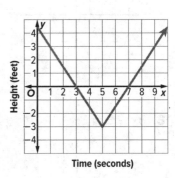

Solve each equation by graphing. Then find the *x*- and *y*-intercepts.

26. $3x + 1 = -5$

27. $9x - 7 = -6x + 8$

 Higher-Order Thinking Skills

28. ANALYZE Do linear equations *sometimes, always,* or *never* have *x*- and *y*-intercepts? Justify your argument.

29. WRITE Describe how to find *x*- and *y*-intercepts by graphing and by using tables.

30. CREATE Write a word problem for the function $y = 60 - 2.5x$. Find the zero of the function. Describe what this value means in the context of your situation.

31. PERSEVERE Describe the steps you use to solve the equation $16 = x + 4 + (2^4 - 6)$ by graphing. Then explain how you can check your solution.

Shapes of Graphs

Explore Line Symmetry

 Online Activity Use graphing technology to complete the Explore.

INQUIRY How can you use the graph of a function to determine whether it is symmetric?

Learn Symmetry and Graphs of Functions

The graphs of some functions exhibit a key feature called symmetry. A figure has **line symmetry** if each half of the figure matches the other side exactly.

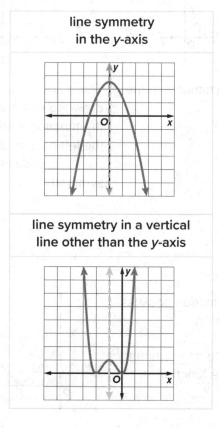

line symmetry
in the *y*-axis

line symmetry in a vertical
line other than the *y*-axis

Today's Goals
- Determine whether functions have line symmetry and, if so, find the line of symmetry.
- Identify extrema and where functions are increasing and decreasing.
- Determine the end behaviors of graphs of functions.

Today's Vocabulary
line symmetry
increasing
decreasing
extrema
relative minimum
relative maximum
end behavior

 **Talk About It!**

Can the graph of a function be symmetric about the *x*-axis? Justify your argument.

Example 1 Line Symmetry

Determine whether each function has line symmetry. Explain.

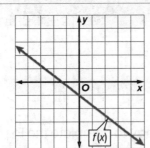

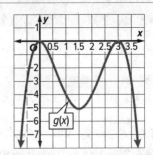

There is _____ that can be drawn to make the right half a mirror image of the left half, so the function _____ display any line symmetry.

This function is symmetric in the line $x =$ _____.

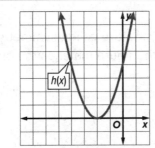

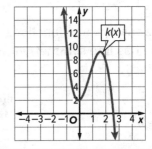

This function is symmetric in the line $x =$ _____.

There is _____ that can be drawn to make the right half of the function a mirror image of the left half, so the function _____ display any line symmetry.

Think About It!

Find the *y*-intercepts of the functions.

Study Tip

Symmetry Remember a graph can be symmetric in the *y*-axis or any other vertical line.

Check

Examine the function.

Part A Does the function possess line symmetry? _____

Part B Describe the line symmetry, if any, of the function.

_____

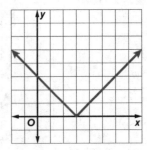

🏃 **Go Online** You can complete an Extra Example online.

🌐 Example 2 Interpret Symmetry

FOUNTAINS A fountain is spraying a stream of water into the air. The solid portion of the graph represents the path of the water, where *x* is the distance in feet from the fountain and *y* is the height in feet of the stream. Find and interpret any symmetry in the graph of the function.

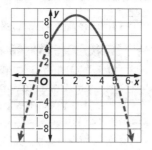

The right half of the graph is the mirror image of the left half in the line $x = $ ____.

In the context of the situation, the symmetry of the graph tells you that the height of the stream of water when it is from 0 to 2 feet away from the fountain is the _____ as the height of the stream of water when it is from 2 to 4 feet away from the fountain.

Check

GOLF The solid portion of the graph represents the path of a golf ball after it is hit off of a platform, where *x* is the distance in feet a golf ball travels and *y* is the height in feet of the golf ball.

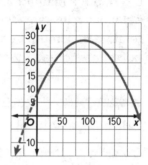

Part A Use the graph to describe any symmetry of the graph of the function. ____

A. symmetric in the *y*-axis

B. symmetric in the line $x = 8$

C. symmetric in the line $x = 28.25$

D. symmetric in the line $x = 90$

Part B Interpret the symmetry in the context of the situation. ____

A. The height of the golf ball when it has traveled a distance of 0 to 8 feet is the same as the height of the golf ball when it has traveled a distance of 8 to 28.25 feet.

B. The height of the golf ball when it has traveled a distance of 0 to 90 feet is the same as the height of the golf ball when it has traveled a distance of 90 to 180 feet.

C. The distance the golf ball has traveled when it is 0 to 8 feet in the air is the same as the distance the golf ball has traveled when it is 8 to 28.25 feet in the air.

D. The distance the golf ball has traveled when it is 0 to 90 feet in the air is the same as the distance the golf ball has traveled when it is 90 to 180 feet in the air.

🌐 Go Online You can complete an Extra Example online.

😮 **Think About It!**

Find the maximum height of the stream of water.

Problem-Solving Tip

Visualize To help visualize a line of symmetry, imagine folding the graph in half. If the graph lines up perfectly, the graph is symmetric about the line you have created with the fold.

Study Tip

Reading Math *Extrema* in this context is the plural form of *extreme point*. The plural forms of *maximum* and *minimum* are *maxima* and *minima*, respectively.

🗨 Think About It!

If *f*(*x*) has a relative maximum when *x* = 2, then is *f*(*x*) increasing or decreasing as *x* approaches 2? as *x* moves past 2?

 Online Activity Use graphing technology to complete the Explore.

@ INQUIRY How do the *y*-values of relatively high and low points on the graph compare to the *y*-values of nearby points?

Learn Extrema of Graphs of Functions

A function is **increasing** where the graph goes up and **decreasing** where the graph goes down when viewed from left to right.

Points that are the locations of relatively high or low function values are called **extrema**. Point A is a **relative minimum** because no other nearby point has a lesser *y*-coordinate. Point B is a **relative maximum** because no other nearby point has a greater *y*-coordinate.

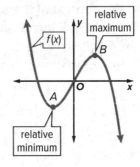

Example 3 Determine Increasing and Decreasing Parts of the Graph of a Function

Determine where *f*(*x*) is increasing and/or decreasing.

When _____, the graph goes up when viewed from left to right. So, the function is increasing for _____.

When _____, the graph goes down when viewed from left to right. So, the function is decreasing for _____.

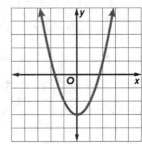

Check

For *x* > 1, *f*(*x*) is _____.

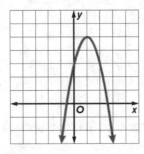

 Go Online You can complete an Extra Example online.

Example 4 Determine Extrema of the Graph of a Function

Determine the extrema of *f(x)*. Then identify each point as a relative maximum or relative minimum.

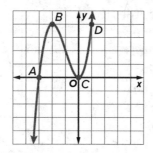

Think About It!

Can a point be a relative minimum or relative maximum and not be an extreme point? Explain.

Extrema: Point _____ and point _____ are the locations of relatively high or low function values. So, they are the extrema of the function.

Relative Minimum: No other points nearby point _____ have a lesser *y*-coordinate. So, point _____ is a relative minimum.

Relative Maximum: No other points nearby point _____ have a greater *y*-coordinate. So, point _____ is a relative maximum.

Check

Which point(s) is(are) a relative minimum? Select all that apply. _____

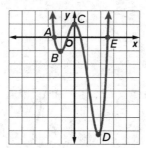

A. *A*

B. *B*

C. *C*

D. *D*

E. *E*

🢒 **Go Online** You can complete an Extra Example online.

Go Online
You can watch a video
to see more about the
comic book store.

Think About It!
Does this function
have a relative
minimum? Explain
your reasoning.

Example 5 Interpret Extrema of the Graph of a Function

COMIC BOOKS A comic book store uses a function to model its profit in thousands of dollars given the price in dollars that it charges for individual issues. Determine whether point *D* is a relative minimum, relative maximum, or neither. Then interpret its meaning in the context of the situation.

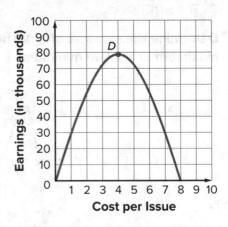

Point *D* is a _____ because all nearby points have a lesser *y*-coordinate.

Point *D* represents the _____ that the comic book store can earn given the price it charges per issue.

Check

AEROBATICS Aerobatics, or stunt flying, is the practice of intentional maneuvers of an aircraft that are not necessary for normal flight. Lincoln Beachey, an inventor of aerobatics, was known for his stunt called the "Dip of Death" in which his plane would plummet toward the ground from 5000 feet until he leveled the plane. His distance from the ground during the stunt can be approximately modeled by the function *f(x)*. Identify any extrema in the context of the situation.

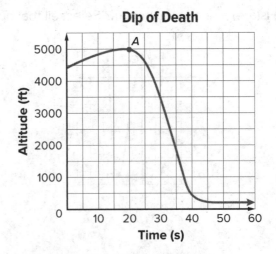

Point A represents that at about 20 seconds, Lincoln Beachey reached a _____ in height at 5000 feet.

After point A, his height is _____ until he levels out the plane.

Go Online You can complete an Extra Example online.

Learn End Behavior of Graphs of Functions

End behavior describes the values of a function at the positive and negative extremes in its domain.

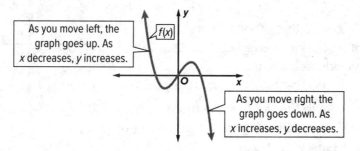

As you move left, the graph goes up. As x decreases, y increases.

f(x)

As you move right, the graph goes down. As x increases, y decreases.

Example 6 Determine End Behavior of the Graph of a Linear Function

Determine the end behavior of f(x).

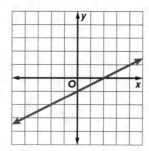

As you move to the left on the graph, the value of y gets increasingly _____. Thus, as x decreases, y _____.

As you move to the right on the graph, the value of y gets increasingly _____. Thus, as x increases, y _____.

Check

Determine the end behavior of f(x).

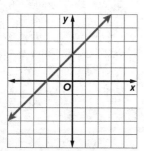

As x increases, y _____.

As x decreases, y _____.

 **Go Online** You can complete an Extra Example online.

Copyright © McGraw-Hill Education

Go Online
You may want to complete the Concept Check to check your understanding.

Watch Out!
End Behavior The end behavior of some graphs can be described as approaching a specific value.

Think About It!
Make a conjecture, or educated guess, about the end behavior of a linear function when the slope is positive or negative.

Study Tip
Conjecture
A conjecture is an educated guess based on known information.

Example 7 Determine End Behavior of the Graph of a Nonlinear Function

Determine the end behavior of $f(x)$.

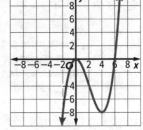

As you move to the left on the graph, the value of y gets increasingly _____. Thus, as x decreases, y _____.

As you move to the right on the graph, the value of y gets increasingly _____. Thus, as x increases, y _____.

Check

Determine the end behavior each function.

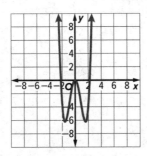

As x increases, y _____.

As x decreases, y _____.

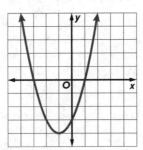

As x increases, y _____.

As x decreases, y _____.

 Go Online You can complete an Extra Example online.

Go Online
to practice what you've learned about interpreting graphs in the Put It All Together over Lessons 3-1 through 3-5.

Practice

🧭 **Go Online** You can complete your homework online.

Example 1

Determine whether each function has line symmetry. Explain.

1.

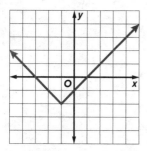

2.

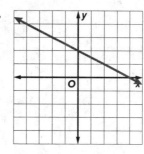

3.

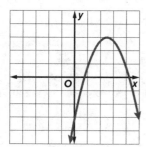

4.

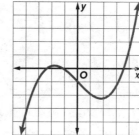

Example 2

5. **GEOMETRY** The solid portion of the graph represents the relationship between the width of a rectangle in centimeters x and the area of the rectangle in centimeters squared y. Find and interpret any symmetry in the graph of the function.

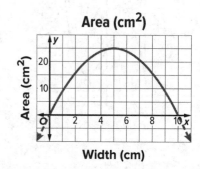

Width (cm)

6. **SPRINKLERS** A sprinkler is spraying a stream of water into the air. The solid portion of the graph represents the path of the water, where x represents the distance in feet from the sprinkler and y represents the height in feet of the water. Find and interpret any symmetry in the graph of the function.

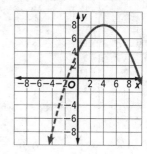

Example 3

Determine where $f(x)$ is increasing and/or decreasing.

7.

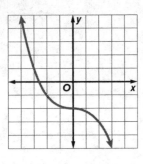

8.

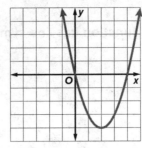

9.

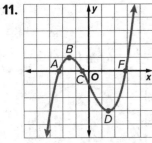

10.

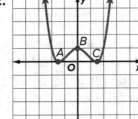

Example 4

Determine the extrema of $f(x)$. Then identify each point as a relative maximum or relative minimum.

11.

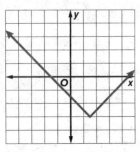

12.

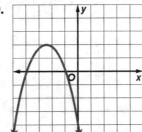

13.

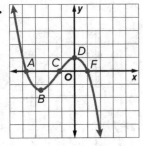

14.

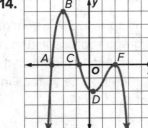

Example 5

15. **GOLF** The height of a golf ball compared to the distance the golf ball is from the tee is shown in the graph. Determine whether point *A* is a *relative minimum, relative maximum,* or *neither*. Describe what this value means in the context of this situation.

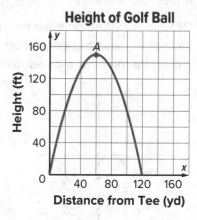

Height of Golf Ball

16. **ROLLERCOASTER** The height of a rollercoaster compared to the distance from start is shown in the graph. Determine whether points *B, C, D,* and *F* are *relative minima, relative maxima,* or *neither*. Describe what each value means in the context of this situation.

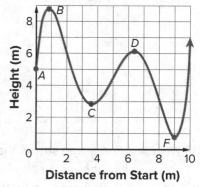

Examples 6 and 7

Determine the end behavior of *f*(*x*).

17.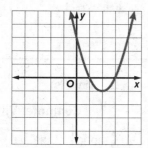

18.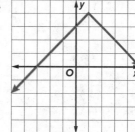

19.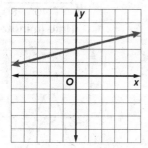

20.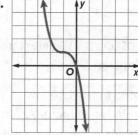

Mixed Exercises

Determine whether each function has line symmetry and where the function is *increasing* and/or *decreasing*. Determine the extrema. Then identify each point as a *relative maximum* or *relative minimum*. Determine the end behavior.

21.

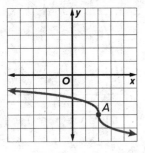

22.

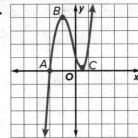

23. ROCKS The graph shows the height of a rock after it is thrown into the air over time. Determine the extrema. Then identify each point as a *relative minimum, relative maximum,* or *neither*. Describe what each value means in the context of this situation.

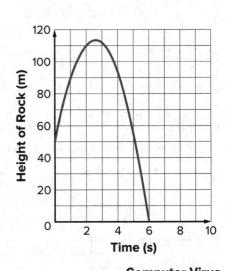

🧠 **Higher-Order Thinking Skills**

24. WRITE The graph shows the number of computers that are affected by a virus over time. Determine and interpret the end behavior.

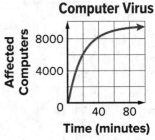

Computer Virus

25. WHICH ONE DOESN'T BELONG? Which statement about the graph is not true? Justify your conclusion.

> The graph is symmetric in the line $x = 0$.
> The graph has one relative minimum at about $(-2.25, -16)$.
> The graph has one relative maximum at about $(0, 9)$.
> As x decreases, y increases. As x increases, y increases.

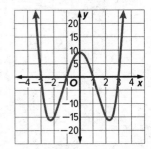

Sketching Graphs and Comparing Functions

Explore Modeling Relationships by Using Functions

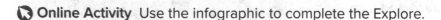 **Online Activity** Use the infographic to complete the Explore.

INQUIRY How can you use key features to approximate the graphs of functions? ×

Learn Sketching Graphs of Functions

You can sketch the graph of a function using its key features. Knowing the domain, range, intercepts, symmetry, end behavior, and extrema of a function as well as intervals where the function is increasing, decreasing, positive, or negative provides a clear idea of what the graph of the function looks like.

Label the graph to identify its key features

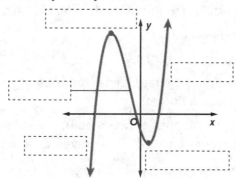

Think About It!

What other key features do you see in the graph of the function? Give two examples.

🌐 Example 1 Sketch the Graph of a Linear Function

CYCLING **In 2015, Christoph Strasser set a new 24-hour cycling record by riding 556 miles in a 24-hour period. The distance he rode over the 24 hours can be represented by a function. Sketch a graph that shows the distance traveled *y* as a function of time *x*.**

Before sketching, consider any possible constraints of the situation. It is not possible for Christoph Strasser to ride for a negative amount of time or ride negative miles. Therefore, the domain and range are restricted to nonnegative *x*- and *y*-values, and the graph exists only in the first quadrant.

y-Intercept: No distance traveled when he has ridden for 0 hours.

The point represents 0 hours ridden and 0 miles traveled. Graph this on the coordinate plane.

Linear or Nonlinear: The graph of the function is linear.

Positive: for time greater than 0

Increasing: for time greater than 0

The graph is a straight line that is positive and increasing for all hours ridden.

End Behavior: As the number of hours he has ridden increases, the number of miles he has traveled increases.

As *x* increases, *y* increases.

🌐 **Go Online** You can complete an Extra Example online.

> ## 💭 Think About It!
>
> What assumption is made when graphing Christoph Strasser's record-setting bike ride? Why is it necessary to make this assumption?

> ## Watch Out!
>
> If information about some key features is not provided, do not assume that it is not important. In this example, the missing information about where the function is negative was not necessary because it did not apply in the context of the situation. However, this is not always the case.

🌐 Example 2 Sketch the Graph of a Symmetric Function

WEATHER **A person's happiness can be affected by temperature. Sketch a nonlinear graph that shows the happiness of a person *y* as a function of temperature *x*. Interpret the key features.**

Positive: between about 25°F and 89°F

Negative: for temperatures less than 25°F and greater than 89°F

Increasing: for temperatures less than about 57°F

Decreasing: for temperatures greater than about 57°F

Relative Maximum: at about 57°F, when a person's happiness is about 85
A relative maximum occurs at 57°F, or $x = 57$, and a happiness of 85, or $y = 85$. This is represented by the point (57, 85), which we can graph on the coordinate plane.

End Behavior: As temperature increases or decreases, a person's happiness decreases.

For temperatures less than 25°F or $x < 25$, the graph is negative. For these temperatures, the graph is also increasing. For temperatures between 25°F and 57°F, the graph is positive and increasing. The graph is positive and decreasing for temperatures between 57°F and 89°F. This interval of the graph is also symmetric to the graph from 25°F to 57°F. This means that the right half of the graph is the mirror image of the left half. For temperatures greater than 89°F, or $x > 89$, the graph is negative, decreasing, and symmetric to the interval of the graph that is less than 25°F. As temperature increases or decreases, a person's happiness decreases. This means that happiness will get increasingly negative to move right and left on the graph.

Symmetry: A person's happiness for temperatures less than 57°F is the same as their happiness for temperatures greater than 57°F.

A person is happiest when it is 57°F. As the temperature gets increasingly cold or hot, a person becomes less happy. When the temperature is below about 25°F or above about 89°F a person is unhappy.

(continued on the next page)

🌐 **Go Online** You can complete an Extra Example online.

🗯 **Think About It!**
Describe the location of the points where the function changes from positive to negative or negative to positive.

🗯 **Think About It!**
What does a negative *y*-value represent in the context of this situation?

One method for sketching a graph given key features is to first graph any given points. Then analyze the key features of small intervals from left to right to make sketching the graph of the function easier.

Personal Happiness

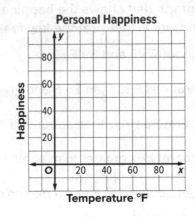

Check

Mariana used the key features to sketch the graph of *x* as a function of *y*. Examine the key features and graph to identify which key features Mariana graphed correctly and incorrectly.

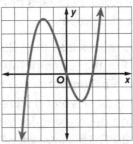

	Correct	Incorrect
Positive: between $x = -3$ and $x = 0$ and for $x > 2$		
Negative: for $x < -3$ and between $x = 0$ and $x = 2$		
Increasing: for about $x < -2$ and between about $x = 1$ and $x = 3$		
Decreasing: for between about $x = -2$ and $x = 1$ and for $x > 3$		
Intercepts: The graph intersects the x-axis at $(-3, 0)$, $(0, 0)$, and $(2, 0)$.		
Relative Minimum: $(1, -2)$		
Relative Maximum: $(-2, 4)$ and $(3, 2)$		
End Behavior: As x increases and decreases, the value of y decreases.		

🌐 Example 3 Sketch the Graph of a Nonlinear Function

AMUSEMENT PARK The number of people in line for a rollercoaster throughout the day can be modeled by a function. Use the key features to sketch a graph of the function. Then interpret the key features if x represents the time in hours since the ride opened at 10:00 A.M. and y represents the number of people in line.

Positive: between $x = -0.5$ and $x = 12$

Negative: for $x < -0.5$ and $x > 12$

Increasing: for $x < 1.4$ and between $x = 5.3$ and $x = 9.9$

Decreasing: for between $x = 1.4$ and $x = 5.3$ and for $x > 9.9$

Intercepts: The graph intersects the x-axis at $(-0.5, 0)$ and $(12, 0)$ and intersects the y-axis at $(0, 220)$.

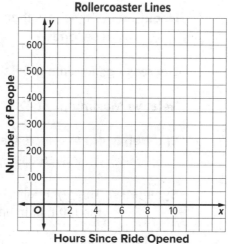

Rollercoaster Lines

Number of People / Hours Since Ride Opened

Relative Minimum: at $(5.3, 133)$

Relative Maximum: at $(1.4, 448)$ and $(9.9, 643)$

End Behavior: As x increases or decreases, the value of y decreases.

The x-intercepts mean that the number of people in line is zero a half hour before the ride opened and ___ hours after it opened. The y-intercept means that ___ people were in line when the ride opened.

The ride experienced a relative low in the number of people in line ___ hours after the ride opened and two relative peaks in the number of people in line ___ hours and ___ hours after it opened.

The number of people in line was negative but increasing until a half hour before the ride opened, positive and increasing from a half hour before the ride opened until ___ hours after it opened and again from ___ hours after the ride opened until ___ hours after it opened, negative and decreasing after the ride had been open for ___ hours, and positive but decreasing from ___ hours after the ride opened until ___ hours after it opened and again from ___ hours after the ride opened until ___ hours after the ride opened.

The graph indicates a period where there is a negative number of people in line. Because it is not possible to have a negative number of people, this graph appears to only model the number of people in line for the ride from a half hour before the ride opened until ___ hours after it opened.

🖱 **Go Online** You can complete an Extra Example online.

Copyright © McGraw-Hill Education

☁ **Think About It!**

Why might there be a relative minimum in the number of people in line around 3:00 P.M., 5.3 hours after the ride opened? Why might there be zero people in line 12 hours after the ride opened at 10:00 P.M.?

Check

MARINE LIFE The path of a dolphin jumping out of the ocean can be modeled with a symmetric function.

Part A Which graph could be used to show the height of a dolphin above the water y as a function of time since it emerged from the water x? _____

Positive: between 0 and 8 seconds

Negative: for time less than 0 seconds and greater than 8 seconds

Decreasing: for time greater than 4 seconds

Relative Maximum: at 4 seconds

End Behavior: As x increases and decreases, the value of y decreases.

Symmetry: The right half of the graph is the mirror image of the left half in approximately the line $x = 4$.

A.
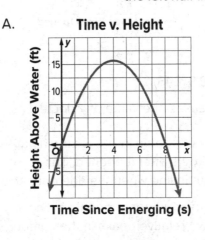

B.
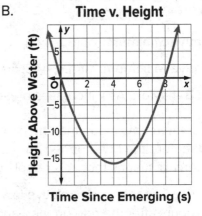

C.
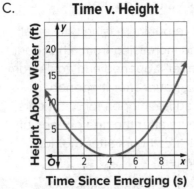

D.
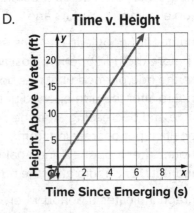

Part B Negative x-values represent _____ _____ and negative y-values represent _____ _____.

🌐 Example 4 Compare Properties of Functions

TENNIS Hawk-Eye is a computer system used in tennis to track the path of the ball. It is used as an officiating aid to locate the landing spot of a tennis ball when players challenge a call. Use the description and graph of a player's forehand and backhand shots to compare the paths of the two shots if *y* is the vertical height and *x* is the distance.

🗨 Talk About It!

Compare the intervals over which each shot is positive and/or negative. Does this make sense in the context of the situation? Explain your reasoning.

Forehand

During the forehand, the ball leaves the player's racquet at a height of 2.8 feet and travels 29 feet, when it reaches a height of about 10 feet. Then, the height of the ball decreases until it hits the ground 58 feet from where it was hit.

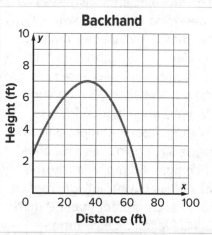

Backhand

	Forehand	Backhand
x-intercept	58	70
y-intercept	2.8	2.5
Extrema	maximum height of 10 feet when *x* = 29.	maximum height of 7 feet when *x* = 35.
Increasing and Decreasing	increases to a height of 10 feet from *x* = 0 to *x* = 29 and then decreases from *x* = 29 to *x* = 58 to a height of 0 feet.	increases to a height of 7 feet from *x* = 0 to *x* = 35 and then decreases from *x* = 35 to *x* = 70 to a height of 0 feet.

_x_-intercept

The tennis ball travels 12 feet farther during the _____ shot.

_y_-intercept

The *y*-intercepts of the two functions mean that the tennis ball is about 0.3 foot higher at the beginning of the _____ shot.

Extrema

The maximum height of the tennis ball is 3 feet higher during the _____ shot.

Increasing and Decreasing

The height of the tennis ball increases over a _____ interval during the forehand shot, but it reaches a _____ maximum height. This means that the tennis ball increases at a _____ rate during the forehand.

🧭 **Go Online** You can complete an Extra Example online.

Copyright © McGraw-Hill Education

Check

CARS Use the description and graph to compare the fuel economy of two cars, where *y* is the fuel efficiency in miles per gallon and *x* is the speed in miles per hour.

Car A

The fuel efficiency increases for speeds up to 25 mph when it reaches a relative maximum efficiency of 53 mpg. The fuel efficiency then decreases for speeds between 25 mph and 41 mph, when it gets down to 37 mpg. Above 41 mph, efficiency increases again until it reaches 48 mpg at 60 mph. Finally, the fuel efficiency rapidly decreases for speeds greater than 60 mph until leveling off at 17 mpg.

Car B

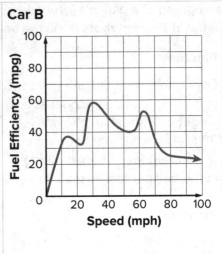

Which statements about the fuel efficiencies of the two cars are true? Select all that apply.

A. Car A has the greatest maximum fuel efficiency.

B. Car B has more relative maximum fuel efficiencies than Car A.

C. As speed increases, Car A levels off at a greater fuel efficiency than Car B.

D. Both cars get 0 mpg when they are traveling at 0 mph.

E. Car A has the least relative minimum fuel efficiency.

F. Both cars increase in fuel efficiency for speeds between 0 mph and about 13 mph.

G. Neither car reaches fuel efficiency below 0 mpg.

Pause and Reflect

Did you struggle with anything in this lesson? If so, how did you deal with it?

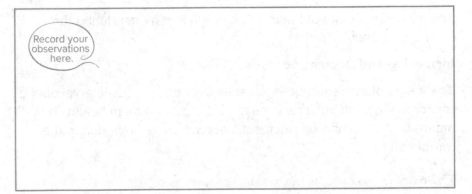

Record your observations here.

Practice

⊙ **Go Online** You can complete your homework online.

Example 1

1. **SAVINGS** David is saving money to buy a new car. The amount he saves can be represented by a function. Sketch a graph that shows the amount in savings *y*, in dollars, as a function of time *x*, in weeks.

   David's Savings for Car

   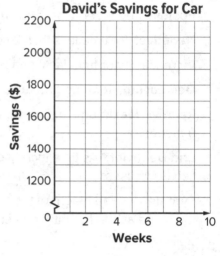

   x-**Intercept:** none

   y-**Intercept:** $1400

   Linear or Nonlinear: The graph of the function is linear.

   Positive: for time greater than 0

   Increasing: for time greater than 0

   End Behavior: As the number of weeks he has saved increases, the amount saved increases.

2. **SWIMMING** Yukio is keeping track of the number of calories she burns while swimming freestyle laps. The number of calories she burns can be represented by a function. Sketch a graph that shows the number of calories burned *y* as a function of time *x*, in hours.

   Calories Burned Swimming

   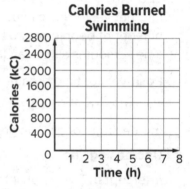

   y-**Intercept:** No calories burned when she has swum for 0 hours.

   Linear or Nonlinear: The graph of the function is linear.

   Positive: for time greater than 0

   Increasing: for time greater than 0

   End Behavior: As the number of hours she has swum increases, the number of calories burned increases.

Example 2

3. **FOOTBALL** The flight of a football thrown by a quarterback can be modeled by an interval of a function. Sketch a nonlinear graph that shows the height of a football *y*, in feet, as a function of time *x*, in seconds.

   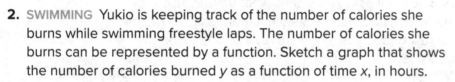

   Positive: between 0 seconds and 5 seconds

   Negative: for time greater than 5 seconds (*represents time after the ball hits the ground*)

   Increasing: for time less than 2 seconds

   Decreasing: for time greater than 2 seconds

   Relative Maximum: at 2 seconds, when the height of the football is 9 feet

   End Behavior: As time increases, the height of the football decreases.

   Symmetry: The height of the football for time between 0 seconds and 2 seconds is the same as the height for time between 2 seconds and 4 seconds.

4. FISH The height of a fish compared to sea level as it jumps out of the ocean water can be represented by a function. Sketch a nonlinear graph that shows the height of a fish y, in inches, as a function of time x, in seconds. Interpret the key features.

Positive: between 2 seconds and 8 seconds

Negative: for time less than 2 seconds and greater than 8 seconds

Increasing: for time less than 5 seconds

Decreasing: for time greater 5 seconds

Relative Maximum: at 5 seconds, when the height of the fish is 9 inches

End Behavior: As time increases or decreases, the height of the fish decreases.

Symmetry: The height of the fish for time less than 5 seconds is the same as the height for time greater than 5 seconds.

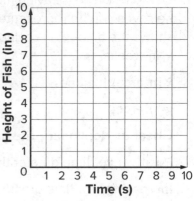

Example 3

5. TECHNOLOGY The results of a poll that asks Americans whether they used the Internet yesterday can be modeled as a function. Sketch a graph that shows the number of people polled that responded yes to the survey y as a function of time x, months since January 2005.

Positive: for time greater than 0 months

Negative: none

Increasing: for all time greater than 0 months

Decreasing: none

Intercepts: The graph intersects the y-axis at about (0, 60).

Extrema: none

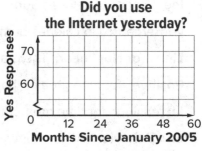

End Behavior: As x increases, y increases. The data represent people, so the maximum it could ever reach is the maximum number of people surveyed.

6. MUSIC The results of a poll that asks Americans whether they have listened to online music can be modeled as a function. Sketch a graph that shows the number of people polled that have listened online y as a function of time x, months since August 2000. Interpret the key features.

Positive: for time greater than 0 months

Negative: none

Increasing: between 0 months and 10 months, for time greater than about 65 months

Decreasing: between 10 months and about 65 months

Intercepts: The graph intersects the y-axis at about (0, 37).

Relative Minimum: at about (65, 31)

Relative Maximum: at about (10, 39)

End Behavior: As x decreases, y decreases. As x increases, y increases. The data represent people, so the maximum it could ever reach is the maximum number of people surveyed.

Example 4

7. **INTERNET** Use the description and graph to compare Internet use at home and Internet use away from home, where *y* is the number of people polled, in thousands, that use the Internet several times a day and *x* is the number of months since March 2004. Use the description and graph to compare Internet use at home and Internet use away from home since March 2004.

Internet Use at Home	**Internet Use Away from Home**
About 10,000 of those polled used the Internet at home in March 2004. The number of users decreased to 7000 at 36 months since March 2004. The number of users continued to increase after 36 months since March 2004.	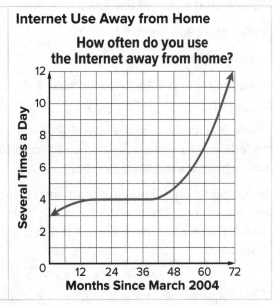

8. **SPENDING** Use the description and graph to compare the amount of U.S. spending on electronics and education, where *y* is the amount spent in billions and *x* is the number of years since 1949. Write statements to compare U.S. spending on electronics and education since 1949.

U.S. Electronic Spending	**U.S. Education Spending**
In 1949, the U.S. spent $0 on electronics. Twenty years after 1949, the U.S. spent about $4 billion on electronics. Thirty years after 1949, the U.S. spent about $3 billion on electronics. Seventy years after 1949, the U.S. spent about $7.5 billion on electronics and spending continues to increase.	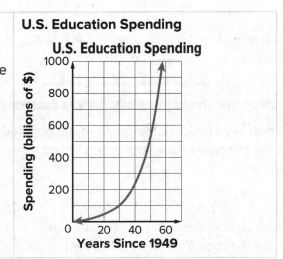

Mixed Exercises

The costume department of a theatre company is making cone-shaped hats for a play set in medieval times. Each hat will be covered with satin over its entire lateral surface area inside and out. The slant height of each hat will remain constant at 20 inches, but the radius of the base will vary to accommodate different head sizes. Using 3.14 for π, the lateral area of the hat can be expressed as a function. Use this information for Exercises 9–11.

9. USE A MODEL Sketch a graph that shows the lateral area of the hat y, in square inches, as a function of the radius of its base x, in inches.

   y-Intercept: No lateral area when the radius of its base is 0 inches.

   Linear or Nonlinear: The graph of the function is linear.

   Positive: for radius of a base greater than 0

   Increasing: for radius of a base greater than 0

   End Behavior: As the radius of the base increases, the lateral area increases.

10. REASONING Write a function y for the lateral area of the hat as a function of the radius of its base, x. (Hint: The formula for the lateral area of a cone is $y = \pi x s$, where s is the slant height.)

11. USE TOOLS Enter the function into your graphing calculator. Press **WINDOW** and enter the following settings: Xmin: −10; Xmax: 10; Ymin: −1000; Ymax: 1000. Then press **GRAPH**. Compare the graph on the calculator to the graph you sketched in Exercise 9.

🧠 Higher-Order Thinking Skills

Aidan buys used bicycles, fixes them up, and sells them. His average cost to buy and fix each bicycle is $47. He also incurred a one-time cost of $840 to purchase tools and a small shed to use as his workshop. He sells bikes for $75 each. Use this information for Exercises 12–15.

12. WRITE Write revenue and cost functions $R(x)$ and $C(x)$ for Aidan's situation, where x is the number of bicycles. How do you include the one-time cost in $C(x)$?

13. WRITE Write a profit function $P(x)$ such that $P(x) = R(x) - C(x)$. In words, what does $P(x)$ represent?

14. PERSEVERE List key features for the profit function $P(x)$. Then use the key features to sketch a graph that shows the profit $P(x)$ as a function of x bicycles.

15. ANALYZE Which key feature of the graph represents Aidan's break-even point (profit = 0)? Explain how to use your graph to find the most accurate value for this feature.

16. CREATE Research the population in your state over a 10-year period. Sketch a graph to model the data. Then list the key feature of the graph.

ⓔ Essential Question

Why are representations of relations and functions useful?

Module Summary

Lesson 3-1

Representing Relations

- A relation is a set of ordered pairs.
- Relations can be shown with ordered pairs, with a table, with a graph, or with a mapping.

Lesson 3-2

Functions

- A function is a relationship between input and output. In a function, there is exactly one output for each input.
- If a vertical line intersects the graph of a relation more than once, then the relation is not a function.
- Function notation is a way of writing an equation so that $y = f(x)$.

Lessons 3-3 through 3-5

Interpreting Graphs

- A discrete function is a set of points that are not connected. A continuous function has points that connect to form a line or curve.
- The x-intercept of a graph is the point where the graph intersects the x-axis. The y-intercept of a graph is the point where the graph intersects the y-axis.
- A figure has line symmetry if each half of the figure matches the other side exactly.
- A function is increasing where the graph goes up and decreasing where the graph goes down when viewed from left to right.

- Points that are the locations of relatively high or low function values are called extrema.
- A point is a relative minimum when no other nearby point has a lesser y-coordinate.
- A point is a relative maximum when no other nearby point has a greater y-coordinate.

Lesson 3-6

Sketching and Using Graphs

- Knowing the intercepts, symmetry, end behavior, and extrema of a function, as well as intervals where the function is increasing, decreasing, positive, or negative, provides a clear idea of what the graph of the function looks like.
- Equations can be solved by graphing related functions.

Study Organizer

📖 Foldables

Use your Foldable to review this module. Working with a partner can be helpful. Ask for clarification of concepts as needed.

Test Practice

1. **MULTI-SELECT** The table and graph show the number of pounds of bananas sold at a local grocery store from 2013 to 2018. (Lesson 3-1)

x	y
0	40
1	60
2	55
3	25
4	30
5	50

Number of Pounds of Bananas Sold

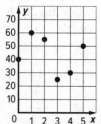

Which of the following statements correctly describes the relation? Select all that apply.

(A) The x-axis has a scale mark of 1 mark = 1 pound of bananas.

(B) The x-axis has a scale mark of 1 mark = 1 year.

(C) The x-axis represents the years since 2013.

(D) The x-axis represents the number of pounds of bananas sold.

(E) The y-axis represents the years since 2013.

(F) The y-axis represents the number of pounds of bananas sold.

(G) The y-axis has a scale mark of 1 mark = 10 years.

2. **MULTIPLE CHOICE** The Hillsborough State Park in Thonotosassa, Florida, charges an admission fee of $4 plus a camping fee of $20 per night. This can be represented by the function $f(x) = 20x + 4$, where $f(x)$ is the total cost and x is the number of nights spent camping. What is the value of $f(5)$, which is the cost of 5 nights camping? (Lesson 3-2)

(A) 80

(B) 100

(C) 104

(D) 120

3. **MULTIPLE CHOICE** If $f(x) = -9x + 8$, then find $f(-2)$. (Lesson 3-2)

(A) −84

(B) −8

(C) 26

(D) 100

4. **MULTIPLE CHOICE** Which function includes the data set below?
{(2, −2), (6, 10), (13, 31)} (Lesson 3-2)

(A) $f(x) = \frac{1}{2}x - 3$

(B) $f(x) = -2x + 2$

(C) $f(x) = 3x - 8$

(D) $f(x) = 4x - 10$

5. **OPEN RESPONSE** Determine whether the relation shown in the table is a function. Explain. (Lesson 3-2)

Domain	−4	2	−4	5
Range	8	11	13	13

6. TABLE ITEM Indicate whether each of the following relations are functions. (Lesson 3-2)

A. 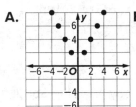 **B.**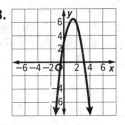

C. 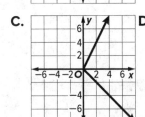 **D.**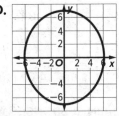

RELATION	FUNCTION?	
	YES	NO
A.		
B.		
C.		
D.		

7. MULTIPLE CHOICE What is the domain of this function? (Lesson 3-3)

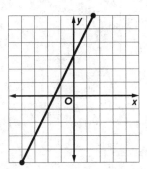

- Ⓐ $-4 < x < 1.5$
- Ⓑ $-4 \leq x \leq 1.5$
- Ⓒ $-6 < y < 6$
- Ⓓ $-6 \leq y \leq 6$

8. OPEN RESPONSE The graph shows the relationship between the number of Fun Pass tickets sold and the total value of the sales. Use the graph to estimate the x- and y-intercepts of the function, where the function is positive and negative, and interpret the meanings in the context of the situation. (Lesson 3-4)

Fun Pass Sales

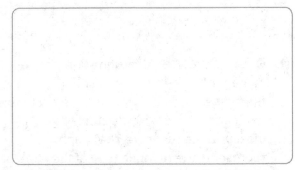

9. OPEN RESPONSE What are the intercepts for the equation $y = -x - 1$ graphed below?

(Lesson 3-4)

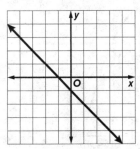

10. OPEN RESPONSE A piece of art is in the shape of an arch. It is modeled by a function where x is the width in feet of the arch and y is the height in feet. The table shows the relationship of x to y. (Lesson 3-4)

x	y
0	0
1	6
2	8
3	6
4	0

Write the y-intercept as an ordered pair and interpret its meaning in the real-world context.

11. OPEN RESPONSE A garden supply store manager found that if she used a certain function she could determine the best price to charge for the shovels she sells to maximize the revenue. The graph represents the revenue ($) y of the store at x price ($) per shovel. Use the graph to find and interpret the symmetry of the function in the context of the situation. (Lesson 3-4)

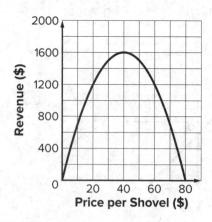

Symmetry: The graph is symmetric in the line $x =$ [____]

Interpret symmetry: The revenue gained when a shovel is sold for $20 is the same as it is when a shovel is sold for $ [____]

12. MULTIPLE CHOICE Suppose the graph of a function is increasing to the left of $x = 2$ and decreasing to the right of $x = 2$. Which describes the point at $x = 2$? (Lesson 3-5)

(A) Unless you know the y-coordinate of the point, you cannot say anything about the point at $x = 2$.

(B) It is an x-intercept.

(C) It is a relative minimum.

(D) It is a relative maximum.

13. OPEN RESPONSE Use the description and graph to compare the population data for Ohio and Florida, where y is the population in millions and x is the number of decades since 1900. Write statements about the populations of Ohio and Florida since 1900. (Lesson 3-6)

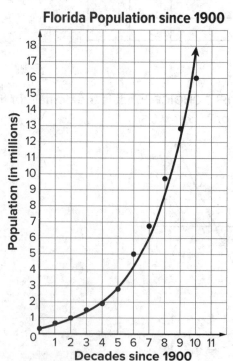

Ohio Population since 1900
In 1900 the population of Ohio was about 4.2 million. Between 1900 and 1950, the population of Ohio nearly doubled to about 8 million. Then between 1950 and 2000, the population of Ohio grew to approximately 11.4 million. Beyond 2000, the population of Ohio continues to gradually increase.

Linear and Nonlinear Functions

e Essential Question

What can a function tell you about the relationship that it represents?

What Will You Learn?

Place a check mark (✓) in each row that corresponds with how much you **already know** about each topic **before** starting this module.

KEY	Before			After		
👎 — I don't know. 👍 — I've heard of it. 👍 — I know it!	👎	👍	👍	👎	👍	👍
graph linear equations by using a table						
graph linear equations by using intercepts						
find rates of change						
determine slopes of linear equations						
write linear equations in slope-intercept form						
graph linear functions in slope-intercept form						
translate, dilate, and reflect linear functions						
identify and find missing terms in arithmetic sequences						
write arithmetic sequences as linear functions						
model and use piecewise functions, step functions, and absolute value functions						
translate absolute value functions						

📖 **Foldables** Make this Foldable to help you organize your notes about functions. Begin with five sheets of grid paper.

1. **Fold** five sheets of grid paper in half from top to bottom.

2. **Cut** along fold. Staple the eight half-sheets together to form a booklet.

3. **Cut** tabs into margin. The top tab is 4 lines wide, the next tab is 8 lines wide, and so on. When you reach the bottom of a sheet, start the next tab at the top of the page.

4. **Label** each tab with a lesson number. Use the extra pages for vocabulary.

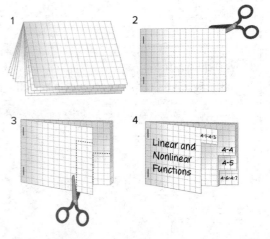

What Vocabulary Will You Learn?

Check the box next to each vocabulary term that you may already know.

- ☐ absolute value function
- ☐ arithmetic sequence
- ☐ common difference
- ☐ constant function
- ☐ dilation
- ☐ family of graphs
- ☐ greatest integer function
- ☐ identity function

- ☐ interval
- ☐ *n*th term of an arithmetic sequence
- ☐ parameter
- ☐ parent function
- ☐ piecewise-defined function
- ☐ piecewise-linear function
- ☐ rate of change

- ☐ reflection
- ☐ sequence
- ☐ slope
- ☐ step function
- ☐ term of a sequence
- ☐ transformation
- ☐ translation
- ☐ vertex

Are You Ready?

Complete the Quick Review to see if you are ready to start this module.
Then complete the Quick Check.

Quick Review

Example 1

Graph A(3, –2) on a coordir

Start at the origin. Since the *x*-coordinate is positive, move 3 units to the right. Then move 2 units down since the *y*-coordinate is negative. Draw a dot and label it *A*.

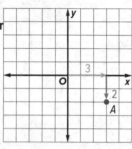

Example 2

Solve $x - 2y = 8$ for *y*.

$x - 2y = 8$	Original expression
$x - x - 2y = 8 - x$	Subtract *x* from each side.
$-2y = 8 - x$	Simplify.
$\dfrac{-2y}{-2} = \dfrac{8-x}{-2}$	Divide each side by -2.
$y = \dfrac{1}{2}x - 4$	Simplify.

Quick Check

Graph and label each point on the coordinate plane.

1. $B(-3, 3)$ **2.** $C(-2, 1)$ **3.** $D(3, 0)$

4. $E(-5, -4)$ **5.** $F(0, -3)$ **6.** $G(2, -1)$

Solve each equation for *y*.

7. $3x + y = 1$ **8.** $8 - y = x$

9. $5x - 2y = 12$ **10.** $3x + 4y = 10$

11. $3 - \dfrac{1}{2}y = 5x$ **12.** $\dfrac{y + 1}{3} = x + 2$

How did you do?

Which exercises did you answer correctly in the Quick Check? Shade those exercise numbers below.

Graphing Linear Functions

Today's Goals
- Graph linear functions by making tables of values.
- Graph linear functions by using the x- and y-intercepts.

Explore Points on a Line

Online Activity Use an interactive tool to complete an Explore.

> ⊗
>
> ⓠ **INQUIRY** How is the graph of a linear equation related to its solutions?

Learn Graphing Linear Functions by Using Tables

A table of values can be used to graph a linear function. Every ordered pair that makes the equation true represents a point on its graph. So, a graph represents all the solutions of an equation.

Linear functions can be represented by equations in two variables.

Example 1 Graph by Making a Table

Graph $-2x - 3 = y$ by making a table.

Step 1 Choose any values of x from the domain and make a table.

Step 2 Substitute each x-value into the equation to find the corresponding y-value. Then, write the x- and y-values as an ordered pair.

x	−2x − 3	y	(x, y)
−4			
−2			
0			
1			
3			

Step 3 Graph the ordered pairs in the table and connect them with a line.

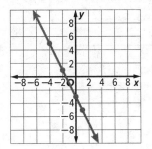

Talk About It!

What values of x might be easiest to use when graphing a linear equation when the x-coefficient is a whole number? Justify your argument.

Study Tip

Exactness Although only two points are needed to graph a linear function, choosing three to five x-values that are spaced out can verify that your graph is correct.

Go Online You can complete an Extra Example online.

Check

Graph $y = 2x + 5$ by using a table.

x	y
−5	
−3	
−1	
0	
2	

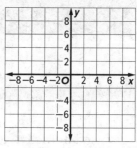

🍃 **Think About It!**

What are some values of x that you might choose in order to graph $y = \frac{1}{7}x - 12$?

Watch Out!

Equivalent Equations Sometimes, the variables are on the same side of the equal sign. Rewrite these equations by solving for y to make it easier to find values for y.

Example 2 Choose Appropriate Domain Values

Graph $y = \frac{1}{4}x + 3$ by making a table.

Step 1 Make a table.

Step 2 Find the y-values.

Step 3 Graph the ordered pairs in the table and connect them with a line.

x	$\frac{1}{4}x + 3$	y	(x, y)
−8			
−4			
0			
4			
8			

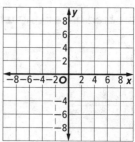

Check

Graph $y = \frac{3}{5}x - 2$ by making a table.

x	y
−10	
−5	
0	
5	
10	

↖ **Go Online** You can complete an Extra Example online.

Example 3 Graph $y = a$

Graph $y = 5$ by making a table.

Step 1 Rewrite the equation.
$$y = 0x + 5$$

Step 2 Make a table.

x	0x + 5	y	(x, y)
−2			
−1			
0			
1			
2			

Step 3 Graph the line.

The graph of $y = 5$ is a horizontal line through $(x, 5)$ for all values of x in the domain.

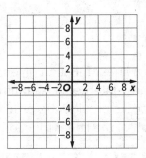

Think About It!

In general, what does the graph of an equation of the form $y = a$, where a is any real number, look like?

Example 4 Graph $x = a$

Graph $x = −2$.

You learned in the previous example that equations of the form $y = a$ have graphs that are horizontal lines. Equations of the form $x = a$ have graphs that are vertical lines.

The graph of $x = −2$ is a vertical line through $(−2, y)$ for all real values of y. Graph ordered pairs that have x-coordinates of $−2$ and connect them with a vertical line.

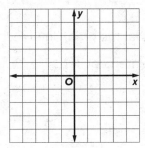

Think About It!

Is $x = a$ a function? Why or why not?

Check

Graph $x = 6$.

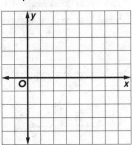

Go Online You can complete an Extra Example online.

Copyright © McGraw-Hill Education

Explore Lines Through Two Points

Go Online You can watch a video to see how to graph linear functions.

Online Activity Use graphing technology to complete an Explore.

> **INQUIRY** How many lines can be formed with two given points?

Think About It!

Why are the *x*- and *y*-intercepts easy to find?

Learn Graphing Linear Functions by Using the Intercepts

You can graph a linear function given only two points on the line. Using the *x*- and *y*-intercepts is common because they are easy to find. The intercepts provide the ordered pairs of two points through which the graph of the linear function passes.

Example 5 Graph by Using Intercepts

Graph $-x + 2y = 8$ by using the *x*- and *y*-intercepts.

To find the *x*-intercept, let _____.

Think About It!

What does a line that only has an *x*-intercept look like? a line that only has a *y*-intercept?

_____ Original equation

_____ Replace *y* with 0.

_____ Simplify.

_____ Divide.

This means that the graph intersects the *x*-axis at _____.

To find the *y*-intercept, let _____.

_____ Original equation

_____ Replace *x* with 0.

_____ Simplify.

_____ Divide.

Study Tip

Tools When drawing lines by hand, it is helpful to use a straightedge or a ruler.

This means that the graph intersects the *y*-axis at _____.

Graph the equation.

Step 1 Graph the *x*-intercept.

Step 2 Graph the *y*-intercept.

Step 3 Draw a line through the points.

Go Online You can complete an Extra Example online.

Check

Graph $4y = -12x + 36$ by using the x- and y- intercepts.

x-intercept: _____

y-Intercept: _____

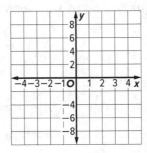

🌐 Example 6 Use Intercepts

PETS Angelina bought a 15-pound bag of food for her dog. The bag contains about 60 cups of food, and she feeds her dog $2\frac{1}{2}$ or $\frac{5}{2}$ cups of food per day. The function $y + \frac{5}{2}x = 60$ represents the amount of food left in the bag y after x days. Graph the amount of dog food left in the bag as a function of time.

Part A

Find the x- and y-intercepts and interpret their meaning in the context of the situation.

To find the x-intercept, let _____ .

_____ Original equation

_____ Replace y with 0.

_____ Simplify.

_____ Multiply each side by $\frac{2}{5}$.

The x-intercept is 24. This means that the graph intersects the x-axis at _____ . So, after 24 days, there is no dog food left in the bag.

To find the y-intercept, let _____ .

_____ Original equation

_____ Replace x with 0.

_____ Simplify.

The y-intercept is 60. This means that the graph intersects the y-axis at _____ . So, after 0 days, there are 60 cups of food in the bag.

🐾 Go Online

You can watch a video to see how to use a graphing calculator with this example.

🐾 Think About It!

Find another point on the graph. What does it mean in the context of the problem?

(continued on the next page)

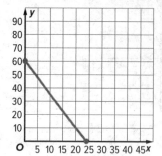
Think About It!

What assumptions did you make about the amount of food Angelina feeds her dog each day?

Go Online

You can watch a video to see how to graph a linear function using a graphing calculator.

Part B
Graph the equation by using the intercepts.

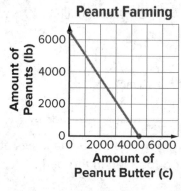

Check

PEANUTS A farm produces about 4362 pounds of peanuts per acre. One cup of peanut butter requires about $\frac{2}{3}$ pound of peanuts. If one acre of peanuts is harvested to make peanut butter, the function $y = -\frac{2}{3}x + 4362$ represents the pounds of peanuts remaining y after x cups of peanut butter are made.

x-intercept: _____

y-intercept: _____

Which graph uses the x- and y-intercepts to correctly graph the equation? _____

A.

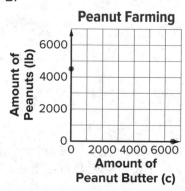

B.

C.

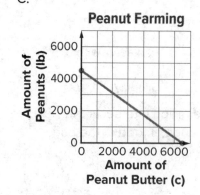

D.

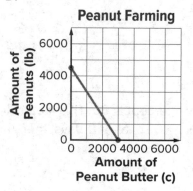

Go Online You can complete an Extra Example online.

Practice

🡢 **Go Online** You can complete your homework online.

Examples 1 through 4

Graph each equation by making a table.

1. $x = -2$

x	y

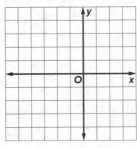

2. $y = -4$

x	y

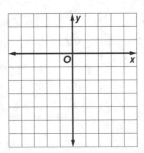

3. $y = -8x$

x	y

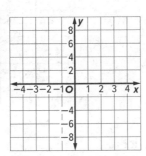

4. $3x = y$

x	y

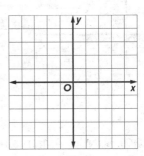

5. $y - 8 = -x$

x	y

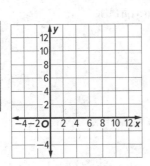

6. $x = 10 - y$

x	y

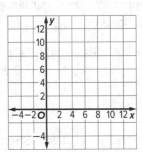

7. $y = \frac{1}{2}x + 1$

x	y

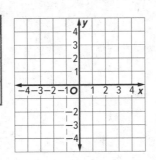

8. $y + 2 = \frac{1}{4}x$

x	y

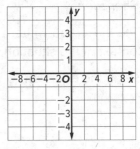

Example 5

Graph each equation by using the *x*-and *y*-intercepts.

9. $y = 4 + 2x$

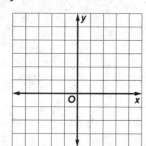

10. $5 - y = -3x$

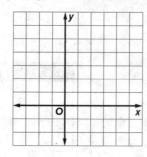

11. $x = 5y + 5$

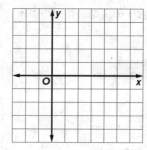

12. $x + y = 4$

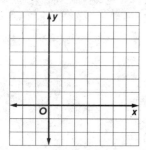

13. $x - y = -3$

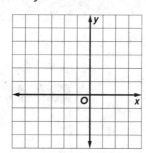

14. $y = 8 - 6x$

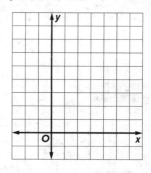

Example 6

15. SCHOOL LUNCH Amanda has $210 in her school lunch account. She spends $35 each week on school lunches. The equation $y = 210 - 35x$ represents the total amount in Amanda's school lunch account y for x weeks of purchasing lunches.

    a. Find the *x*- and *y*-intercepts and interpret their meaning in the context of the situation.

    b. Graph the equation by using the intercepts.

16. SHIPPING The *OOCL Shenzhen,* one of the world's largest container ships, carries 8063 TEUs (1280-cubic-feet containers). Workers can unload a ship at a rate of 1 TEU every minute. The equation $y = 8063 - 60x$ represents the number of TEUs on the ship y after x hours of the workers unloading the containers from the *Shenzhen.*

    a. Find the *x*- and *y*-intercepts and interpret their meaning in the context of the situation.

    b. Graph the equation by using the intercepts.

Mixed Exercises

Graph each equation.

17. $1.25x + 7.5 = y$

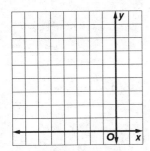

18. $2x - 3 = 4y + 6$

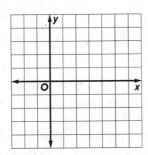

19. $3y - 7 = 4x + 1$

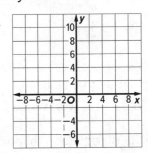

Find the x-intercept and y-intercept of the graph of each equation.

20. $5x + 3y = 15$

21. $2x - 7y = 14$

22. $2x - 3y = 5$

23. $6x + 2y = 8$

24. $y = \frac{1}{4}x - 3$

25. $y = \frac{2}{3}x + 1$

26. **HEIGHT** The height of a woman can be predicted by the equation $h = 81.2 + 3.34r$, where h is her height in centimeters and r is the length of her radius bone in centimeters.

    a. Is this a linear function? Explain.

    b. What are the r- and h-intercepts of the equation? Do they make sense in the situation? Explain.

    c. Graph the equation by using the intercepts.

    d. Use the graph to find the approximate height of a woman whose radius bone is 25 centimeters long.

27. **TOWING** Pick-M-Up Towing Company charges $40 to hook a car and $1.70 for each mile that it is towed. Write an equation that represents the total cost y for x miles towed. Graph the equation. Find the y-intercept, and interpret its meaning in the context of the situation.

28. **USE A MODEL** Elias has $18 to spend on peanuts and pretzels for a party. Peanuts cost $3 per pound and pretzels cost $2 per pound. Write an equation that relates the number of pounds of pretzels y and the number of pounds of peanuts x. Graph the equation. Find the x- and y-intercepts. What does each intercept represent in terms of context?

29. REASONING One football season, a football team won 4 more games than they lost. The function $y = x + 4$ represents the number of games won y and the number of games lost x. Find the x- and y-intercepts. Are the x- and y-intercepts reasonable in this situation? Explain.

30. WRITE Consider real-world situations that can be modeled by linear functions.
   a. Write a real-world situation that can be modeled by a linear function.

   b. Write an equation to model your real-world situation. Be sure to define variables. Then find the x- and y-intercepts. What does each intercept represent in your context?

   c. Graph your equation by making a table. Include a title for the graph as well as labels and titles for each axis. Explain how you labeled the x- and y-axes. State a reasonable domain for this situation. What does the domain represent?

31. FIND THE ERROR Geroy claims that every line has both an x- and a y-intercept. Is he correct? Explain your reasoning.

32. WHICH ONE DOESN'T BELONG? Which equation does not belong with the other equations? Justify your conclusion.

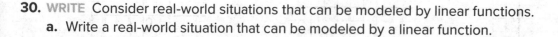

$y = 2 - 3x$	$5x = y - 4$	$y = 2x + 5$	$y - 4 = 0$

33. ANALYZE Robert sketched a graph of a linear equation $2x + y = 4$. What are the x- and y-intercepts of the graph? Explain how Robert could have graphed this equation using the x- and y-intercepts.

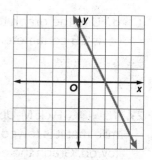

34. ANALYZE Compare and contrast the graph of $y = 2x + 1$ with the domain $\{1, 2, 3, 4\}$ and $y = 2x + 1$ with the domain all real numbers.

CREATE Give an example of a linear equation in the form $Ax + By = C$ for each condition. Then describe the graph of the equation.

35. $A = 0$ **36.** $B = 0$ **37.** $C = 0$

Rate of Change and Slope

Learn Rate of Change of a Linear Function

The **rate of change** is how a quantity is changing with respect to a change in another quantity.

If x is the independent variable and y is the dependent variable, then

$$\text{rate of change} = \frac{\text{change in } y}{\text{change in } x}.$$

Example 1 Find the Rate of Change

COOKING **Find the rate of change of the function by using two points from the table.**

Amount of Flour x (cups)	Pancakes y
2	12
4	24
6	36

$$\text{rate of change} = \frac{\text{change in } y}{\text{change in } x}$$

$$= \frac{\text{change in pancakes}}{\text{change in flour}}$$

$$= \frac{24 - 12}{4 - 2}$$

$$= \underline{\quad\quad} \text{ or } \underline{\quad\quad}$$

The rate is _____ or _____. This means that you could make _____ pancakes for each cup of flour.

Check

Find the rate of change.

$$\underline{\quad\quad} \frac{\text{dollars}}{\text{gallons}}$$

Amount of Gasoline Purchased (Gallons)	Cost (Dollars)
4.75	15.77
6	19.92
7.25	24.07
8.5	28.22

Go Online You can complete an Extra Example online.

Today's Goals
- Calculate and interpret rate of change.
- Calculate and interpret slope.

Today's Vocabulary
rate of change

slope

💭 Think About It!

Suppose you found a new recipe that makes 6 pancakes when using 2 cups of flour, 12 pancakes when using 4 cups of flour, and 18 pancakes when using 6 cups of flour. How does this change the rate you found for the original recipe?

Study Tip

Placement Be sure that the dependent variable is in the numerator and the independent variable is in the denominator. In this example, the number of pancakes you can make *depends* on the amount of flour you can use.

How is a greater increase or decrease of funds represented graphically?

Study Tip

Assumptions In this example, we assumed that the rate of change for the budget was constant between each 5-year period. Although the budget might have varied from year to year, analyzing in larger periods of time allows us to see trends within data.

🌐 **Example 2** Compare Rates of Change

STUDENT COUNCIL **The Jackson High School Student Council budget varies based on the fundraising of the previous year.**

Part A Find the rate of change for 2000–2005 and describe its meaning in the context of the situation.

$$\frac{\text{change in budget}}{\text{change in time}} =$$

$$\frac{1675 - 1350}{2005 - 2000} = \underline{\hspace{1cm}}, \text{ or } \underline{\hspace{1cm}}$$

This means that the student council's budget increased by

$\underline{\hspace{1cm}}$ over the $\underline{\hspace{1cm}}$ -year

period, with a rate of change of $\$\underline{\hspace{1cm}}$ per year.

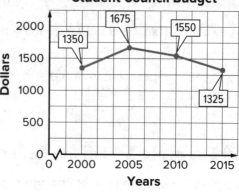

Jackson High School
Student Council Budget

Part B Find the rate of change for 2010–2015 and describe its meaning in the context of the situation.

$$\frac{\text{change in budget}}{\text{change in time}} = \frac{1325 - 1550}{2015 - 2010} = \underline{\hspace{1cm}}, \text{ or } \underline{\hspace{1cm}}$$

This means that the student council's budget was reduced by $\$\underline{\hspace{1cm}}$

over the $\underline{\hspace{1cm}}$-year period, with a rate of change of $-\$\underline{\hspace{1cm}}$ per year.

Check

TICKETS The graph shows the average ticket prices for the Miami Dolphins football team.

Part A Find the rate of change in ticket prices between 2009–2010.

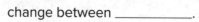
$$\frac{\text{dollars}}{\text{year}}$$

Part B The ticket prices have the greatest rate of

change between $\underline{\hspace{2cm}}$.

Part C Between $\underline{\hspace{2cm}}$ and $\underline{\hspace{2cm}}$, the rate of change is negative.

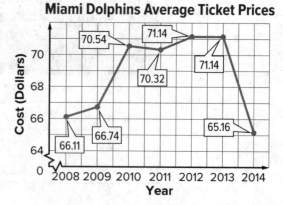

Miami Dolphins Average Ticket Prices

🌐 **Go Online** You can complete an Extra Example online.

Example 3 Constant Rate of Change

Determine whether the function is linear. If it is, state the rate of change.

Find the changes in the *x*-values and the changes in the *y*-values.

Notice that the rate of change for each pair of points shown is _____

The rates of change are _____, so the function is _____. The rate of change is _____.

x	y
11	−5
8	−3
5	−1
2	1
−1	3

Example 4 Rate of Change

Determine whether the function is linear. If it is, state the rate of change.

Find the changes in the *x*-values and the changes in the *y*-values.

The rates of change are _____. Between some pairs of points the rate of change is _____, and between the other pairs it is _____. Therefore, this is

_____.

x	y
22	−4
29	−1
36	1
43	4
50	6

Check

Complete the table so that the function is linear.

x	y
	−2.25
	1
11	
10.5	7.5
10	10.75
9.5	

🅑 **Go Online** You can complete an Extra Example online.

Study Tip

Linear Versus Not Linear Remember that the word *linear* means that the graph of the function is a straight line. For the graph of a function to be a line, it has to be increasing or decreasing at a constant rate.

Explore Investigating Slope

 Online Activity Use graphing technology to complete an Explore.

@ **INQUIRY** How does slope help to describe a line?

Learn Slope of a Line

The **slope** of a line is the rate of change in the y-coordinates (rise) for the corresponding change in the x-coordinates (run) for points on the line.

Key Concept • Slope	
Words	The slope of a nonvertical line is the ratio of the rise to the run.
Symbols	The slope m of a nonvertical line through any two points (x_1, y_1) and (x_2, y_2) can be found as follows. $m = \dfrac{y_2 - y_1}{x_2 - x_1}$
Example	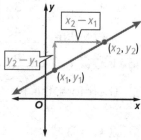

The slope of a line can show how a quantity changes over time. When finding the slope of a line that represents a real-world situation, it is often referred to as the *rate of change*.

Example 5 Positive Slope

Find the slope of a line that passes through (−3, 4) and (1, 7).

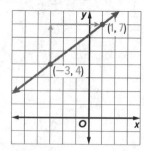

$$m = \frac{y_2 - y_1}{x_2 - x_1}$$

$$= \frac{7 - 4}{1 - (-3)}$$

$$= \underline{}$$

Check

Determine the slope of a line passing through the given points. If the slope is undefined, write *undefined*. Enter your answer as a decimal if necessary.

(−1, 8) and (7, 10) _____

 Go Online You can complete an Extra Example online.

Example 6 Negative Slope

Find the slope of a line that passes through (−1, 3) and (4, 1).

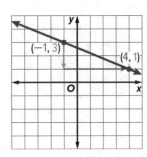

$$m = \frac{y_2 - y_1}{x_2 - x_1}$$

$$= \frac{1 - 3}{4 - (-1)}$$

$$= \underline{\hspace{3em}}$$

Check

Determine the slope of a line passing through the given points. If the slope is undefined, write *undefined*. Enter your answer as a decimal if necessary.

a. (5, −4) and (0, 1) _____

Example 7 Slopes of Horizontal Lines

Find the slope of a line that passes through (−2, −5) and (4, −5).

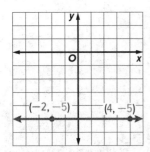

$$m = \frac{y_2 - y_1}{x_2 - x_1}$$

$$= \frac{-5 - (-5)}{4 - (-2)}$$

$$= \underline{\hspace{2em}} \text{ or } \underline{\hspace{2em}}$$

Example 8 Slopes of Vertical Lines

Find the slope of a line that passes through (−3, 4) and (−3, −2).

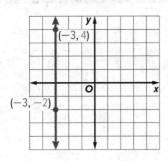

$$m = \frac{y_2 - y_1}{x_2 - x_1}$$

$$= \frac{-2 - 4}{-3 - (-3)}$$

$$= \underline{\hspace{2em}} \text{ or } \underline{\hspace{3em}}$$

🅑 **Go Online** You can complete an Extra Example online.

Study Tip

Positive and Negative Slope To know whether a line has a positive or negative slope, read the graph of the line just like you would read a sentence, from left to right. If the line "goes uphill," then the slope is positive. If the line "goes downhill," then the slope is negative.

💬 **Talk About It!**

Why is the slope for vertical lines always undefined? Justify your argument.

Study Tip

Converting Slope
When solving for an unknown coordinate, like the previous example, converting a slope from a decimal or mixed number to an improper fraction might make the problem easier to solve. For example, a slope of $1.\overline{333}$ can be rewritten as $\frac{4}{3}$.

🌧 Think About It!

If a crab is walking along the ocean floor 112 meters away from the shoreline to 114 meters away from the shoreline, how far does it descend?

Example 9 Find Coordinates Given the Slope

Find the value of r so that the line passing through $(-4, 5)$ and $(4, r)$ has a slope of $\frac{3}{4}$.

$$m = \frac{y_2 - y_1}{x_2 - x_1}$$ Use the Slope Formula.

$$\frac{3}{4} = \frac{r - 5}{4 - (-4)}$$ $(-4, 5) = (x_1, y_1)$ and $(4, r) = (x_2, y_2)$

$$\frac{3}{4} = \frac{r - 5}{8}$$ Subtract.

$$\underline{\hspace{1.2cm}} = \underline{\hspace{1.5cm}}$$ Multiply each side by 8.

$$\underline{\hspace{1.2cm}} = \underline{\hspace{1.5cm}}$$ Simplify.

$$\underline{\hspace{1.2cm}} = \underline{\hspace{1.5cm}}$$ Add 5 to each side.

$$\underline{\hspace{1.2cm}} = \underline{\hspace{0.7cm}}$$ Simplify.

Check

Find the value of r so that the line passing through $(-3, r)$ and $(7, -6)$ has a slope of $2\frac{2}{5}$.

$r = \underline{\hspace{1.5cm}}$

Example 10 Use Slope

OCEANS

What is the slope of the continental slope at Cape Hatteras?

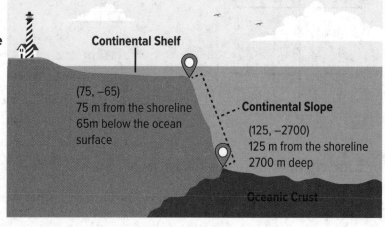

The continental slope at Cape Hatteras has a slope of _____.

$$m = \frac{y_2 - y_1}{x_2 - x_1}$$ Use the Slope Formula

$$= \frac{-2700 - (-65)}{125 - 75}$$ $(75, -65) = (x_1, y_1)$ and $(125, -2700) = (x_2, y_2)$

$$= \underline{\hspace{1.5cm}} \text{ or } \underline{\hspace{1cm}}$$ Simplify.

The continental slope at Cape Hatteras has a slope of _____.

🅝 **Go Online** You can complete an Extra Example online.

Practice

Go Online You can complete your homework online.

Example 1

Find the rate of change of the function by using two points from the table.

1.

x	y
5	2
10	3
15	4
20	5

2.

x	y
1	15
2	9
3	3
4	−3

3. **POPULATION DENSITY** The table shows the population density for the state of Texas in various years. Find the average annual rate of change in the population density from 2000 to 2009.

4. **BAND** In 2012, there were approximately 275 students in the Delaware High School band. In 2018, that number increased to 305. Find the annual rate of change in the number of students in the band.

Population Density	
Year	People Per Square Mile
1930	22.1
1960	36.4
1980	54.3
2000	79.6
2009	96.7

Source: Bureau of the Census, U.S. Dept. of Commerce

Example 2

5. **TEMPERATURE** The graph shows the temperature in a city during different hours of one day.

   a. Find the rate of change in temperature between 6 A.M. and 7 A.M. and describe its meaning in the context of the situation.

   b. Find the rate of change in temperature from 1 P.M. and 2 P.M. and describe its meaning in the context of the situation.

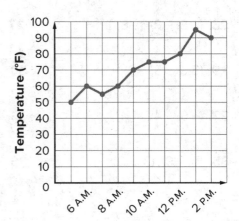

6. **COAL EXPORTS** The graph shows the annual coal exports from U.S. mines in millions of short tons.

   a. Find the rate of change in coal exports between 2000 and 2002 and describe its meaning in the context of the situation.

   b. Find the rate of change in coal exports between 2005 and 2006 and describe its meaning in the context of the situation.

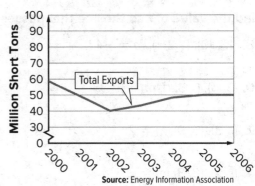

Source: Energy Information Association

Examples 3 and 4

Determine whether the function is linear. If it is, state the rate of change.

7.

x	4	2	0	−2	−4
y	−1	1	3	5	7

8.

x	−7	−5	−3	−1	0
y	11	14	17	20	23

9.

x	−0.2	0	0.2	0.4	0.6
y	0.7	0.4	0.1	0.3	0.6

10.

x	$\frac{1}{2}$	$\frac{3}{2}$	$\frac{5}{2}$	$\frac{7}{2}$	$\frac{9}{2}$
y	$\frac{1}{2}$	1	$\frac{3}{2}$	2	$\frac{5}{2}$

Examples 5 through 8

Find the slope of the line that passes through each pair of points.

11. $(4, 3), (-1, 6)$

12. $(8, -2), (1, 1)$

13. $(2, 2), (-2, -2)$

14. $(6, -10), (6, 14)$

15. $(5, -4), (9, -4)$

16. $(11, 7), (-6, 2)$

17. $(-3, 5), (3, 6)$

18. $(-3, 2), (7, 2)$

19. $(8, 10), (-4, -6)$

20. $(-12, 15), (18, -13)$

21. $(-8, 6), (-8, 4)$

22. $(-8, -15), (-2, 5)$

23. $(2, 5), (3, 6)$

24. $(6, 1), (-6, 1)$

25. $(4, 6), (4, 8)$

26. $(-5, -8), (-8, 1)$

27. $(2, 5), (-3, -5)$

28. $(9, 8), (7, -8)$

29. $(5, 2), (5, -2)$

30. $(10, 0), (-2, 4)$

31. $(17, 18), (18, 17)$

32. $(-6, -4), (4, 1)$

33. $(-3, 10), (-3, 7)$

34. $(2, -1), (-8, -2)$

35. $(5, -9), (3, -2)$

36. $(12, 6), (3, -5)$

37. $(-4, 5), (-8, -5)$

Example 9

Find the value of r so the line that passes through each pair of points has the given slope.

38. $(12, 10), (-2, r), m = -4$

39. $(r, -5), (3, 13), m = 8$

40. $(3, 5), (-3, r), m = \frac{3}{4}$

41. $(-2, 8), (r, 4), m = -\frac{1}{2}$

42. $(r, 3), (5, 9), m = 2$

43. $(5, 9), (r, -3), m = -4$

44. $(r, 2), (6, 3), m = \frac{1}{2}$

45. $(r, 4), (7, 1), m = \frac{3}{4}$

Example 10

46. ROAD SIGNS Roadway signs such as the one shown are used to warn drivers of an upcoming steep down grade. What is the grade, or slope, of the hill described on the sign?

47. HOME MAINTENANCE Grading the soil around the foundation of a house can reduce interior home damage from water runoff. For every 6 inches in height, the soil should extend 10 feet from the foundation. What is the slope of the soil grade?

48. USE A SOURCE Research the Americans with Disabilities Act (ADA) regulation for the slope of a wheelchair ramp. What is the maximum slope of an ADA regulation ramp? Use the slope to determine the length and height of an ADA regulation ramp.

49. DIVERS A boat is located at sea level. A scuba diver is 80 feet along the surface of the water from the boat and 30 feet below the water surface. A fish is 20 feet along the horizontal plane from the scuba diver and 10 feet below the scuba diver. What is the slope between the scuba diver and fish?

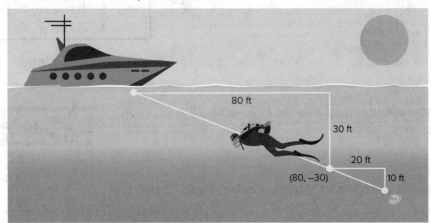

Mixed Exercises

STRUCTURE **Find the slope of the line that passes through each pair of points.**

50.

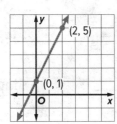

51.

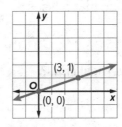

52.

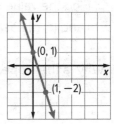

53. (6, −7), (4, −8)

54. (0, 5), (5, 5)

55. (−2, 6), (−5, 9)

56. (5, 8), (−4, 6)

57. (9, 4), (5, −3)

58. (1, 4), (3, −1)

59. REASONING Find the value of *r* that gives the line passing through (3, 2) and (*r*, −4) a slope that is undefined.

60. REASONING Find the value of *r* that gives the line passing through (−5, 2) and (3, *r*) a slope of 0.

61. CREATE Draw a line on a coordinate plane so that you can determine at least two points on the graph. Describe how you would determine the slope of the graph and justify the slope you found.

62. ARGUMENTS The graph shows median prices for small cottages on a lake since 2005. A real estate agent says that since 2005, the rate of change for house prices is $10,000 each year. Do you agree? Use the graph to justify your answer.

Cottage Prices Since 2005

🌑 **Higher-Order Thinking Skills**

63. CREATE Use what you know about rate of change to describe the function represented by the table.

Time (wk)	Height of Plant (in.)
4	9.0
6	13.5
8	18.0

64. WRITE Explain how the rate of change and slope are related and how to find the slope of a line.

65. FIND THE ERROR Fern is finding the slope of the line that passes through (−2, 8) and (4, 6). Determine in which step she made an error. Explain your reasoning.

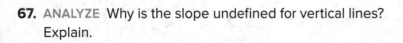

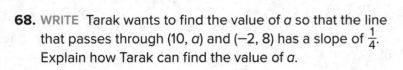

$$m = \frac{6-8}{-2-4} \quad \text{Step 1}$$

$$= \frac{-2}{-6} \quad \text{Step 2}$$

$$= \frac{1}{3} \quad \text{Step 3}$$

66. PERSEVERE Find the value of *d* so that the line that passes through (*a*, *b*) and (*c*, *d*) has a slope of $\frac{1}{2}$.

67. ANALYZE Why is the slope undefined for vertical lines? Explain.

68. WRITE Tarak wants to find the value of *a* so that the line that passes through (10, *a*) and (−2, 8) has a slope of $\frac{1}{4}$. Explain how Tarak can find the value of *a*.

Slope-Intercept Form

Learn Writing Linear Equations in Slope-Intercept Form

An equation of the form $y = mx + b$, where m is the slope and b is the y-intercept, is written in slope-intercept form. When an equation is not in slope-intercept form, it might be easier to rewrite it before graphing. An equation can be rewritten in slope-intercept form by using the properties of equality.

Key Concept • Slope-Intercept Form	
Words	The slope-intercept form of a linear equation is $y = mx + b$, where m is the slope and b is the y-intercept.
Example	$y = mx + b$ $y = -2x + 7$

Example 1 Write Linear Equations in Slope-Intercept Form

Write an equation in slope-intercept form for the line with a slope of $\frac{4}{7}$ and a y-intercept of 5.

Write the equation in slope-intercept form.

$y = mx + b$ Slope-intercept form.

$y = (\underline{\quad})x + 5$ $m = \frac{4}{7}, b = 5$

$y = \underline{\qquad}$ Simplify.

Check

Write an equation for the line with a slope of −5 and a y-intercept of 12. _____

Today's Goals
• Rewrite linear equations in slope-intercept form.
• Graph and interpret linear functions.

Today's Vocabulary
parameter

constant function

 Think About It!
Explain why the y-intercept of a linear equation can be written as $(0, b)$, where b is the y-intercept.

Go Online You can complete an Extra Example online.

Think About It!

Can $x = 5$ be rewritten in slope-intercept form? Justify your argument.

Example 2 Rewrite Linear Equations in Slope-Intercept Form

Write $-22x + 8y = 4$ in slope-intercept form.

$-22x + 8y = 4$	Original equation
_____	Add 22x to each side.
_____	Simplify.
_____	Divide each side by 8.
_____	Simplify.

Check

What is the slope intercept form of $-16x - 4y = -56$? _____

Example 3 Write Linear Equations

JOBS The number of job openings in the United States during a recent year increased by an average of 0.06 million per month since May. In May, there were about 4.61 million job openings in the United States. Write an equation in slope-intercept form to represent the number of job openings in the United States in the months since May.

Use the given information to write an equation in slope-intercept form.

- You are given that there were _____ million job openings in May.

- Let _____ and _____ _____.

- Because the number of job openings is 4.61 million when _____, _____, and because the number of job openings has increased by _____ million each month, _____.

- So, the equation _____ represents the number of job openings in the United States since May.

Think About It!

When $x = 2$, describe the meaning of the equation in the context of the situation.

Check

SOCIAL MEDIA In the first quarter of 2012, there were 183 million users of a popular social media site in North America. The number of users increased by an average of 9 million per year since 2012. Write an equation that represents the number of users in millions of the social media site in North America after 2012.

_____

 Go Online You can complete an Extra Example online.

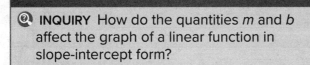

Online Activity Use graphing technology to complete an Explore.

INQUIRY How do the quantities *m* and *b* affect the graph of a linear function in slope-intercept form?

×

Learn Graphing Linear Functions in Slope-Intercept Form

The slope-intercept form of a linear equation is $y = mx + b$, where *m* is the slope and *b* is the *y*-intercept. The variables *m* and *b* are called **parameters** of the equation because changing either value changes the graph of the function.

A **constant function** is a linear function of the form $y = b$. Constant functions where $b \neq 0$ do not cross the *x*-axis. The graphs of constant functions have a slope of 0. The domain of a constant function is all real numbers, and the range is *b*.

Example 4 Graph Linear Functions in Slope-Intercept Form

Graph a linear function with a slope of $-\frac{3}{2}$ and a *y*-intercept of 4.

Write the equation in slope-intercept form and graph the function.

$y = mx + b$

$y = (\underline{\quad\quad}) x \underline{\quad}$

$y = \underline{\quad\quad\quad\quad}$

Study Tip

Negative Slope When counting rise and run, a negative sign may be associated with the value in the numerator or denominator. In this case, we associated the negative sign with the numerator. If we had associated it with the denominator, we would have moved up 3 and left 2 to the point $(-2, 7)$. Notice that this point is also on the line. The resulting line will be the same whether the negative sign is associated with the numerator or denominator.

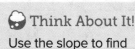 **Think About It!**

Use the slope to find another point on the graph. Explain how you found the point.

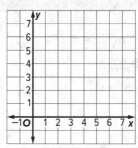
Check

Graph a linear function with a slope of −2 and a y-intercept of 7.

Example 5 Graph Linear Functions

Graph 12x − 3y = 18.

Rewrite the equation in slope-intercept form.

$$12x - 3y = 18$$ Original equation

_____   Subtract 12x from each side.

_____   Simplify.

_____   Divide each side by −3.

_____   Simplify.

Graph the function.

Plot the y-intercept (0, −6).

The slope is $\frac{rise}{run}$ = 4. From (0, −6), move up 4 units and right 1 unit. Plot the point (1, −2).

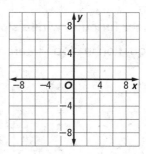

Draw a line through the points (0, −6) and (1, −2).

🐦 **Go Online** You can complete an Extra Example online.

Example 6 Graph Constant Functions

Graph $y = 2$.

Step 1 Plot $(0, 2)$.

Step 2 The slope of $y = 2$ is 0.

Step 3 Draw a line through all the points that have a y-coordinate of 2.

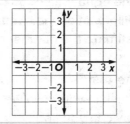

Check

Graph $y = 1$.

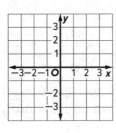

Match each graph with its function.

_____ $y = 8$	_____ $3x + 7y = -28$	_____ $y = \frac{3}{7}x - 4$
_____ $y = -4$	_____ $y = -3x + 8$	_____ $3x - y = 8$

A.

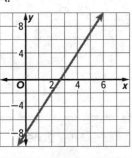

B.

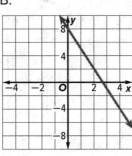

C.

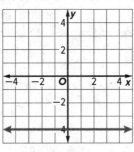

D.

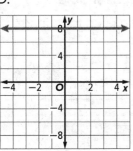

E.

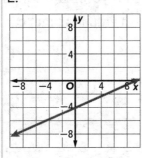

F.

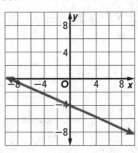

 Go Online You can complete an Extra Example online.

Watch Out!

Slope A line with zero slope is not the same as a line with no slope. A line with zero slope is horizontal, and a line with no slope is vertical.

SHOPPING **The number of online shoppers in the United States can be modeled by the equation $-5.88x + y = 172.3$, where y represents the number of millions of online shoppers in the United States x years after 2010. Estimate the number of people shopping online in 2020.**

1. **What is the task?**
Describe the task in your own words. Then list any questions that you may have. How can you find answers to your questions?

_____

_____

2. **How will you approach the task? What have you learned that you can use to help you complete the task?**

_____

_____

3. **What is your solution?**
Use your strategy to solve the problem. Graph the function.

In 2020, there will be approximately
_____ online shoppers in the
United States.

Online Shoppers in the United States

4. **How can you know that your solution is reasonable?**
🖊 **Write About It!** Write an argument that can be used to defend your solution.

_____

_____

_____

_____

_____

💭 **Think About It!**
Estimate the year when the number of online shoppers in the United States will reach 271 million.

🧭 **Go Online**
to learn about intervals in linear growth patterns in Expand 4-3.

🧭 **Go Online** You can complete an Extra Example online.

Copyright © McGraw-Hill Education

Practice

Go Online You can complete your homework online.

Example 1

Write an equation of a line in slope-intercept form with the given slope and *y*-intercept.

1. slope: 5, *y*-intercept: −3

2. slope: −2, *y*-intercept: 7

3. slope: −6, *y*-intercept: −2

4. slope: 7, *y*-intercept: 1

5. slope: 3, *y*-intercept: 2

6. slope: −4, *y*-intercept: −9

7. slope: 1, *y*-intercept: −12

8. slope: 0, *y*-intercept: 8

Example 2

Write each equation in slope-intercept form.

9. $-10x + 2y = 12$

10. $4y + 12x = 16$

11. $-5x + 15y = -30$

12. $6x - 3y = -18$

13. $-2x - 8y = 24$

14. $-4x - 10y = -7$

Example 3

15. **SAVINGS** Wade's grandmother gave him $100 for his birthday. Wade wants to save his money to buy a portable game console. Each month, he adds $25 to his savings. Write an equation in slope-intercept form to represent Wade's savings *y* after *x* months.

16. **FITNESS CLASSES** Toshelle wants to take strength training classes at the community center. She has to pay a one-time enrollment fee of $25 to join the community center, and then $45 for each class she wants to take. Write an equation in slope-intercept form for the cost of taking *x* classes.

17. **EARNINGS** Macario works part time at a clothing store in the mall. He is paid $9 per hour plus 12% commission on the items he sells in the store. Write an equation in slope-intercept form to represent Macario's hourly wage *y*.

18. **ENERGY** From 2002 to 2005, U.S. consumption of renewable energy increased an average of 0.17 quadrillion BTUs per year. About 6.07 quadrillion BTUs of renewable power were produced in the year 2002. Write an equation in slope-intercept form to find the amount of renewable power *P* in quadrillion BTUs produced in year *y* between 2002 and 2005.

Example 4

Graph a linear function with the given slope and *y*-intercept.

19. slope: 5, *y*-intercept: 8

20. slope: 3, *y*-intercept: 10

21. slope: −4, *y*-intercept: 6

22. slope: −2, *y*-intercept: 8

Examples 5 and 6

Graph each function.

23. $5x + 2y = 8$

24. $4x + 9y = 27$

25. $y = 7$

26. $y = -\frac{2}{3}$

27. $21 = 7y$

28. $3y - 6 = 2x$

Example 7

29. STREAMING An online company charges $13 per month for the basic plan. They offer premium channels for an additional $8 per month.

a. Write an equation in slope-intercept form for the total cost c of the basic plan with p premium channels in one month.

b. Graph the function.

c. What would the monthly cost be for a basic plan plus 3 premium channels?

30. CAR CARE Suppose regular gasoline costs $2.76 per gallon. You can purchase a car wash at the gas station for $3.

a. Write an equation in slope-intercept form for the total cost y of purchasing a car wash and x gallons of gasoline.

b. Graph the function.

c. Find the cost of purchasing a car wash and 8 gallons of gasoline.

Mixed Exercises

Write an equation of a line in slope-intercept form with the given slope and *y*-intercept.

31. slope: $\frac{1}{2}$, y-intercept: -3

32. slope: $\frac{2}{3}$, y-intercept: -5

Graph a function with the given slope and *y*-intercept.

33. slope: 3, y-intercept: -4

34. slope: 4, y-intercept: -6

Graph each function.

35. $-3x + y = 6$

36. $-5x + y = 1$

Write an equation in slope-intercept form for each graph shown.

37.

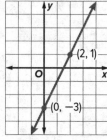

38.

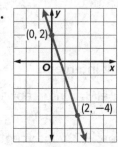

39.

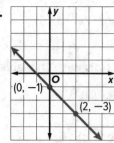

40. MOVIES MovieMania, an online movie rental Web site charges a one-time fee of $6.85 and $2.99 per movie rental. Let m represent the number of movies you watch and let C represent the total cost to watch the movies.

  a. Write an equation that relates the total cost to the number of movies you watch from MovieMania.

  b. Graph the function.

  c. Explain how to use the graph to estimate the cost of watching 13 movies at MovieMania.

  d. SuperFlix has no sign-up fee, just a flat rate per movie. If renting 13 movies at MovieMania costs the same as renting 9 movies at SuperFlix, what does SuperFlix charge per movie? Explain your reasoning.

  e. Write an equation that relates the total cost to the number of movies you watch from SuperFlix. Round to the nearest whole number.

41. FACTORY A factory uses a heater in part of its manufacturing process. The product cannot be heated too quickly, nor can it be cooled too quickly after the heating portion of the process is complete.

  a. The heater is digitally controlled to raise the temperature inside the chamber by 10°F each minute until it reaches the set temperature. Write an equation to represent the temperature, T, inside the chamber after x minutes if the starting temperature is 80°F.

  b. Graph the function.

  c. The heating process takes 22 minutes. Use your graph to find the temperature in the chamber at this point.

  d. After the heater reaches the temperature determined in **part c**, the temperature is kept constant for 20 minutes before cooling begins. Fans within the heater control the cooling so that the temperature inside the chamber decreases by 5°F each minute. Write an equation to represent the temperature, T, inside the chamber x minutes after the cooling begins.

42. SAVINGS When Santo was born, his uncle started saving money to help pay for a car when Santo became a teenager. Santo's uncle initially saved $2000. Each year, his uncle saved an additional $200.

  a. Write an equation that represents the amount, in dollars, Santo's uncle saved y after x years.

  b. Graph the function.

  c. Santo starts shopping for a car when he turns 16. The car he wants to buy costs $6000. Does he have enough money in the account to buy the car? Explain.

43. STRUCTURE Jazmin is participating in a 25.5-kilometer charity walk. She walks at a rate of 4.25 km per hour. Jazmin walks at the same pace for the entire event.

   a. Write an equation in slope-intercept form for the remaining distance y in kilometers of walking for x hours.

   b. Graph the function.

   c. What do the x- and y-intercepts represent in this situation?

   d. After Jazmin has walked 17 kilometers, how much longer will it take her to complete the walk? Explain how you can use your graph to answer the question.

Higher-Order Thinking Skills

For Exercises 44 and 45, refer to the equation $y = -\frac{4}{5}x + \frac{2}{5}$ **where** $-2 \leq x \leq 5$.

44. ANALYZE Complete the table to help you graph the function $y = -\frac{4}{5}x + \frac{2}{5}$ over the interval. Identify any values of x where maximum or minimum values of y occur.

x	$-\frac{4}{5}x + \frac{2}{5}$	y	(x, y)
−2			
0			
5			

45. WRITE A student says you can find the solution to $-\frac{4}{5}x + \frac{2}{5} = 0$ using the graph. Do you agree? Explain your reasoning. Include the solution to the equation in your response.

46. PERSEVERE Consider three points that lie on the same line, $(3, 7)$, $(-6, 1)$, and $(9, p)$. Find the value of p and explain your reasoning.

47. CREATE Linear equations are useful in predicting future events. Create a linear equation that models a real-world situation. Make a prediction from your equation.

Transformations of Linear Functions

Copyright © McGraw-Hill Education

Explore Transforming Linear Functions

Online Activity Use graphing technology to complete an Explore.

> **INQUIRY** How does performing an operation on a linear function change its graph?

Learn Translations of Linear Functions

A **family of graphs** includes graphs and equations of graphs that have at least one characteristic in common. The **parent function** is the simplest of the functions in a family.

The family of linear functions includes all lines, with the parent function $f(x) = x$, also called the **identity function**. A **transformation** moves the graph on the coordinate plane, which can create new linear functions.

One type of transformation is a translation. A **translation** is a transformation in which a figure is slid from one position to another without being turned. A linear function can be slid up, down, left, right, or in two directions.

Vertical Translations

When a constant k is added to a linear function $f(x)$, the result is a vertical translation. The y-intercept of $f(x)$ is translated up or down.

Key Concept • Vertical Translations of Linear Functions

The graph of $g(x) = x + k$ is the graph of $f(x) = x$ translated vertically.

If $k > 0$, the graph of $f(x)$ is translated k units up.

If $k < 0$, the graph of $f(x)$ is translated $|k|$ units down.

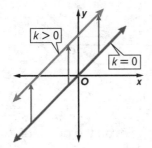

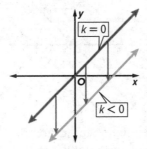

Every point on the graph of $f(x)$ moves k units up.

Every point on the graph of $f(x)$ moves $|k|$ units down.

Today's Goals
- Apply translations to linear functions.
- Apply dilations to linear functions.
- Apply reflections to linear functions.

Today's Vocabulary
family of graphs

parent function

identity function

transformation

translation

dilation

reflection

Study Tip

Slope When translating a linear function, the graph of the function moves from one location to another, but the slope remains the same.

Watch Out!

Translations of $f(x)$ When a translation is the only transformation performed on the identity function, adding a constant before or after evaluating the function has the same effect on the graph. However, when more than one type of transformation is applied, this will not be the case.

🗨 **Think About It!**

What do you notice about the *y*-intercepts of vertically translated functions compared to the *y*-intercept of the parent function?

Example 1 Vertical Translations of Linear Functions

Describe the translation in $g(x) = x - 2$ as it relates to the graph of the parent function.

Graph the parent graph for linear functions.

Because $f(x) = x$, $g(x) = f(x) + k$ where _____.

x	$f(x)$	$f(x) - 2$	$(x, g(x))$
-2	-2	-4	$(-2, -4)$
0	0	-2	$(0, -2)$
1	1	-1	$(1, -1)$

$g(x) = x - 2 \rightarrow$ _____

The constant k is not grouped with x, so k affects the _____, or _____. The value of k is less than 0, so the graph of $f(x) = x$ is translated _____ units down, or 2 units down.

$g(x) = x - 2$ is the translation of the graph of the parent function 2 units down.

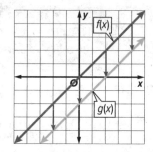

Check

Describe the translation in $g(x) = x - 1$ as it relates to the graph of the parent function.

The graph of $g(x) = x - 1$ is a translation of the graph of the parent function 1 unit _____.

🔿 **Go Online**

You can watch a video to see how to describe translations of functions.

Horizontal Translations

When a constant h is subtracted from the *x*-value before the function $f(x)$ is performed, the result is a horizontal translation. The *x*-intercept of $f(x)$ is translated right or left.

Key Concept • Horizontal Translations of Linear Functions
The graph of $g(x) = (x - h)$ is the graph of $f(x) = x$ translated horizontally.

If $h > 0$, the graph of $f(x)$ is translated h units right.

If $h < 0$, the graph of $f(x)$ is translated $|h|$ units left.

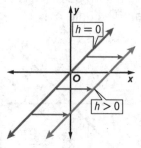

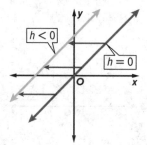

Every point on the graph of $f(x)$ moves h units right.

Every point on the graph of $f(x)$ moves $|h|$ units left.

🔿 **Go Online**

You may want to complete the Concept Check to check your understanding.

🔿 **Go Online** You can complete an Extra Example online.

Example 2 Horizontal Translations of Linear Functions

Describe the translation in $g(x) = (x + 5)$ as it relates to the graph of the parent function.

Graph the parent graph for linear functions.

Because $f(x) = x$, _____
where $h = -5$.

$g(x) = (x + 5) \rightarrow$ _____

x	$x + 5$	$f(x + 5)$	$(x, g(x))$
-2	3	3	$(-2, 3)$
0	5	5	$(0, 5)$
1	6	6	$(1, 6)$

The constant h is grouped with x, so k affects the _____, or _____. The value of h is less than 0, so the graph of $f(x) = x$ is translated _____ units left, or 5 units left.

$g(x) = (x + 5)$ is the translation of the graph of the parent function 5 units left.

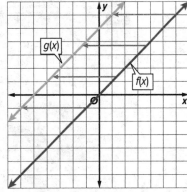

Check

Describe the translation in $g(x) = (x + 12)$ as it relates to the graph of the parent function.

The graph of $g(x) = (x + 12)$ is a translation of the graph of the parent function 12 units _____.

Think About It!

What do you notice about the x-intercepts of horizontally translated functions compared to the x-intercept of the parent function?

Example 3 Multiple Translations of Linear Functions

Describe the translation in $g(x) = (x - 6) + 3$ as it relates to the graph of the parent function.

Graph the parent graph for linear functions.

Because $f(x) = x$,

_____

where _____ and _____.

$g(x) = (x - 6) + 3 \rightarrow$ _____

x	$x - 6$	$f(x - 6)$	$f(x - 6) + 3$	$(x, g(x))$
-2	-8	-8	-5	$(-2, -5)$
0	-6	-6	-3	$(0, -3)$
1	-5	-5	-2	$(1, -2)$

The value of h is grouped with x and is greater than 0, so the graph of $f(x) = x$ is translated _____.

The value of k is not grouped with x and is greater than 0, so the graph of $f(x) = x$ is translated _____.

$g(x) = (x - 6) + 3$ is the translation of the graph of the parent function 6 units right and 3 units up.

Think About It!

Eleni described the graph of $g(x) = (x - 6) + 3$ as the graph of the parent function translated down 3 units. Is she correct? Explain your reasoning.

Go Online You can complete an Extra Example online.

🌐 Example 4 Translations of Linear Functions

TICKETS A Web site sells tickets to concerts and sporting events. The total price of the tickets to a certain game can be modeled by $f(t) = 12t$, where t represents the number of tickets purchased. The Web site then charges a standard service fee of \$4 per order. The total price of an order can be modeled by $g(t) = 12t + 4$. Describe the translation of $g(t)$ as it relates to $f(t)$.

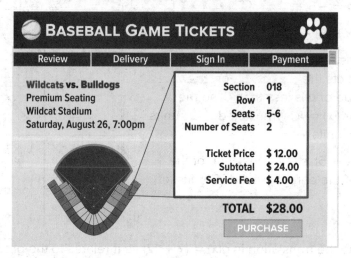

Complete the steps to describe the translation of $g(t)$ as it relates to $f(t)$. Because $f(t) = 12t$, $g(t) = f(t) + k$, where $k = 4$. $g(t) = 12t + 4 \rightarrow f(t) +$ _____

The constant k is added to $f(t)$ after the total price of the tickets has been evaluated and is greater than 0, so the graph will be shifted 4 units up. $g(t) = 12t + 4$ is the translation of the graph of $f(t)$ _____ units _____.

Graph the parent function and the translated function.

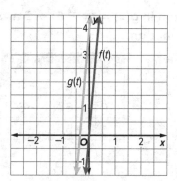

Check

RETAIL Jerome is buying paint for a mural. The total cost of the paint can be modeled by the function $f(p) = 6.99p$. He has a coupon for \$5.95 off his purchase at the art supply store, so the final cost of his purchase can be modeled by $g(p) = 6.99p - 5.95$. Describe the translation in $g(p)$ as it relates to $f(p)$.

_____

 Go Online You can complete an Extra Example online.

Learn Dilations of Linear Functions

A **dilation** stretches or compresses the graph of a function.

When a linear function $f(x)$ is multiplied by a positive constant a, the result $a \cdot f(x)$ is a vertical dilation.

Key Concept • Vertical Dilations of Linear Functions

The graph of $g(x) = ax$ is the graph of $f(x) = x$ stretched or compressed vertically.

If $|a| > 1$, the graph of $f(x)$ is stretched vertically away from the x-axis.	If $0 < |a| < 1$, the graph of $f(x)$ is compressed vertically toward the x-axis.
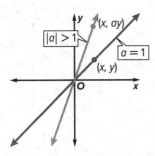	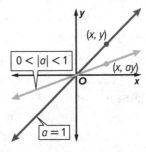
The slope of the graph of $a \cdot f(x)$ is steeper than that of the graph of $f(x)$.	The slope of the graph of $a \cdot f(x)$ is less steep than that of the graph of $f(x)$.

When x is multiplied by a positive constant a before a linear function $f(a \cdot x)$ is evaluated, the result is a horizontal dilation.

Key Concept • Horizontal Dilations of Linear Functions

The graph of $g(x) = (a \cdot x)$ is the graph of $f(x) = x$ stretched or compressed horizontally.

If $|a| > 1$, the graph of $f(x)$ is compressed horizontally toward the y-axis.	If $0 < |a| < 1$, the graph of $f(x)$ is stretched horizontally away from the y-axis.
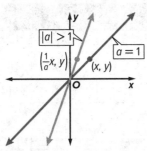	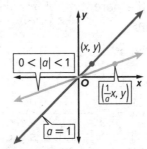
The slope of the graph of $f(a \cdot x)$ is steeper than that of the graph of $f(x)$.	The slope of the graph of $f(a \cdot x)$ is less steep than that of the graph of $f(x)$.

 Go Online You can complete an Extra Example online.

Watch Out!

Dilations of $f(x) = x$
When a dilation is the only transformation performed on the identity function, multiplying by a constant before or after evaluating the function has the same effect on the graph. However, when more than one type of transformation is applied, this will not be the case.

Go Online
You can watch a video to see how to describe dilations of functions.

Example 5 Vertical Dilations of Linear Functions

Describe the dilation in $g(x) = 2(x)$ as it relates to the graph of the parent function.

Graph the parent graph for linear functions.

Since $f(x) = x$, _____

where _____.

$g(x) = 2(x) \rightarrow$ _____

x	$f(x)$	$2f(x)$	$(x, g(x))$
-2	-2	-4	$(-2,-4)$
0	0	0	$(0, 0)$
1	1	2	$(1, 2)$

The positive constant a is not grouped with x, and $|a|$ is greater than 1, so the graph of $f(x) = x$ is _____ by a factor of a, or ____.

$g(x) = 2(x)$ is a vertical stretch of the graph of the parent function. The slope of the graph of $g(x)$ is steeper than that of $f(x)$.

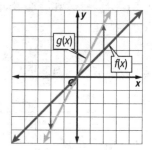

Check

Describe the transformation in $g(x) = 6(x)$ as it relates to the graph of the parent function.

The graph of $g(x) = 6(x)$ is a _____ of the graph of the parent function.

The slope of the graph $g(x)$ is _____ than that of the parent function.

Example 6 Horizontal Dilations of Linear Functions

Describe the dilation in $g(x) = \left(\frac{1}{4}x\right)$ as it relates to the graph of the parent function.

Graph the parent graph for linear functions.

Since $f(x) = x$, _____

where _____.

$g(x) = \left(\frac{1}{4}x\right) \rightarrow$ _____

x	$\frac{1}{4}x$	$f\left(\frac{1}{4}x\right)$	$(x, g(x))$
-4	-1	-1	$(-4, -1)$
0	0	0	$(0, 0)$
4	1	1	$(4, 1)$

The positive constant a is grouped with x, and $|a|$ is between 0 and 1, so the graph of $f(x) = x$ is _____ by a factor of $\frac{1}{a}$, or 4.

$g(x) = \left(\frac{1}{4}x\right)$ is a horizontal stretch of the graph of the parent function. The slope of the graph of $g(x)$ is less steep than that of $f(x)$.

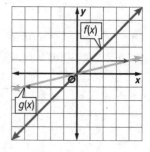

 Go Online You can complete an Extra Example online.

 Think About It!

What do you notice about the slope of the vertical dilation $g(x)$ compared to the slope of $f(x)$?

_____

_____

_____

_____

How does this relate to the constant a in the vertical dilation?

_____

_____

_____

_____

_____

_____

_____

 Think About It!

What do you notice about the slope of the horizontal dilation $g(x)$ compared to the slope of $f(x)$?

_____

_____

_____

How does this relate to the constant a in the horizontal dilation?

_____

_____

_____

Learn Reflections of Linear Functions

A **reflection** is a transformation in which a figure, line, or curve, is flipped across a line. When a linear function $f(x)$ is multiplied by -1 before or after the function has been evaluated, the result is a reflection across the x- or y-axis. Every x- or y-coordinate of $f(x)$ is multiplied by -1.

Key Concept • Reflections of Linear Functions

The graph of $-f(x)$ is the reflection of the graph of $f(x) = x$ across the x-axis.

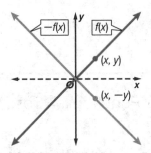

Every y-coordinate of $-f(x)$ is the corresponding y-coordinate of $f(x)$ multiplied by -1.

The graph of $f(-x)$ is the reflection of the graph of $f(x) = x$ across the y-axis.

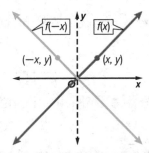

Every x-coordinate of $f(-x)$ is the corresponding x-coordinate of $f(x)$ multiplied by -1.

Example 7 Reflections of Linear Functions Across the x-Axis

Describe how the graph of $g(x) = -\frac{1}{2}(x)$ is related to the graph of the parent function.

Graph the parent graph for linear functions.

Since $f(x) = x$, _____

where _____.

$g(x) = -\frac{1}{2}(x) \rightarrow$ _____

x	$f(x)$	$-\frac{1}{2}f(x)$	$(x, g(x))$
-2	-2	1	$(-2, 1)$
0	0	0	$(0, 0)$
4	4	-2	$(4, -2)$

The constant a is not grouped with x, and $|a|$ is less than 1, so the graph of $f(x) = x$ is

_____.

The negative is not grouped with x, so the graph is also reflected across the _____.

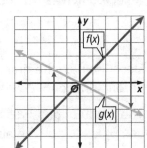

The graph of $g(x) = -\frac{1}{2}(x)$ is the graph of the parent function vertically compressed and reflected across the x-axis.

Go Online You can complete an Extra Example online.

Go Online
You can watch a video to see how to describe reflections of functions.

Watch Out!

Reflections of $f(x) = x$
When a reflection is the only transformation performed on the identity function, multiplying by -1 before or after evaluating the function appears to have the same effect on the graph. However, when more than one type of transformation is applied, this will not be the case.

Talk About It!
In the example, the slope of $g(x)$ is negative. Will this always be the case when multiplying a linear function by -1? Justify your argument.

Check

How can you tell whether multiplying −1 by the parent function will result in a reflection across the x-axis? _____

A. If the constant is not grouped with x, the result will be a reflection across the x-axis.

B. If the constant is grouped with x, the result will be a reflection across the x-axis.

C. If the constant is greater than 0, the result will be a reflection across the x-axis.

D. If the constant is less than 0, the result will be a reflection across the x-axis.

Example 8 Reflections of Linear Functions Across the y-Axis

Describe how the graph of $g(x) = (−3x)$ is related to the graph of the parent function.

Graph the parent graph for linear functions.

Since $f(x) = x,$ _____ where _____.

$g(x) = −3x \rightarrow$ _____

x	−3x	f(−3x)	(x, g(x))
−1	3	3	(−1, 3)
0	0	0	(0, 0)
1	−3	−3	(1, −3)

The constant a is grouped with x, and $|a|$ is greater than 1, so the graph of $f(x) = x$ is _____.

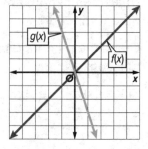

The negative is grouped with x, so the graph is also reflected across the _____.

The graph of $g(x) = (−3x)$ is the graph of the parent function horizontally compressed and reflected across the y-axis.

Check

Describe how the graph of $g(x) = (−10x)$ is related to the graph of the parent function.

The graph of $g(x) = (−10x)$ is the graph of the parent function compressed horizontally and reflected across the _____.

Go Online
You can watch a video to see how to graph transformations of a linear function using a graphing calculator.

 Go Online You can complete an Extra Example online.

Practice

Go Online You can complete your homework online.

Examples 1 through 3

Describe the translation in each function as it relates to the graph of the parent function.

1. $g(x) = x + 11$

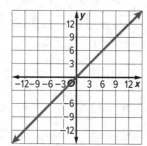

2. $g(x) = x - 8$

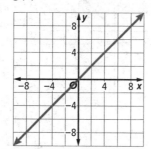

3. $g(x) = (x - 7)$

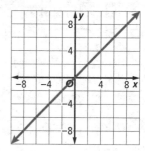

4. $g(x) = (x + 12)$

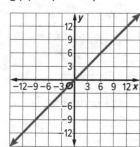

5. $g(x) = (x + 10) - 1$

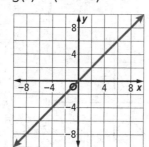

6. $g(x) = (x - 9) + 5$

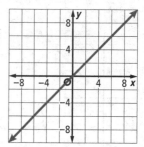

Example 4

7. **BOWLING** The cost for Nobu to go bowling is $4 per game plus an additional flat fee of $3.50 for the rental of bowling shoes. The cost can be modeled by the function $f(x) = 4x + 3.5$, where x represents the number of games bowled. Describe the graph of $g(x)$ as it relates to $f(x)$ if Nobu does not rent bowling shoes.

8. **SAVINGS** Natalie has $250 in her savings account, into which she deposits $10 of her allowance each week. The balance of her savings account can be modeled by the function $f(w) = 250 + 10w$, where w represents the number of weeks. Write a function $g(w)$ to represent the balance of Natalie's savings account if she withdraws $40 to purchase a new pair of shoes. Describe the translation of $f(w)$ that results in $g(w)$.

9. **BOAT RENTAL** The cost to rent a paddle boat at the county park is $8 per hour plus a nonrefundable deposit of $10. The cost can be modeled by the function $f(h) = 8h + 10$, where h represents the number of hours the boat is rented. Describe the graph of $g(h)$ as it relates to $f(h)$ if the nonrefundable deposit increases to $15.

Describe the dilation in each function as it relates to the graph of the parent function.

10. $g(x) = 5(x)$

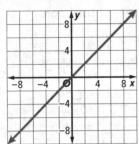

11. $g(x) = \frac{1}{3}(x)$

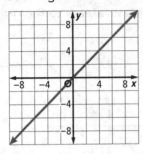

12. $g(x) = 1.5(x)$

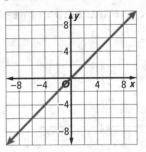

13. $g(x) = (3x)$

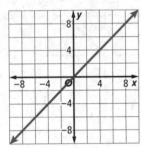

14. $g(x) = \left(\frac{3}{4}x\right)$

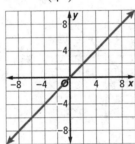

15. $g(x) = (0.4x)$

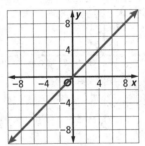

Example 7

Describe how the graph of each function is related to the graph of the parent function.

16. $g(x) = -4(x)$

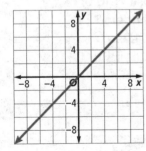

17. $g(x) = -8(x)$

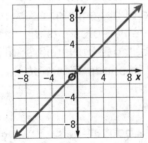

18. $g(x) = -\frac{2}{3}(x)$

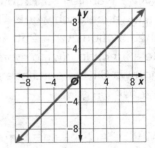

Example 8

Describe how the graph of each function is related to the graph of the parent function.

19. $g(x) = \left(-\frac{4}{5}x\right)$

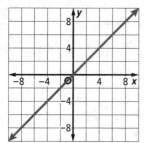

20. $g(x) = (-6x)$

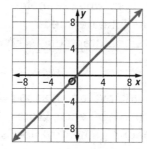

21. $g(x) = (-1.5x)$

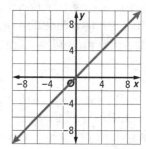

Mixed Exercises

Describe the transformation in each function as it relates to the graph of the parent function.

22. $g(x) = x + 4$

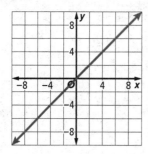

23. $g(x) = (x - 2) - 8$

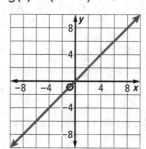

24. $g(x) = \left(-\frac{5}{8}x\right)$

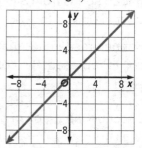

25. $g(x) = \frac{1}{5}(x)$

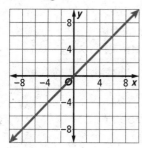

26. $g(x) = -3(x)$

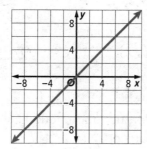

27. $g(x) = (2.5x)$

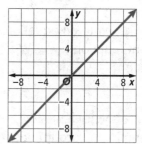

REASONING Write a function $g(x)$ to represent the translated graph.

28. $f(x) = 3x + 7$ translated 4 units up.   **29.** $f(x) = x - 5$ translated 2 units down.

30. **PERIMETER** The function $f(s) = 4s$ represents the perimeter of a square with side length s. Write a function $g(s)$ to represent the perimeter of a square with side lengths that are twice as great. Describe the graph of $g(s)$ compared to $f(s)$.

31. **GAMES** The function $f(x) = 0.50x$ gives the average cost in dollars for x cell phone game downloads that cost an average of $0.50 each. Write a function $g(x)$ to represent the cost in dollars for x cell phone game downloads that cost $1.50 each. Describe the graph of $g(x)$ compared to $f(x)$.

32. **TRAINER** The function $f(x) = 90x$ gives the cost of working out with a personal trainer, where $90 is the trainer's hourly rate, and x represents the number of hours spent working out with the trainer. Describe the dilation, $g(x)$ of the function $f(x)$, if the trainer increases her hourly rate to $100.

33. **DOWNLOADS** Hannah wants to download songs. She researches the price to download songs from Site F. Hannah wrote the function $f(x) = x$, which represents the cost in dollars for x songs downloaded that cost $1.00 each.

   a. Hannah researches the price to download songs from Site G. Write a function $g(x)$ to represent the cost in dollars for x songs downloaded that cost $1.29 each.

   b. Describe the graph of $g(x)$ compared to the graph of $f(x)$.

34. **PERSEVERE** For any linear function, replacing $f(x)$ with $f(x + k)$ results in the graph of $f(x)$ being shifted k units to the right for $k < 0$ and shifted k units to the left for $k > 0$. Does shifting the graph horizontally k units have the same effect as shifting the graph vertically $-k$ units? Justify your answer. Include graphs in your response.

35. **CREATE** Write an equation that is a vertical compression by a factor of a of the parent function $y = x$. What can you say about the horizontal dilation of the function?

36. **WHICH ONE DOESN'T BELONG** Consider the four functions. Which one does not belong in this group? Justify your conclusion.

| $f(x) = 2(x + 1) - 3$ | $f(x) = \frac{1}{2}x - 4$ | $f(x) = -3x + 10$ | $f(x) = 5(x - 7) + 3$ |

Arithmetic Sequences

Learn Arithmetic Sequences

A **sequence** is a set of numbers that are ordered in a specific way. Each number within a sequence is called a **term of a sequence**.

In an **arithmetic sequence**, each term after the first is found by adding a constant, the **common difference** d, to the previous term.

Words	An arithmetic sequence is a numerical pattern that increases or decreases at a constant rate called the common difference.
Examples	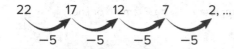 The common difference is −5. 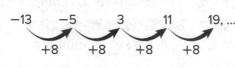 The common difference is 8.

Today's Goals
• Construct arithmetic sequences.
• Apply the arithmetic sequence formula.

Today's Vocabulary
sequence

term of sequence

arithmetic sequence

common difference

nth term of an arithmetic function

Example 1 Identify Arithmetic Sequences

Determine whether the sequence is an arithmetic sequence. Justify your reasoning.

17, 14, 10, 7, 3

Check the difference between terms.

This sequence _____ a common difference between its terms. This _____ an arithmetic sequence.

Check

Determine whether the sequence is an arithmetic sequence. Justify your reasoning.

82, 73, 64, 55, . . .

_____

> **Think About It!**
> How are arithmetic sequences and number patterns alike and different?

🔵 **Go Online** You can complete an Extra Example online.

Talk About It!

Why would it be useful to develop a rule to find terms of a sequence? Explain.

Example 2 Find the Next Term

Determine the next three terms in the sequence.

11, 7, 3, –1

Find the common difference between terms. _____

Add the common difference to the last term of the sequence to find the next terms.

$-1 + ($____$) = $ _____   _____ $+ ($____$) = $ _____   _____ $+ ($____$) = $ _____

Check

Determine the next three terms in the sequence.

31, 18, 5, _____, _____, _____

Go Online You can complete an Extra Example online.

Explore Common Differences

Online Activity Use a real-world situation to complete the Explore.

INQUIRY How can you tell if a set of numbers models a linear function?

Watch Out!

Subscripts Subscripts are used to indicate a specific term. For example, a_8 means the 8th term of the sequence. It does not mean $a \times 8$.

Learn Arithmetic Sequences as Linear Functions

Each term of an arithmetic sequence can be expressed in terms of the first term a_1 and the common difference d.

Key Concept • nth Term of an Arithmetic Sequence

The **nth term of an arithmetic sequence** with the first term a_1 and common difference d is given by $a_n = a_1 + (n - 1)d$, where n is a positive integer.

The graph of an arithmetic sequence includes points that lie along a line. Because there is a constant difference between each pair of points, the function is linear. For the equation of an arithmetic sequence, $a_n = a_1 + (n - 1)d$

Think About It!

Why is the domain of a sequence counting numbers instead of all real numbers?

- n is the independent variable,
- a_n is the dependent variable, and
- d is the slope.

The function of an arithmetic sequence is written as $f(n) = a_1 + (n - 1)d$, where n is a counting number.

Go Online You can complete an Extra Example online.

Example 3 Find the *n*th Term

Use the arithmetic sequence −4, −1, 2, 5, . . . to complete the following.

Part A Write an equation.

_____

Part B Find the 16th term of the sequence.

Use the equation from Part A to find the 16th term in the arithmetic sequence.

_____   Equation from Part *A*

_____   Substitute 16 for *n*.

_____   Multiply.

_____   Simplify.

Check

RUNNING Randi has been training for a marathon, and it is important for her to keep a constant pace. She recorded her time each mile for the first several miles that she ran.

- At 1 mile, her time was 10 minutes and 30 seconds.
- At 2 miles, her time was 21 minutes.
- At 3 miles, her time was 31 minutes and 30 seconds.
- At 4 miles, her time was 42 minutes.

Part A Write a function to represent her sequence of data. Use *n* as the variable.

_____

Part B How long will it take her to run a whole marathon? Round your answer to the nearest thousandth if necessary. (Hint: a marathon is 26.2 miles.)

_____

🌐 Example 4 Apply Arithmetic Sequences as Linear Functions

MONEY Laniqua opened a savings account to save for a trip to Spain. With the cost of plane tickets, food, hotel, and other expenses, she needs to save $1600. She opened the account with $525. Every month, she adds the same amount to her account using the money she earns at her after school job. From her bank statement, Laniqua can write a function that represents the balance of her savings account.

🐦 **Go Online** You can complete an Extra Example online.

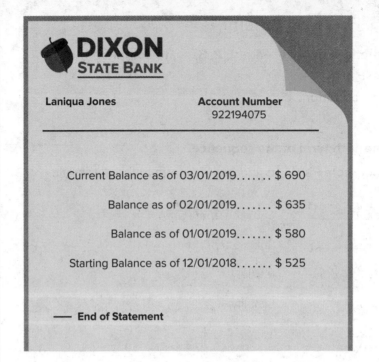

DIXON STATE BANK

Laniqua Jones

Account Number
922194075

Current Balance as of 03/01/2019....... $ 690

Balance as of 02/01/2019....... $ 635

Balance as of 01/01/2019....... $ 580

Starting Balance as of 12/01/2018....... $ 525

—— **End of Statement**

Use a Source

Find the cost of a flight from the airport closest to you to Madrid, the capital of Spain. How many months would Laniqua need to save to afford the ticket?

Study Tip

Graphing You might not need to create a table of the sequence first. However, it might serve as a reminder that an arithmetic sequence is a series of points, not a line.

Part A Create a function to represent the sequence.

First, find the _____.

525   580   635   690

The common difference is ____.

The balance after 1 month is $580, so let $a_1 = 580$. Notice that the starting balance is $525. You can think of this starting point as $a_0 = 580$.

_____ Formula for the nth term

_____ $a_1 = 580$ and $d = 55$

_____ Simplify.

_____

Part B Graph the function and determine its domain.

n	$f(n)$
0	____
1	____
2	____
3	____
4	____
5	____
6	____

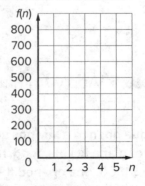

The domain is the number of months since Laniqua opened her savings account. The domain is {0, 1, 2, 3, 4, 5, ...}.

🔎 **Go Online** You can complete an Extra Example online.

Practice

Go Online You can complete your homework online.

Example 1

ARGUMENTS **Determine whether each sequence is an arithmetic sequence. Justify your reasoning.**

1. $-3, 1, 5, 9, \ldots$

2. $\frac{1}{2}, \frac{3}{4}, \frac{5}{8}, \frac{7}{16}, \ldots$

3. $-10, -7, -4, 1, \ldots$

4. $-12.3, -9.7, -7.1, -4.5, \ldots$

5. $4, 7, 9, 12, \ldots$

6. $15, 13, 11, 9, \ldots$

7. $7, 10, 13, 16, \ldots$

8. $-6, -5, -3, -1, \ldots$

Example 2

Find the common difference of each arithmetic sequence. Then find the next three terms.

9. $0.02, 1.08, 2.14, 3.2, \ldots$

10. $6, 12, 18, 24, \ldots$

11. $21, 19, 17, 15, \ldots$

12. $-\frac{1}{2}, 0, \frac{1}{2}, 1, \ldots$

13. $2\frac{1}{3}, 2\frac{2}{3}, 3, 3\frac{1}{3}, \ldots$

14. $\frac{7}{12}, 1\frac{1}{3}, 2\frac{1}{12}, 2\frac{5}{6}, \ldots$

15. $3, 7, 11, 15, \ldots$

16. $22, 19.5, 17, 14.5, \ldots$

17. $-13, -11, -9, -7, \ldots$

18. $-2, -5, -8, -11, \ldots$

Example 3

Use the given arithmetic sequence to write an equation and then find the 7th term of the sequence.

19. $-3, -8, -13, -18, \ldots$

20. $-2, 3, 8, 13, \ldots$

21. $-11, -15, -19, -23, \ldots$

22. $-0.75, -0.5, -0.25, 0, \ldots$

Example 4

23. **SPORTS** Wanda is the manager for the soccer team. One of her duties is to hand out cups of water at practice. Each cup of water is 4 ounces. She begins practice with a 128-ounce cooler of water.

    a. Create a function to represent the arithmetic sequence.

    b. Graph the function.

    c. How much water is remaining after Wanda hands out the 14th cup?

24. **THEATER** A theater has 20 seats in the first row, 22 in the second row, 24 in the third row, and so on for 25 rows.

    a. Create a function to represent the arithmetic sequence.

    b. Graph the function.

    c. How many seats are in the last row?

25. **POSTAGE** The price to send a large envelope first class mail is 88 cents for the first ounce and 17 cents for each additional ounce. The table shows the cost for weights up to 5 ounces.

Weight (ounces)	1	2	3	4	5
Postage (dollars)	0.88	1.05	1.22	1.39	1.56

Source: United States Postal Service

    a. Create a function to represent the arithmetic sequence.

    b. Graph the function.

    c. How much did a large envelope weigh that cost $2.07 to send?

26. **VIDEO DOWNLOADING** Brian is downloading episodes of his favorite TV show to play on his personal media device. The cost to download 1 episode is $1.99. The cost to download 2 episodes is $3.98. The cost to download 3 episodes is $5.97.

    a. Create a function to represent the arithmetic sequence.

    b. Graph the function.

    c. What is the cost to download 9 episodes?

27. USE A MODEL Chapa is beginning an exercise program that calls for 30 push-ups each day for the first week. Each week thereafter, she has to increase her push-ups by 2.

    a. Write a function to represent the arithmetic sequence.

    b. Graph the function.

    c. Which week of her program will be the first one in which she will do at least 50 push-ups a day?

Mixed Exercises

CONSTRUCT ARGUMENTS **Determine whether each sequence is an arithmetic sequence. Justify your argument.**

28. −9, −12, −15, −18, ...

29. 10, 15, 25, 40, ...

30. −10, −5, 0, 5, ...

31. −5, −3, −1, 1, ...

Write an equation for the nth term of each arithmetic sequence. Then graph the first five terms of the sequence.

32. 7, 13, 19, 25, ...

33. 30, 26, 22, 18, ...

34. −7, −4, −1, 2, ...

35. SAVINGS Fabiana decides to save the money she's earning from her after-school job for college. She makes an initial contribution of $3000 and each month deposits an additional $500. After one month, she will have contributed $3500.

    a. Write an equation for the nth term of the sequence.

    b. How much money will Fabiana have contributed after 24 months?

36. NUMBER THEORY One of the most famous sequences in mathematics is the Fibonacci sequence. It is named after Leonardo de Pisa (1170–1250) or Filius Bonacci, alias Leonardo Fibonacci. The first several numbers in the Fibonacci sequence are shown.

1, 1, 2, 3, 5, 8, 13, 21, 34, 55, 89, . . .

Does this represent an arithmetic sequence? Why or why not?

37. STRUCTURE Use the arithmetic sequence 2, 5, 8, 11, ...

    a. Write an equation for the nth term of the sequence.

    b. What is the 20th term in the sequence?

38. CREATE Write a sequence that is an arithmetic sequence. State the common difference, and find a_6.

39. CREATE Write a sequence that is not an arithmetic sequence. Determine whether the sequence has a pattern, and if so describe the pattern.

40. REASONING Determine if the sequence 1, 1, 1, 1, . . . is an arithmetic sequence. Explain your reasoning.

41. CREATE Create an arithmetic sequence with a common difference of −10.

42. PERSEVERE Find the value of x that makes $x + 8$, $4x + 6$, and $3x$ the first three terms of an arithmetic sequence.

43. CREATE For each arithmetic sequence described, write a formula for the nth term of a sequence that satisfies the description.

    a. first term is negative, common difference is negative

    b. second term is −5, common difference is 7

    c. $a_2 = 8$, $a_3 = 6$

Andre and Sam are both reading the same novel. Andre reads 30 pages each day. Sam created the table at the right. Refer to this information for Exercises 44–46.

Sam's Reading Progress	
Day	Pages Left to Read
1	430
2	410
3	390
4	370

44. ANALYZE Write arithmetic sequences to represent each boy's daily progress. Then write the function for the nth term of each sequence.

45. PERSEVERE Enter both functions from Exercise 44 into your calculator. Use the table to determine if there is a day when the number of pages Andre has read is equal to the number of pages Sam has left to read. If so, which day is it? Explain how you used the table feature to help you solve the problem.

46. ANALYZE Graph both functions on your calculator, then sketch the graph in the coordinate plane at the right. How can you use the graph to answer the question from Exercise 45?

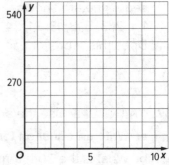

Piecewise and Step Functions

Learn Graphing Piecewise-Defined Functions

Some functions cannot be described by a single expression because they are defined differently depending on the interval of *x*. These functions are **piecewise-defined functions**. A **piecewise-linear function** has a graph that is composed of some number of linear pieces.

Example 1 Graph a Piecewise-Defined Function

To graph a piecewise-defined function, graph each "piece" separately.

Graph $f(x) = \begin{cases} 2x + 4 \text{ if } x \le 1 \\ -x + 3 \text{ if } x > 1 \end{cases}$. **State the domain and range.**

First, graph $f(x) = 2x + 4$ if $x \le 1$.

- Create a table for $f(x) = 2x + 4$ using values of $x > 1$.
- Because *x* is *less than or equal to* 1, place a _____ at _____ to indicate that the endpoint is included in the graph.
- Then, plot the points and draw the graph beginning at (1, 6).

x	y
1	
0	
−1	
−2	
−3	

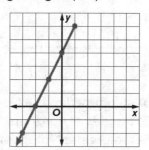

Next, graph $f(x) = -x + 3$ if $x > 1$.

- Create a table for $f(x) = -x + 3$ using values of $x > 1$.
- Because *x* is *greater than but not equal to* 1, place a _____ at _____ to indicate that the endpoint is not included in the graph.
- Then, plot the points and draw the graph beginning at (1, 2).

x	y
1	
2	
3	
4	
5	

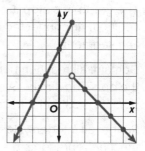

The domain is all real numbers. The range is $y \le$ _____.

⚫ **Go Online** You can complete an Extra Example online.

Copyright © McGraw-Hill Education

Today's Goals
- Identify and graph piecewise-defined functions.
- Identify and graph step functions.

Today's Vocabulary
piecewise-defined function

piecewise-linear function

step function

greatest integer function

🍄 Think About It!

What would be an advantage of graphing the entire expression and removing the portion that is not in the interval?

_____

_____

_____

_____

_____

_____

_____

_____

_____

_____

▶ Go Online
An alternate method is available for this example.

Circles and Dots Do not forget to examine the endpoint(s) of each piece to determine whether there should be a circle or a dot. > and < mean that a circle should be used, while ≥ and ≤ mean that a dot should be used.

Study Tip

Piecewise-Defined Functions When graphing piecewise-defined functions, there should be a dot or line that contains each member of the domain.

Go Online
You can watch a video to see how to graph a piecewise-defined function on a graphing calculator

Check

Part A Graph $f(x) = \begin{cases} -x + 1 \text{ if } x \leq -2 \\ -3x - 2 \text{ if } x > -2 \end{cases}$.

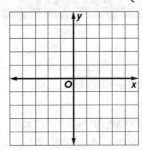

Part B Find the domain and range of the function.

Explore Age as a Function

Online Activity Use a real-world situation to complete an Explore.

> **INQUIRY** When can real-world data be described using a step function?

Learn Graphing Step Functions

A **step function** is a type of piecewise-linear function with a graph that is a series of horizontal line segments. One example of a step function is the **greatest integer function,** written as $f(x) = [\![x]\!]$ in which $f(x)$ is the greatest integer less than or equal to x.

Key Concept • Greatest Integer Function

Type of graph: disjointed line segments

The graph of a step function is a series of disconnected horizontal line segments.

Domain: all real numbers; Because the dots and circles overlap, the domain is all real numbers.

Range: all integers; Because the function represents the greatest integer less than or equal to x, the range is all integers.

Parent function: $f(x) = [\![x]\!]$

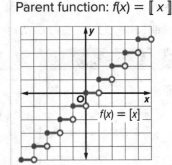

$f(x) = [x]$

Go Online You can complete an Extra Example online.

Example 2 Graph a Greatest Integer Function

Graph $f(x) = [\![x + 1]\!]$. State the domain and range.

First, make a table. Select a few values that are between integers.

x	x + 1	$[\![x + 1]\!]$	
−2	−1	−1	−1, −0.75, and −0.25 are greater than or equal to −1 but less than 0. So, −1 is the greatest integer that is not greater than −1, −0.75, or −0.25.
−1.75	−0.75	−1	
−1.25	−0.25	−1	
−1	0	0	0, 0.5, and 0.75 are greater than or equal to 0 but less than 1. So, 0 is the greatest integer that is not greater than 0, 0.5, or 0.75.
−0.5	0.5	0	
−0.25	0.75	0	
0	1	1	1, 1.25, and 1.5 are greater than or equal to 1 but less than 2. So, 1 is the greatest integer that is not greater than 1, 1.25, or 1.5.
0.25	1.25	1	
0.5	1.5	1	
1	2	2	2, 2.25, and 2.75 are greater than or equal to 2 but less than 3. So, 2 is the greatest integer that is not greater than 2, 2.25, or 2.75.
1.25	2.25	2	
1.75	2.75	2	

On the graph, dots represent included points. Circles represent points that are not included. The domain is all real numbers. The range is all integers. Note that this is the graph of $f(x) = [\![x]\!]$ shifted 1 unit to the left.

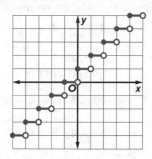

Check

Graph $f(x) = [\![x - 2]\!]$ by making a table.

x	x − 2	$[\![x - 2]\!]$
−1	−3	
−0.75	−2.75	
−0.25	−2.25	−3
0	−2	−2
0.25	−1.75	−2
0.5	−1.5	
1	−1	
1.25	−0.75	−1
1.5	−0.5	
2	0	
2.25	0.25	

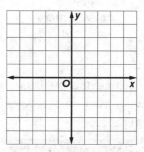

Go Online You can complete an Extra Example online.

Talk About It!

What do you notice about the symmetry, extrema, and end behavior of the function?

Watch Out!

Greatest Integer Function When finding the value of a greatest integer function, do not round to the nearest integer. Instead, always round nonintegers down to the greatest integer that is not greater than the number.

Example 3 Graph a Step Function

SAFETY **A state requires a ratio of 1 lifeguard to 60 swimmers in a swimming pool. This means that 1 lifeguard can watch up to and including 60 swimmers. Make a table and draw a graph that shows the number of lifeguards that must be on duty f(x) based on the number of swimmers in the pool x.**

The number of lifeguards that must be on duty can be represented by a step function.

- If the number of swimmers is greater than 0 but fewer than or equal to 60, only 1 lifeguard must be on duty.

- If the number of swimmers is greater than 60 but fewer than or equal to 120, there must be 2 lifeguards on duty.

- If the number of swimmers is greater than 180 but fewer than or equal to 240, there must be 4 lifeguards on duty.

x	$f(x)$
$0 < x \leq 60$	1
$60 < x \leq 120$	2
$120 < x \leq 180$	
$180 < x \leq 240$	4
$240 < x \leq 300$	
$300 < x \leq 360$	
$360 < x \leq 420$	

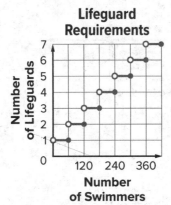

Lifeguard Requirements

Number of Lifeguards / Number of Swimmers

The circles mean that when there are more than a multiple of 60 swimmers,

_____.

The dots represent the _____ number of swimmers that can be in the pool for that particular number of _____ on duty.

Check

PETS At Luciana's pet boarding facility, it costs $35 per day to board a dog. Every fraction of a day is rounded up to the next day. Graph the function representing this situation by making a table.

Days	Cost ($)
$0 < x \leq 1$	
$1 < x \leq 2$	
$2 < x \leq 3$	
$3 < x \leq 4$	
$4 < x \leq 5$	
$5 < x \leq 6$	

Dog Boarding

Cost ($) / Days

🐾 **Go Online** You can complete an Extra Example online.

Practice

Go Online You can complete your homework online.

Example 1

Graph each function. State the domain and range.

1. $f(x) = \begin{cases} \frac{1}{2}x - 1 \text{ if } x > 3 \\ -2x + 3 \text{ if } x \leq 3 \end{cases}$

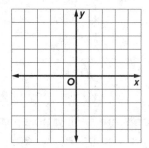

2. $f(x) = \begin{cases} 2x - 5 \text{ if } x > 1 \\ 4x - 3 \text{ if } x \leq 1 \end{cases}$

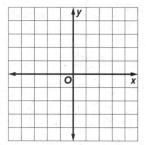

3. $f(x) = \begin{cases} 2x + 3 \text{ if } x \geq -3 \\ -\frac{1}{3}x + 1 \text{ if } x < -3 \end{cases}$

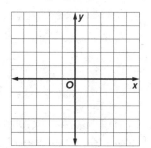

4. $f(x) = \begin{cases} 3x + 4 \text{ if } x \geq 1 \\ x + 3 \text{ if } x < 1 \end{cases}$

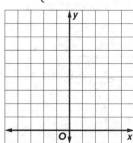

5. $f(x) = \begin{cases} 3x + 2 \text{ if } x > -1 \\ -\frac{1}{2}x - 3 \text{ if } x \leq -1 \end{cases}$

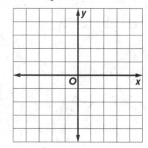

6. $f(x) = \begin{cases} 2x + 1 \text{ if } x < -2 \\ -3x - 1 \text{ if } x \geq -2 \end{cases}$

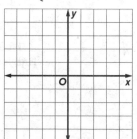

Example 2

Graph each function. State the domain and range.

7. $f(x) = 3 [\![x]\!]$

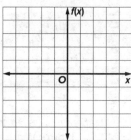

8. $f(x) = [\![-x]\!]$

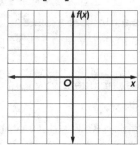

9. $g(x) = -2 [\![x]\!]$

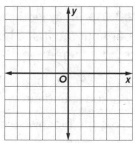

10. $g(x) = [\![x]\!] + 3$

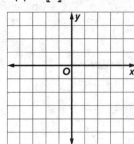

11. $h(x) = [\![x]\!] - 1$

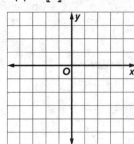

12. $h(x) = \frac{1}{2}[\![x]\!] + 1$

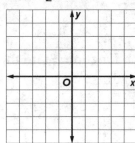

Example 3

13. BABYSITTING Ariel charges $8 per hour as a babysitter. She rounds every fraction of an hour up to the next half-hour. Draw a graph to represent Ariel's total earnings y after x hours.

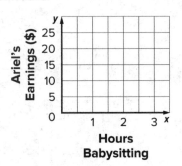

14 FUNDRAISING Students are selling boxes of cookies at a fundraiser. The boxes of cookies can only be ordered by the case, with 12 boxes per case. Draw a graph to represent the number of cases needed y when x boxes of cookies are sold.

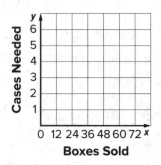

Mixed Exercises

15. PRECISION A package delivery service determines rates for express shipping by the weight of a package, with every fraction of a pound rounded up to the next pound. The table shows the cost of express shipping packages that weigh no more than 5 pounds. Write a piecewise-linear function representing the cost to ship a package that weighs no more than 5 pounds. State the domain and range.

Weight (pounds)	Rate (dollars)
1	16.20
2	19.30
3	22.40
4	25.50
5	28.60

16. EARNINGS Kelly works in a hospital as a medical assistant. She earns $8 per hour the first 8 hours she works in a day and $11.50 per hour each hour thereafter.

a. Organize the information into a table. Include a row for hours worked x, and a row for daily earnings $f(x)$.

b. Write the piecewise equation describing Kelly's daily earnings $f(x)$ for x hours.

c. Draw a graph to represent Kelly's daily earnings.

17. REASONING Write a piecewise function that represents the graph.

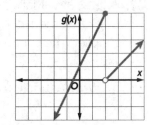

18. STRUCTURE Suppose $f(x) = 2[\![x - 1]\!]$.

a. Find $f(1.5)$.

b. Find $f(2.2)$.

c. Find $f(9.7)$.

d. Find $f(-1.25)$.

19. RENTAL CARS Mr. Aronsohn wants to rent a car on vacation. The rate the car rental company charges is $19 per day. If any fraction of a day is counted as a whole day, how much would it cost for Mr. Aronsohn to rent a car for 6.4 days?

20. USE A MODEL A roadside fruit and vegetable stand determines rates for selling produce, with every fraction of a pound rounded up to the next pound. The table shows the cost of tomatoes by weight in pounds.

a. Write a piecewise-linear function representing the cost of purchasing 0 to 5 pounds of tomatoes, where C is the cost in dollars and p is the number of pounds.

Weight (pounds)	Rate (dollars)
1	3.50
2	7.00
3	10.50
4	14.00
5	17.50

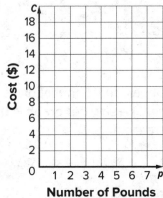

b. Graph the function.

c. State the domain and range.

d. What would be the cost of purchasing 4.3 pounds of tomatoes at the roadside stand?

21. ELECTRONIC REPAIRS Tech Repairs charges $25 for an electronic device repair that takes up to one hour. For each additional hour of labor, there is a charge of $50. The repair shop charges for the next full hour for any part of an hour.

a. Complete the table to organize the information. Include a row for hours of repair x, and a row for total cost $f(x)$.

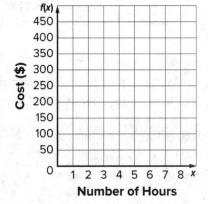

x	0	2	4	6	8
$f(x)$					

b. Write a step function to represent the total cost for every hour x of repair.

c. Graph the function.

d. Devesh was charged $125 to repair his tablet. How long did the repair take to complete?

22. INVENTORY Malik owns a bakery. Every week he orders chocolate chips from a supplier. The supplier's pricing is shown in the table.

Chocolate Chip Pricing	
$4 per pound	Up to 3 pounds
$1.50 per pound	For each pound over 3 pounds

   a. Write a function to represent the cost of chocolate chips.

   b. Malik's budget for chocolate chips for the week is $25. How many whole pounds of chocolate chips can he order?

23. CREATE Write a piecewise-defined function with three linear pieces. Then graph the function.

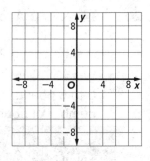

24. FIND THE ERROR Amy graphed a function that gives the height of a car on a roller coaster as a function of time. She said her graph is the graph of a step function. Is this possible? Explain your reasoning.

25. WRITE What is the difference between a step function and a piecewise-defined function?

26. ANALYZE Does the piecewise relation $y = \begin{cases} -2x + 4 \text{ if } x \geq 2 \\ -\frac{1}{2}x - 1 \text{ if } x \leq 4 \end{cases}$ represent a function? Justify your argument.

ANALYZE Refer to the graph for Exercises 27–31.

27. Write a piecewise function to represent the graph.

28. What is the domain?

29. What is the range?

30. Find $f(8.5)$.

31. Find $f(1.2)$.

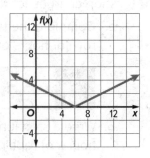

Absolute Value Functions

Explore Parameters of an Absolute Value Function

 Online Activity Use graphing technology to complete the Explore.

> **⊘** **INQUIRY** How does performing an operation on an absolute value function change its graph?

Learn Graphing Absolute Value Functions

The **absolute value function** is a type of piecewise-linear function. An absolute value function is written as $f(x) = a|x - h| + k$, where a, h, and k are constants and $f(x) \geq 0$ for all values of x.

The **vertex** is either the lowest point or the highest point of a function. For the parent function, $y = |x|$, the vertex is at the origin.

Key Concept • Absolute Value Function	
Parent Function	$f(x) = |x|$, defined as $f(x) = \begin{cases} x \text{ if } x \geq 0 \\ -x \text{ if } x < 0 \end{cases}$
Type of Graph	V-Shaped
Domain:	all real numbers
Range:	all nonnegative real numbers

Learn Translations of Absolute Value Functions

Key Concept • Vertical Translations of Absolute Value Functions

If $k > 0$, the graph of $f(x) = |x|$ is translated k units up.
If $k < 0$, the graph of $f(x) = |x|$ is translated $|k|$ units down.

Key Concept • Horizontal Translations of Linear Functions

If $h > 0$, the graph of $f(x) = |x|$ is translated h units right.
If $h < 0$, the graph of $f(x) = |x|$ is translated h units left.

Example 1 Vertical Translations of Absolute Value Functions

Describe the translation in $g(x) = |x| - 3$ as it relates to the graph of the parent function.

Graph the parent function, $f(x) = |x|$, for absolute values.

The constant, k, is outside the absolute value signs, so k affects the y-values. The graph will be a vertical translation.

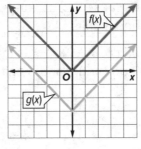

(continued on the next page)

Today's Goals
- Graph absolute value functions.
- Apply translations to absolute value functions.
- Apply dilations to absolute value functions.
- Apply reflections to absolute value functions.
- Interpret constants within equations of absolute value functions.

Today's Vocabulary
absolute value function

vertex

🤔 Think About It!
Why does adding a positive value of k shift the graph k units up?

Go Online
You can watch a video to see how to describe translations of functions.

Study Tip
Horizontal Shifts
Remember that the general form of an absolute value function is $y = a|x - h| + k$. So, $y = |x + 7|$ is actually $y = |x - (-7)|$ in the function's general form.

Since $f(x) = |x|$, $g(x) = f(x) + k$ where $k = -3$.
$g(x) = |x| - 3 \longrightarrow g(x) = f(x) + (-3)$

The value of k is less than 0, so the graph will be translated $|k|$ units down, or 3 units down.

$g(x) = |x| - 3$ is a translation of the graph of the parent function 3 units down.

Example 2 Horizontal Translations of Absolute Value Functions

Describe the translation in $j(x) = |x - 4|$ as it relates to the parent function.

Graph the parent function, $f(x) = |x|$, for absolute values.

The constant, h, is inside the absolute value signs, so h affects the input or, x-values. The graph will be a horizontal translation.

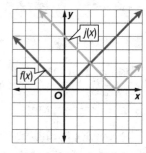

Since $f(x) = |x|$, $j(x) = f(x - h)$, where $h = 4$.
$j(x) = |x - 4| \longrightarrow j(x) = f(x - 4)$

The value of h is greater than 0, so the graph will be translated h units right, or 4 units right.

$j(x) = |x - 4|$ is the translation of the graph of the parent function 4 units right.

Example 3 Multiple Translations of Absolute Value Functions

Describe the translation in $g(x) = |x - 2| + 3$ as it relates to the graph of the parent function.

The equation has both h and k values. The input and output will be affected by the constants. The graph of $f(x) = |x|$ is vertically and horizontally translated.

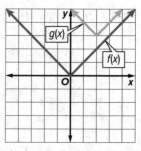

Since $f(x) = |x|$, $g(x) = f(x - h) + k$ where $h = 2$ and $k = 3$.

Because _____ and _____, the graph is translated 2 units _____ and 3 units _____.

$g(x) = |x - 2| + 3$ is the translation of the graph of the parent function 2 units right and 3 units_____.

 Go Online You can complete an Extra Example online.

Think About It!

Since the vertex of the parent function is at the origin, what is a quick way to determine where the vertex is of $q(x) = |x - h| + k$?

Emilio says that the graph of $g(x) = |x + 1| - 1$ is the same graph as $f(x) = |x|$. Is he correct? Why or why not?

Example 4 Identify Absolute Value Functions from Graphs

Use the graph of the function to write its equation.

The graph is the translation of the parent graph 1 unit to the right.

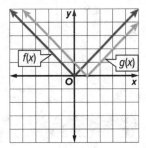

$g(x) = |x - h|$ General equation for a horizontal translation

$g(x) = |x - 1|$ The vertex is 1 unit to the right of the origin.

Example 5 Identify Absolute Value Functions from Graphs (Multiple Translations)

Use the graph of the function to write its equation.

The graph is a translation of the parent graph 2 units to the left and 5 units down.

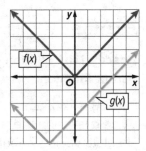

$g(x) = |x - h| + k$ General equation for translations

$g(x) = |x - (-2)| + k$ The vertex is 2 units left of the origin.

$g(x) = |x - (-2)| + (-5)$ The vertex is 5 units down from the origin.

$g(x) = |x + 2| - 5$ Simplify.

Learn Dilations of Absolute Value Functions

Multiplying by a constant a after evaluating an absolute value function creates a vertical change, either a stretch or compression.

Key Concept • Vertical Dilations of Absolute Value Functions
If $|a| > 1$, the graph of $f(x) = |x|$ is stretched vertically.
If $0 < |a| < 1$, the graph of $f(x) = |x|$ is compressed vertically.

When an input is multiplied by a constant a before for the absolute value is evaluated, a horizontal change occurs.

Key Concept • Horizontal Dilations of Absolute Value Functions
If $|a| > 1$, the graph of $f(x) = |x|$ is compressed horizontally.
If $0 < |a| < 1$, the graph of $f(x) = |x|$ is stretched horizontally.

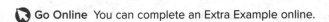 **Go Online** You can complete an Extra Example online.

Talk About It!

How is the value of a in an absolute value function related to slope? Explain.

Copyright © McGraw-Hill Education

Lesson 4-7 • Absolute Value Functions 269

Example 6 Dilations of the Form $a|x|$ When $a > 1$

Describe the dilation in $g(x) = \frac{5}{2}|x|$ as it relates to the graph of the parent function.

Since $f(x) = |x|$, $g(x) = a \cdot f(x)$, where $a = \frac{5}{2}$.

$g(x) = \frac{5}{2}|x| \longrightarrow g(x) = \frac{5}{2} \cdot f(x)$

$g(x) = \frac{5}{2}|x|$ is a vertical stretch of the graph of the parent graph.

x	$\|x\|$	$\frac{5}{2}\|x\|$	$(x, g(x))$
-4	$\|-4\| = 4$	10	$(-4, 10)$
-2	$\|-2\| = 2$	5	$(-2, 5)$
0	$\|0\| = 0$	0	$(0, 0)$
2	$\|2\| = 2$	5	$(2, 5)$
4	$\|4\| = 4$	10	$(4, 10)$

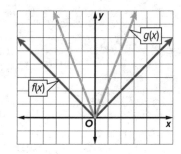

Go Online
You can watch a video to see how to describe dilations of functions.

Think About It!
How are $a|x|$ and $|ax|$ evaluated differently?

Example 7 Dilations of the Form $|ax|$

Describe the dilation in $p(x) = |2x|$ as it relates to the graph of the parent function.

For $p(x) = |2x|$, _____.
Since a is inside the absolute value symbols, the input is first multiplied by a. Then, the absolute value of ax is evaluated.

x	$\|2x\|$	$p(x)$	$(x, p(x))$
-4	$\|2(-4)\| = \|-8\|$	8	$(-4, 8)$
-2	$\|2(-2)\| = \|-4\|$	4	$(-2, 4)$
0	$\|2(0)\| = \|0\|$	0	$(0, 0)$
2	$\|2(2)\| = \|4\|$	4	$(2, 4)$
4	$\|2(4)\| = \|8\|$	8	$(4, 8)$

Plot the points from the table.

Since $f(x) = |x|$, _____ where _____.

$p(x) = |2x| \rightarrow$ _____

$p(x) = |2x|$ is a _____ of the graph of the parent graph.

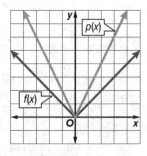

Watch Out!

Differences in Dilations
Although $a|x|$ and $|ax|$ appear to have the same effect on a function, they are evaluated differently and that difference is more apparent when a function is dilated and translated horizontally. For a function with multiple transformations, it is best to first create a table.

Check

Match each description of the dilation with its equation.

stretched vertically

compressed vertically

stretched horizontally

compressed horizontally

$j(x) = \left|\frac{4}{3}x\right|$

$q(x) = \left|\frac{1}{5}x\right|$

$p(x) = 6|x|$

$g(x) = \frac{5}{7}|x|$

Go Online You can complete an Extra Example online.

Example 8 | Dilations When $0 < a < 1$

Describe the dilation in $j(x) = \frac{1}{3}|x|$ as it relates to the graph of the parent function.

For $j(x) = \frac{1}{3}|x|$, _____.
Because a is outside the absolute value signs, the absolute value of the input is evaluated first. Then, the function is multiplied by a.

x	$|x|$	$\frac{1}{3}|x|$	$(x, j(x))$
-6	$|-6| = 6$	2	$(-6, 2)$
-3	$|-3| = 3$	1	$(-3, 1)$
0	$|0| = 0$	0	$(0, 0)$
3	$|3| = 3$	1	$(3, 1)$
6	$|6| = 6$	2	$(6, 2)$

Plot the points from the table.

Because $f(x) = |x|$, _____

where _____.

$j(x) = \frac{1}{3}|x| \rightarrow$ _____

$j(x) = \frac{1}{3}|x|$ is a _____ of the graph of the parent function.

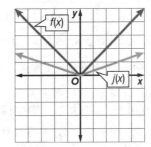

Check

Write an equation for each graph shown.

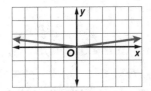 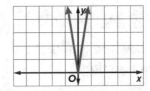

Learn Reflections of Absolute Value Functions

The graph of $-a|x|$ appears to be flipped upside down compared to $a|x|$, and they are symmetric about the x-axis.

> **Key Concept • Reflections of Absolute Value Functions Across the x-axis**
>
> The graph of $-af(x)$ is the reflection of the graph of $af(x) = a|x|$ across the x-axis.

When the only transformation occurring is a reflection or a dilation and reflection, the graphs of $f(ax)$ and $f(-ax)$ appear the same.

> **Key Concept • Reflections of Absolute Value Functions Across the y-axis**
>
> The graph of $f(-ax)$ is the reflection of the graph of $f(ax) = |ax|$ across the y-axis.

Go Online
You can watch a video to see how to describe reflections of functions.

Think About It!
Why would $g(x) = |-2x|$ and $j(x) = |2x|$ appear to be the same graphs?

Go Online You can complete an Extra Example online.

Example 9 Graphs of Reflections with Transformations

Describe how the graph of $j(x) = -|x + 3| + 5$ is related to the graph of the parent function.

x	$|x + 3|$	$-|x + 3|$	$-|x + 3| + 5$	$(x, j(x))$
-5	$|-5 + 3| = |-2| = 2$	-2	$-2 + 5 = 3$	$(-5, 3)$
-4	$|-4 + 3| = |-1| = 1$	-1	$-1 + 5 = 4$	$(-4, 4)$
-3	$|-3 + 3| = |0| = 0$	0	$0 + 5 = 5$	$(-3, 5)$
-2	$|-2 + 3| = |1| = 1$	-1	$-1 + 5 = 4$	$(-2, 4)$
-1	$|-1 + 3| = |2| = 2$	-2	$-2 + 5 = 3$	$(-1, 3)$

First, the absolute value of $x + 3$ is evaluated.
Then, the function is multiplied by $-1 \cdot a$.
Finally, 5 is added to the function.

Plot the points from the table.

Because $f(x) = |x|$, _____

where _____ .

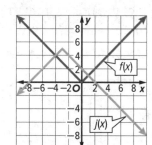

$j(x) = -|x + 3| + 5 \rightarrow$ _____

$j(x) = -|x + 3| + 5$ is the graph of the parent function reflected across the _____ and translated 3 units _____ and 5 units _____ .

Example 10 Graphs of $y = -a|x|$

Describe how the graph of $q(x) = -\frac{3}{4}|x|$ is related to the graph of the parent function.

First, the absolute value of x is evaluated. Then, the function is multiplied by $-1 \cdot a$.

Plot the points from the table.

x	$|x|$	$-\frac{3}{4}|x|$	$(x, q(x))$
-8	$|-8| = 8$	-6	$(-8, -6)$
-4	$|-4| = 4$	-3	$(-4, -3)$
0	$|0| = 0$	0	$(0, 0)$
4	$|4| = 4$	3	$(4, 3)$
8	$|8| = 8$	6	$(8, 6)$

Because $f(x) = |x|$, _____

where _____ .

$q(x) = -\frac{3}{4}|x| \rightarrow$ _____

$q(x) = -\frac{3}{4}|x|$ is the graph of the parent function reflected across the _____ and

_____ .

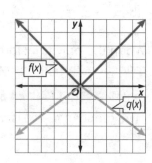

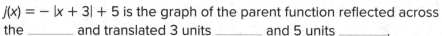

 Go Online You can complete an Extra Example online.

Example 11 Graphs of $y = |-ax|$

Describe how the graph of $g(x) = |-4x|$ is related to the graph of the parent function.

First, the input is multiplied by $-1 \cdot a$. Then the absolute value of $-ax$ is evaluated.

Because $f(x) = |x|$, _____

where _____.

$g(x) = |-4x| \rightarrow$ _____

$g(x) = |-4x|$ is the graph of the parent function reflected across the _____ and _____.

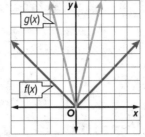

Learn Transformations of Absolute Value Functions

You can use the equation of a function to understand the behavior of the function. Because the constants a, h, and k affect the function in different ways, they can help develop an accurate graph of the function.

Concept Summary Transformations of Graphs of Absolute Value Functions

$$g(x) = a|x - h| + k$$

Horizontal Translation, h

If _____, the graph of $f(x) = |x|$ is translated ____ units _____.

If $h < 0$, the graph of $f(x) = |x|$ is translated $|h|$ units left.

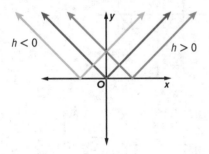

Vertical Translation, k

If _____, the graph of $f(x) = |x|$ is translated ____ units _____.

If $k < 0$, the graph of $f(x) = |x|$ is translated $|k|$ units down.

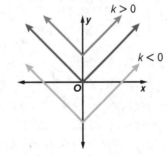

Reflection, a

If $a > 0$, the graph opens up.

If _____, the graph opens down.

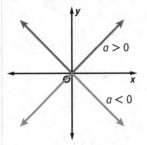

Dilation, a

If _____, the graph of $f(x) = |x|$ is stretched vertically.

If $0 < |a| < 1$, the graph is compressed vertically.

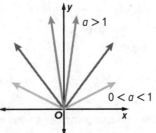

Go Online You can complete an Extra Example online.

Think About It!

Describe how the graph of $y = |-ax|$ is related to the parent function.

Why does there appear to be no reflection for the graph of $y = |-ax|$?

Choose the phrase that best describes how each parameter affects the graph of $g(x) = -5|x - 2| + 3$ in relation to the parent function.

−5 _____

2 _____

3 _____

Translates right	Translates left
Translates up	Translates down
Stretches vertically only	Reflects and compresses vertically
Compresses vertically only	Reflects and stretches vertically

Example 12 Graph an Absolute Value Function with Multiple Translations

Graph $g(x) = |x + 1| - 4$. State the domain and range.

$a = 1$	The graph is not reflected or dilated in relation to the parent function.	
$h = -1$	The graph is _____ 1 unit left from the parent function.	
$k = -4$	The graph is translated _____ units down from the parent function.	

The graph of $g(x) = |x + 1| - 4$ is the graph of the parent function translated 1 unit left and 4 units down without dilation or reflection. The domain is _____. The range is _____.

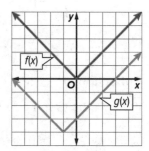

Watch Out!

Dilations and Translations
Don't assume that $j(x) = 2|x - 5| + 1$ and $p(x) = |2x - 5| + 1$ are the same graph. Functions are evaluated differently depending on whether a is inside or outside the absolute value symbols. It might be best to create a table to generate an accurate graph.

Go Online
You can watch a video to see how to graph a transformed absolute value function.

Go Online You can complete an Extra Example online.

Example 13 Graph an Absolute Value Function with Translations and Dilation

Graph $j(x) = |3x - 6|$. State the domain and range.

Because a is inside the absolute value symbols, the effect of h on the translation changes.

Evaluate the function for several values of x to find points on the graph.

x	$(x, j(x))$
0	(0, 6)
1	(1, 3)
2	(2, 0)
3	(3, 3)
4	(4, 6)

The graph of $j(x) = |3x - 6|$ is the graph of the parent function compressed horizontally and translated 2 units right.

The domain is _____. The range is _____

_____.

Example 14 Graph an Absolute Value Function with Translations and Reflection

Graph $p(x) = -|x - 3| + 5$. State the domain and range.

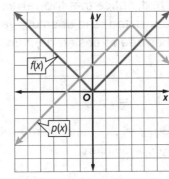

In $p(x) = -|x - 3| + 5$, the parent function is reflected across the x-axis because the absolute value is being multiplied by -1.

The function is then translated 3 units right.

Finally, the function is translated 5 units up.

$p(x) = -|x - 3| + 5$ is the graph of the parent function translated 3 units _____ and 5 units _____ and reflected across the _____.

The domain is _____. The range is _____

_____.

Go Online You can complete an Extra Example online.

💭 **Think About It!**

How is the vertical translation k of an absolute value function related to its range?

Copyright © McGraw-Hill Education

🌐 Example 15 Apply Graphs of Absolute Value Functions

BUILDINGS **Determine an absolute value function that models the shape of The Palace of Peace and Reconciliation.**

To write the equation for the absolute value function, we must determine the values of a, h, and k in $f(x) = a|x - h| + k$ from the graph.

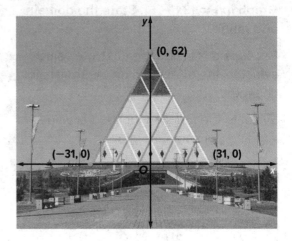

If we consider the absolute value as a piecewise function, we can find the slope of one side of the graph to determine the value of a.

Because this function opens downward, the graph is a reflection of the parent graph across the x-axis. So we know that the a-value in the equation should be negative.

$m = \dfrac{y_2 - y_1}{x_2 - x_1}$ The Slope Formula

$= \underline{\hspace{1.5cm}}$ $(0, 62) = (x_1, y_1)$ and $(31, 0) = (x_2, y_2)$

$= \underline{\hspace{1cm}}$ or $\underline{\hspace{1cm}}$

Next, notice that the vertex is not located at the origin. It has been translated. The absolute value function is not shifted left or right, but has been translated 62 units up from the origin.

$y = \underline{\hspace{3cm}}$ $a = -2, h = 0, k = 62$

$y = \underline{\hspace{3cm}}$ Simplify.

So, $y = \underline{\hspace{2.5cm}}$ models the shape of The Palace of Peace and Reconciliation.

Check

GLASS PRODUCTION Certain types of glass heat and cool at a nearly constant rate when they are melted to create new glass products. Use the graph to determine the equation that represents this process.

$y = \underline{\hspace{1cm}} |x - \underline{\hspace{1cm}}| + \underline{\hspace{1.5cm}}$

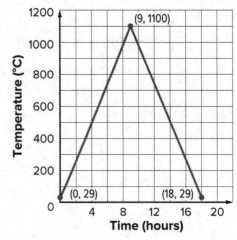

🌐 **Go Online** You can complete an Extra Example online.

🡕 **Go Online to** practice what you've learned about graphing special functions in the Put It All Together over Lessons 4–6 through 4–7.

Practice

🡒 **Go Online** You can complete your homework online.

Examples 1 through 3

Describe the translation in $g(x)$ as it relates to the graph of the parent function.

1. $g(x) = |x| - 5$

2. $g(x) = |x + 6|$

3. $g(x) = |x - 2| + 7$

4. $g(x) = |x + 1| - 3$

5. $g(x) = |x| + 1$

6. $g(x) = |x - 8|$

Examples 4 and 5

Use the graph of the function to write its equation.

7.
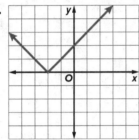

8.
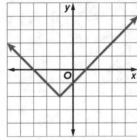

9.
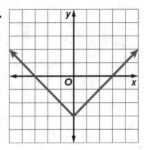

10.
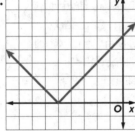

11.
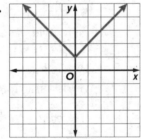

12.
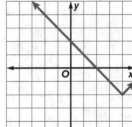

Examples 6 through 8

Describe the dilation in $g(x)$ as it relates to the graph of the parent function.

13. $g(x) = \frac{2}{5}|x|$

14. $g(x) = |0.7x|$

15. $g(x) = 1.3|x|$

16. $g(x) = |3x|$

17. $g(x) = \left|\frac{1}{6}x\right|$

18. $g(x) = \frac{5}{4}|x|$

Examples 9 through 11

Describe how the graph of g(x) is related to the graph of the parent function.

19. $g(x) = -3|x|$

20. $g(x) = -|x| - 2$

21. $g(x) = \left|-\frac{1}{4}x\right|$

22. $g(x) = -|x - 7| + 3$

23. $g(x) = |-2x|$

24. $g(x) = -\frac{2}{3}|x|$

Examples 12 through 14

USE TOOLS Graph each function. State the domain and range.

25. $g(x) = |x + 2| + 3$

26. $g(x) = |2x - 2| + 1$

27. $f(x) = \left|\frac{1}{2}x - 2\right|$

28. $f(x) = |2x - 1|$

29. $f(x) = \frac{1}{2}|x| + 2$

30. $h(x) = -2|x - 3| + 2$

31. $f(x) = -4|x + 2| - 3$

32. $g(x) = -\frac{2}{3}|x + 6| - 1$

33. $h(x) = -\frac{3}{4}|x - 8| + 1$

Example 15

Determine an absolute value function that models each situation.

34. ESCALATORS An escalator travels at a constant speed. The graph models the escalator's distance, in floors, from the second floor x seconds after leaving the ground floor.

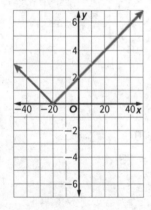

35. TRAVEL The graph models the distance, in miles, a car traveling from Chicago, Illinois is from Annapolis, Maryland, where x is the number of hours since the car departed from Chicago, Illinois.

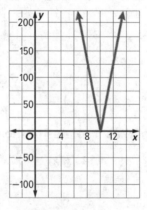

Mixed Exercises

MODELING Graph each function. State the domain and range. Describe how each graph is related to its parent graph.

36. $f(x) = -4|x - 2| + 3$

37. $f(x) = |2x|$

38. $f(x) = |2x + 5|$

278 Module 4 · Linear and Nonlinear Functions

Use the graph of the function to write its equation.

39.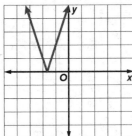

40.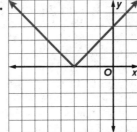

41.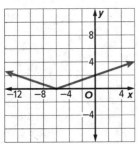

42.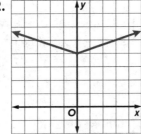

43. SUNFLOWER SEEDS A company produces and sells bags of sunflower seeds. A medium-sized bag of sunflower seeds must contain 16 ounces of seeds. If the amount of sunflower seeds s in the medium-sized bag differs from the desired 16 ounces by more than x, the bag cannot be delivered to companies to be sold. Write an equation that can be used to find the highest and lowest amounts of sunflower seeds in a medium-sized bag.

44. REASONING The function $y = \frac{5}{4}|x - 5|$ models a car's distance in miles from a parking lot after x minutes. Graph the function. After how many minutes will the car reach the parking lot?

45. STATE YOUR ASSUMPTION A track coach set up an agility drill for members of the track team. According to the coach, 21.7 seconds is the target time to complete the agility drill. If the time differs from the desired 21.7 seconds by more than x, the track coach may require members of the track team to change their training. Write an equation that can be used to find the fastest and slowest times members of the track team can complete the agility drill so that their training does not have to change. If $x = 3.2$, what can you assume about the range of times the coach wants the members of the track team to complete the agility drill? Solve your equation for $x = 3.2$ and use the results to justify your assumption.

46. SCUBA DIVING The function $y = 3|x - 12| - 36$ models a scuba diver's elevation in feet compared to sea level after x minutes. Graph the function. How far below sea level is the scuba diver at the deepest point in their dive?

47. MANUFACTURING A manufacturing company produces boxes of cereal. A small box of cereal must have 12 ounces. If the amount of cereal b in a small box differs from the desired 12 ounces by more than x, the box cannot be shipped for selling. Write an equation that can be used to find the highest and lowest amounts of cereal in a small box.

48. STRUCTURE Amelia is competing in a bicycle race. The race is along a circular path. She is 6 miles from the start line. She is approaching the start line at a speed of 0.2 mile per minute. After Amelia reaches the start line, she continues at the same speed, taking another lap around the track.

  a. Organize the information into a table. Include a row for time in minutes x, and a row for distance from start line $f(x)$.

  b. Draw a graph to represent Amelia's distance from the start line.

49. WRITE Use transformations to describe how the graph of $h(x) = -|x + 2| - 3$ is related to the graph of the parent absolute value function.

50. ANALYZE On a straight highway, the town of Garvey is located at mile marker 200. A car is located at mile marker x and is traveling at an average speed of 50 miles per hour.

  a. Write a function $T(x)$ that gives the time, in hours, it will take the car to reach Garvey. Then graph the function on the coordinate plane.

  b. Does the graph have a maximum or minimum? If so, name it and describe what it represents in the context of the problem.

51. PERSEVERE Write the equation $y = |x - 3| + 2$ as a piecewise-defined function. Then graph the piecewise function.

52. CREATE Write an absolute value function, $f(x)$, that has a domain of all real numbers and a range that is greater than or equal to 4. Be sure your function also includes a dilation of the parent function. Describe how your function relates to the parent absolute value graph. Then graph your function.

@ Essential Question

What can a function tell you about the relationship that it represents?

Module Summary

Lessons 4-1 through 4-3

Graphing Linear Functions, Rate of Change, and Slope

- The graph of an equation represents all of its solutions.

- The x-value of the y-intercept is 0. The y-value of the x-intercept is 0.

- The rate of change is how a quantity is changing with respect to a change in another quantity. If x is the independent variable and y is the dependent variable, then rate of change $= \frac{\text{change in } y}{\text{change in } x}$.

- The slope m of a nonvertical line through any two points can be found using $m = \frac{y_2 - y_1}{x_2 - x_1}$.

- A line with positive slope slopes upward from left to right. A line with negative slope slopes downward from left to right. A horizontal line has a slope of 0. The slope of a vertical line is undefined.

Lesson 4-4

Transformations of Linear Functions

- When a constant k is added to a linear function $f(x)$, the result is a vertical translation.

- When a linear function $f(x)$ is multiplied by a constant a, the result $a \cdot f(x)$ is a vertical dilation.

- When a linear function $f(x)$ is multiplied by -1 before or after the function has been evaluated, the result is a reflection across the x- or y-axis.

Lesson 4-5

Arithmetic Sequences

- An arithmetic sequence is a numerical pattern that increases or decreases at a constant rate called the common difference.

- The nth term of an arithmetic sequence with the first term a_1 and common difference d is given by $a_n = a_1 + (n - 1)d$, where n is a positive integer.

Lessons 4-6, 4-7

Special Functions

- A piecewise-linear function has a graph that is composed of a number of linear pieces.

- A step function is a type of piecewise-linear function with a graph that is a series of horizontal line segments.

- An absolute value function is V-shaped.

Study Organizer

 Foldables

Use your Foldable to review this module. Working with a partner can be helpful. Ask for clarification of concepts as needed.

Test Practice

1. GRAPH Jalyn made a table of how much money she will earn from babysitting. (Lesson 4-1)

Hours Babysitting	Money Earned
1	5
2	10
3	15
4	20

Use the table to graph the function in the coordinate grid.

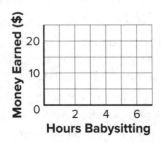

2. TABLE ITEM What are the missing values in the table that show the points on the graph of $f(x) = 2x - 4$? (Lesson 4-1)

x	−2	0	2	4	6
f(x)	−8	−4			

3. OPEN RESPONSE Mr. Hernandez is draining his pool to have it cleaned. At 8:00 A.M., it had 2000 gallons of water and at 11:00 A.M. it had 500 gallons left to drain. What is the rate of change in the amount of water in the pool? (Lesson 4-2)

4. MULTIPLE CHOICE Find the slope of the graphed line. (Lesson 4-2)

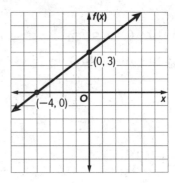

Ⓐ $-\dfrac{4}{3}$

Ⓑ $-\dfrac{3}{4}$

Ⓒ $\dfrac{3}{4}$

Ⓓ $\dfrac{4}{3}$

5. MULTIPLE CHOICE Determine the slope of the line that passes through the points (4, 10) and (2, 10). (Lesson 4-2)

Ⓐ −1

Ⓑ 0

Ⓒ 1

Ⓓ undefined

6. GRAPH Graph the equation of a line with a slope of −3 and a y-intercept of 2. (Lesson 4-3)

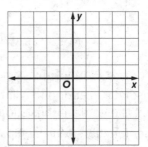

7. MULTIPLE CHOICE What is the slope of the line that passes through (3, 4) and (−7, 4)? (Lesson 4-3)

Ⓐ 0

Ⓑ undefined

Ⓒ −2

Ⓓ −10

8. **MULTIPLE CHOICE** A teacher buys 100 pencils to keep in her classroom at the beginning of the school year. She allows the students to borrow pencils, but they are not always returned. On average, she loses about 8 pencils a month. Write an equation in slope-intercept form that represents the number of pencils she has left y after a number of x months. (Lesson 4-3)

Ⓐ $y = -8x - 100$

Ⓑ $y = -8x + 100$

Ⓒ $y = 8x + 100$

Ⓓ $y = 8x - 100$

9. **OPEN RESPONSE** Name the transformation that changes the slope, or the steepness of a graph. (Lesson 4-4)

10. **OPEN RESPONSE** Describe the dilation of $g(x) = \frac{1}{2}(x)$ as it relates to the graph of the parent function, $f(x) = x$. (Lesson 4-4)

11. **MULTIPLE CHOICE** Arjun begins the calendar year with $40 in his bank account. Each week he receives an allowance of $20, half of which he deposits into his bank account. The situation describes an arithmetic sequence. Which function represents the amount in Arjun's account after n weeks? (Lesson 4-5)

Ⓐ $f(n) = 20n + 40$

Ⓑ $f(n) = 40n + 20$

Ⓒ $f(n) = 40 + 10n$

Ⓓ $f(n) = 10 + 40n$

12. **OPEN RESPONSE** What number can be used to complete the equation below that describes the nth term of the arithmetic sequence $-2, -1.5, -1, -0.5, 0, 0.5, ...$? (Lesson 4-5)

$a_n = 0.5n -$ ____

13. **OPEN RESPONSE** Write and graph a function to represent the sequence 1, 10, 19, 28, ... (Lesson 4-5)

14. **OPEN RESPONSE** Christa has a box of chocolate candies. The number of chocolates in each row forms an arithmetic sequence, as shown in the table. (Lesson 4-5)

Row	1	2	3	4
Number of Chocolates	3	6	9	12

Write an arithmetic function that can be used to find the number of chocolates in each row.

15. TABLE ITEM Daniel earns $9 per hour at his job for the first 40 hours he works each week. However, his pay rate increases to $13.50 per hour thereafter. This situation can be represented with the function

$$f(x) = \begin{cases} 9x, \text{ if } x \le 40 \\ 360 + 13.5(x - 40), \text{ if } x > 40 \end{cases}$$

Use this function to complete the table with the correct values. (Lesson 4-6)

Hours Worked, x	Money Earned, f(x)
30	
35	315
40	
45	427.5
50	

16. GRAPH Graph the function $f(x) = 2[[x]]$.

(Lesson 4-6)

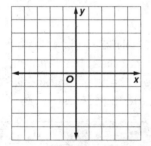

17. MULTIPLE CHOICE Which of the following describes the effect a dilation has upon the graph of the absolute value parent function? (Lesson 4-7)

- (A) Flipped across axis
- (B) Stretch or compression
- (C) Rotated about the origin
- (D) Shifted horizontally or vertically

18. MULTI-SELECT Describe the transformation(s) of the function graphed below in relation to the absolute value parent function. Select all that apply. (Lesson 4-7)

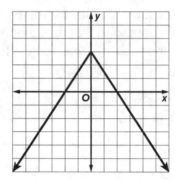

- (A) Reflected across x-axis
- (B) Vertical stretch
- (C) Vertical compression
- (D) Reflected across y-axis
- (E) Translated right 3
- (F) Translated up 3

19. OPEN RESPONSE Describe the graph of $g(x) = |x| + 5$ in relation to the graph of the absolute value parent function. (Lesson 4-7)

20. OPEN RESPONSE Across which axis is the graph of $h(x) = -5|x|$ reflected? (Lesson 4-7)

21. OPEN RESPONSE Use the graph of the function to write its equation. (Lesson 4-7)

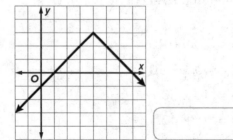

Creating Linear Equations

e Essential Question

What can a function tell you about the relationship that it represents?

What Will You Learn?

Place a check mark (✓) in each row that corresponds with how much you already know about each topic **before** starting this module.

KEY

👎 — I don't know. 👊 — I've heard of it. 👍 — I know it!

	Before			After		
	👎	👊	👍	👎	👊	👍
write linear equations in slope-intercept form when given the slope and the coordinates of a point						
write linear equations in slope-intercept form when given the coordinates of two points on the line						
write linear equations in standard form						
write linear equations in point-slope form						
write equations of parallel and perpendicular lines						
examine scatter plots to describe relationships between quantities						
make and evaluate predictions by fitting linear functions to sets of data						
distinguish between correlation and causation						
write equations of best-fit lines						
plot and analyze residuals						
find inverses of linear relations and functions						

📔 **Foldables** Make this Foldable to help you organize your notes about linear equations. Begin with one sheet of 11″ × 17″ paper.

1. Fold each end of the paper in about 2 inches.

2. Fold along the width and the length. Unfold. Cut along the fold line from the top to the center.

3. Fold the top flaps down. Then fold in half and turn to form a folder. Staple the flaps down to form pockets.

4. Label the front with the chapter title.

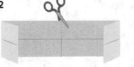

What Vocabulary Will You Learn?

Check the box next to each vocabulary term that you may already know.

- ☐ best-fit line
- ☐ bivariate data
- ☐ causation
- ☐ correlation coefficient
- ☐ inverse functions
- ☐ inverse relations

- ☐ line of fit
- ☐ linear extrapolation
- ☐ linear interpolation
- ☐ linear regression
- ☐ negative correlation
- ☐ no correlation

- ☐ parallel lines
- ☐ perpendicular lines
- ☐ positive correlation
- ☐ residual
- ☐ scatter plot
- ☐ trend

Are You Ready?

Complete the Quick Review to see if you are ready to start this module.
Then complete the Quick Check.

Quick Review

Example 1

Solve $5x + 15y = 9$ for x.

$5x + 15y = 9$	Original equation
$5x + 15y - 15y = 9 - 15y$	Subtract 15y from each side.
$5x = 9 - 15y$	Simplify.
$\frac{5x}{5} = \frac{9 - 15y}{5}$	Divide each side by 5.
$x = \frac{9}{5} - 3y$	Simplify.

Example 2

Write the ordered pair for A.

Step 1 Begin at point A.

Step 2 Follow along a vertical line to the x-axis. The x-coordinate is −4.

Step 3 Follow along a horizontal line to the y-axis. The y-coordinate is 2.

The ordered pair for point A is (−4, 2).

Quick Check

Solve each equation for the given variable.

1. $x + y = 5$ for y

2. $2x - 4y = 6$ for x

3. $y - 2 = x + 3$ for y

4. $4x - 3y = 12$ for x

Write the ordered pair for each point.

5. A

6. B

7. C

8. D

9. E

10. F

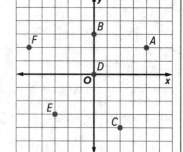

How Did You Do?

Which exercises did you answer correctly in the Quick Check? Shade those exercise numbers below.

① ② ③ ④ ⑤ ⑥ ⑦ ⑧ ⑨ ⑩

Writing Equations in Slope-Intercept Form

Explore Slope-Intercept Form

Online Activity Use graphing technology to complete the Explore.

> **INQUIRY** How does changing the coordinates of two points on a line affect the slope of the line?

Learn Creating Linear Equations in Slope-Intercept Form Given the Slope and a Point

If you are given the slope of a line and the coordinates of any point on that line, you can create an equation for that line.

Key Concept • Creating Equations in Slope-Intercept Form Given the Slope and a Point	
Step 1	Determine whether the given point is the y-intercept. If not, substitute the given information into the slope-intercept form equation to find the y-intercept.
Step 2	Use the given slope and y-intercept you found in Step 1 to write the equation of the line in slope-intercept form.

Example 1 Write an Equation Given the Slope and a Point

Write an equation of the line that passes through (−8, 6) and has a slope of $-\frac{3}{4}$.

Step 1 Find the y-intercept.

$y = mx + b$ Slope-intercept form

$6 = -\frac{3}{4}(-8) + b$ $m = $ _____, $x = $ _____, and $y = $ _____

$6 = $ _____ $+ b$ Simplify.

_____ $= b$ Subtract 6 from each side.

Step 2 Write the equation in slope-intercept form.

$y = mx + b$ Slope-intercept form

$y = -\frac{3}{4}x + 0$ $m = $ _____ and $b = $ _____

$y = $ _____   Simplify.

Go Online You can complete an Extra Example online.

Today's Goals
- Write an equation of a line in slope-intercept form given the slope and one point.
- Write an equation of a line in slope-intercept form given two points.

Think About It!
How can you determine whether the given point is the y-intercept of the line?

Think About It!
What does it mean if $b = 0$ when an equation is written in slope-intercept form?

Study Tip
Slope-Intercept Form Remember, you need two things to write an equation in slope-intercept form: the slope and the y-intercept.

🌐 **Example 2** Write an Equation in Slope-Intercept Form

BAKING Marissa is baking a recipe that calls for her to turn down the temperature on her oven for part of the baking time. Write an equation to represent the situation if the temperature in her oven drops 25°F every 30 seconds, and after 2 minutes the temperature is 350°F.

Step 1 Determine a point on the line and the slope.

After 2 minutes the temperature is _____°F.

Let $x =$ _____ and $y =$ the temperature in °F.

So, the point (_____, _____) is on the line.

The temperature drops _____°F every _____.

The change in x is 30 seconds, or _____ minute.

"Drops" means a negative change, so the change in y is _____°F.

$\text{Slope} = \dfrac{\text{change in } y}{\text{change in } x} =$ _____ or −50°F per minute.

So, the slope is _____.

Step 2 Find the y-intercept.

$y = mx + b$	Slope-intercept form
_____ = _____(2) + b	$m = -50$, $x = 2$, and $y = 350$
$350 =$ _____ + b	Simplify.
_____ = b	Add 100 to each side.

This means that the temperature of the oven was _____°F when it was turned off.

Step 3 Write the equation in slope-intercept form.

$y = mx + b$	Slope-intercept form
$y =$ _____	$m = -50$ and $b = 450$

Check

MEMBERSHIP The total monthly cost of Ayzha's gym membership increases by $5 per class she attends. After signing up for 4 classes one month, her total cost is $49.99. Which equation represents Ayzha's total monthly cost y after attending x classes? _____

A. $y = -5x + 29.99$

B. $y = -5x + 69.99$

C. $y = 5x + 29.99$

D. $y = 5x + 69.99$

🅒 **Go Online** You can complete an Extra Example online.

Study Tip

Slope When determining the slope, words like "drops" and "decreasing" represent a negative slope, and words like "growing" and "increasing" represent a positive slope.

☁️ **Think About It!**

Find the domain of your equation, and describe the meaning in the context of the situation.

Learn Creating Linear Equations in Slope-Intercept Form Given Two Points

If you are given the coordinates of any two points on a line, you can create an equation for that line.

Key Concept • Creating Equations in Slope-Intercept Form Given Two Points	
Step 1	Use the given points to find the slope of the line containing the points.
Step 2	Use the slope from Step 1 and either of the given points to find the y-intercept of the line.
Step 3	Use the slope you found in Step 1 and the y-intercept you found in Step 2 to write the equation of the line in slope-intercept form.

☁ Talk About It!

Will your equation for the line be different depending on the point you choose in Step 2? Justify your argument.

Example 3 Write Equations Given Two Points

Write an equation of the line that passes through (1.2, −0.7) and (−3.4, 1.6).

Step 1 Find the slope.

$$m = \frac{y_2 - y_1}{x_2 - x_1}$$ Slope Formula

$$m = \frac{1.6 - (-0.7)}{-3.4 - 1.2}$$ $(x_1, y_1) = (1.2, -0.7), (x_2, y_2) = (-3.4, 1.6)$

$$m = \frac{2.3}{-4.6}$$ Simplify.

$$m = \underline{\hspace{1cm}}$$ Simplify.

Step 2 Use either point to find the y-intercept.

$y = mx + b$ Slope-intercept form

$1.6 = -0.5(-3.4) + b$ $m = \underline{\hspace{1cm}}, x = \underline{\hspace{1cm}},$ and $y = \underline{\hspace{1cm}}$

$1.6 = 1.7 + b$ Simplify.

$-0.1 = b$ Subtract 1.7 from each side.

Step 3 Write the equation in slope-intercept form.

$y = mx + b$ Slope-intercept form

$y = -0.5x - 0.1$ $m = \underline{\hspace{1cm}}$ and $b = \underline{\hspace{1cm}}$

Check

Write an equation of the line that passes through (−5, −3) and (−7, −12).

_____

🧭 **Go Online** You can complete an Extra Example online.

Watch Out!

Subtraction If the (x_1, y_1) coordinates are negative, be sure to account for both the negative signs and the subtraction symbols in the Slope Formula. Remember, the result of subtracting a negative number is the same as adding its opposite.

Copyright © McGraw-Hill Education

Use the table to make an estimate of the number of students enrolled in public high schools in 2030. Then, use the equation to predict the number of students enrolled. How does your estimate compare to the number of students that you calculated?

_____

_____

_____

_____

_____

_____

Problem-Solving Tip

Use a Graph You can also estimate and make predictions using a graph. Plot two points from the table, connect them with a line, and then estimate using the graph.

Study Tip

Units The number of students enrolled is in thousands. While it is impossible to have one-quarter of a student, 14,708.25 thousand students really means 14,708,250 students. So, this solution is within the constraints of the situation.

Apply Example 4 Write an Equation Given Real-World Data

SCHOOLS **The number of students enrolled in public high schools in the United States has risen slightly since 2010. Write an equation that could be used to predict the number of students enrolled in public high schools if enrollment continues to grow at the same rate.**

Year	Students (in thousands)
2011	14,749
2012	14,753
2013	14,754
2014	14,826
2015	14,912

1. What is the task?
Describe the task in your own words. Then list any questions that you may have. How can you find answers to your questions?

_____

_____

_____

2. How will you approach the task? What have you learned that you can use to help you complete the task?

_____

_____

_____

_____

3. What is your solution?
Use your strategy to solve the problem.

Find the slope.

$m =$ _____

Find the y-intercept.

$b =$ _____

Write an equation to predict the number of students enrolled in public high schools if enrollment continues to grow at the same rate.

_____

4. How can you know that your solution is reasonable?
Write About It! Write an argument that can be used to defend your solution.

_____

_____

🌐 **Go Online** You can complete an Extra Example online.

Practice

⬤ Go Online You can complete your homework online.

Example 1

Write an equation of the line that passes through the given point and has the given slope.

1. (4, 2); slope $\frac{1}{2}$

2. (3, −2); slope $\frac{1}{3}$

3. (6, 4); slope $-\frac{3}{4}$

4. (−5, 4); slope −3

5. (4, 3); slope $\frac{1}{2}$

6. (1, −5); slope $-\frac{3}{2}$

Example 2

7. EXERCISE Carlos is jogging at a constant speed. He starts a timer when he is 12 feet from his starting position. After 3 seconds, Carlos is 21 feet from his starting position. Write a linear equation to represent the distance d of Carlos from his starting position after t seconds.

8. JOBS Mr. Kimball sells computer software. He earns a base salary of $41,250 and 8% commission on his sales. Write an equation to represent Mr. Kimball's total pay p after selling d dollars of software.

9. USE A MODEL In 2006, the average ticket price for a National Football League game was $62.38. Since then the cost has increased an average of $2.54 per year. Write a linear equation to represent the cost C of an NFL ticket y years after 2006.

10. TYPING Nebi has already typed 250 words. He then starts a timer and finds that he types 150 words in 3 minutes. If Nebi types at a constant rate, write a linear equation to represent the number of words w Nebi types m minutes after starting the timer.

Example 3

Write an equation of the line that passes through each pair of points.

11. (0, −4), (5, −4)

12. (−4, −2), (4, 0)

13. (−2, −3), (4, 5)

14. (0, 1), (5, 3)

15. (−3, 0), (1, −6)

16. (1, 0), (5, −1)

17. (9, 2), (−2, 6)

18. (−6, 5), (−6, −4)

19. (5, −2), (7, −1)

20. (5, −3), (2, 5)

21. $\left(\frac{5}{4}, 1\right), \left(-\frac{1}{4}, \frac{3}{4}\right)$

22. $\left(\frac{5}{12}, -1\right), \left(-\frac{3}{4}, \frac{1}{6}\right)$

Example 4

23. **GUITAR** Lydia wants to purchase guitar lessons. She sees a sign that gives the prices for 7 guitar lessons and 11 guitar lessons. Write a linear equation to find the total cost C for d lessons.

GUITAR LESSONS

7 Lessons = $82
11 Lessons = $122

24. **CENSUS** The population of Laredo, Texas, was about 215,500 in 2007. It was about 123,000 in 1990. If we assume that the population growth is constant, write a linear equation with an integer slope to represent *p*, Laredo's population *t* years after 1990.

25. **WEATHER** A meteorologist finds that the temperature at the 6000-foot level of a mountain is 76°F and the temperature at the 12,000-foot level of the mountain is 49°F. Write a linear equation to represent the temperature *T* at an elevation of *x*, where *x* is in thousands of feet.

26. **FUNDRAISING** Natalia and her friends held a bake sale to benefit a local charity. The friends sold 15 cakes on the first day and 22 cakes on the second day of the bake sale. They collected $60 on the first day and $88 on the second day. Write an equation to represent the amount *R* Natalia and her friends raised after selling *c* cakes.

Mixed Exercises

Write an equation of each line.

27.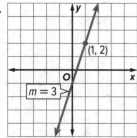

28.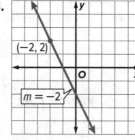

29.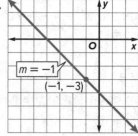

30.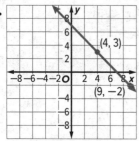

31.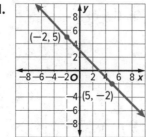

32.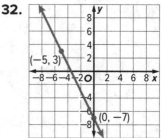

Determine whether the given point is on the line. Explain your reasoning.

33. $(3, -1)$; $y = \frac{1}{3}x + 5$

34. $(6, -2)$; $y = \frac{1}{2}x - 5$

35. $(15, -13)$; $y = -\frac{1}{5}x - 10$

36. $(3, 3)$; $y = -\frac{2}{3}x + 1$

Determine another point on a line given two points on the line.

37. $(2, -4)$, $(4, -2)$

38. $(0, 5)$, $(4, 1)$

39. $(-3, 1)$, $(-1, -3)$

40. $(0, 4)$, $(2, 5)$

41. $(-2, 9)$, $(2, -1)$

42. $(3, 0)$, $(12, 3)$

For Exercises 43–45, determine which equation best represents each situation. Explain the meaning of each variable.

A. $y = \frac{1}{25}x + 300$

B. $y = 25x + 300$

C. $y = 300x + 25$

43. PLANES Plane tickets cost $300 each plus a fee of $25 to select seats per order.

44. SAVINGS Larry has $300. He saves $25 each week.

45. OIL The current oil level in a tank is 25 feet. The rate that oil is being poured into the tank is $\frac{1}{25}$ inch per hour.

46. USE A MODEL The table of ordered pairs shows the coordinates of the two points on the graph of a function. Write an equation that describes the function.

x	y
-2	2
4	-1

47. USE A SOURCE The table shows how women's shoe sizes in the United Kingdom compare to women's shoe sizes in the United States.

Women's Shoe Sizes							
U.K.	3	3.5	4	4.5	5	5.5	6
U.S.	5.5	6	6.5	7	7.5	8	8.5

Source: DanceSport UK

  a. Write a linear equation to determine the U.S. size y if you are given the U.K. size x.

  b. What would be the U.S. shoe size for a woman who wears a U.K. size 7.5?

  c. Research women's shoe sizes in Australia compared to women's shoe sizes in the United States. Write a linear equation to determine the U.S. size y if you are given the Australia size x.

48. REASONING Shikita borrowed money from her brother and paid back a set amount each week. The table shows how much she owed in a given week.

Week	3	6	8	10	13
Amount Owed	$32.50	$25.00	$20.00	$15.00	$7.50

a. Let x represent the number of weeks and y represent the amount owed. Write an equation in slope-intercept form to model the amount Shikita owed each week.

b. Describe the graph of the equation you found in **part a.** What does its shape tell you about the problem?

49. REASONING Koby tracked the weight of his puppy for 6 months. Her growth is shown in the table where x = age in months and y = weight in pounds.

x	y
2	16
3	23.5
4	31
5	38.5
6	46

a. Write an equation in slope-intercept form to model the growth of Koby's puppy.

b. What is the y-intercept? What does the y-intercept mean in the context of the problem?

c. What is the slope? What does the slope mean in the context of the problem?

Higher-Order Thinking Skills

50. PERSEVERE Write the equation of each line in slope-intercept form.

a. slope: $\frac{4}{5}$, y-intercept: -8

b. $(-3, 0)$, $(3, -16)$

c. What point do the graphs of both equations have in common, and what does this tell you about their graphs?

51. FIND THE ERROR Tess and Jacinta are writing an equation of the line through $(3, -2)$ and $(6, 4)$. Is either of them correct? Explain your reasoning.

52. WRITE Linear equations are useful in predicting future events. Describe some factors in real-world situations that might affect the reliability of the graph in making any predictions.

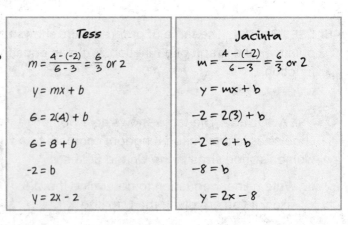

Tess
$$m = \frac{4 - (-2)}{6 - 3} = \frac{6}{3} \text{ or } 2$$
$$y = mx + b$$
$$6 = 2(4) + b$$
$$6 = 8 + b$$
$$-2 = b$$
$$y = 2x - 2$$

Jacinta
$$m = \frac{4 - (-2)}{6 - 3} = \frac{6}{3} \text{ or } 2$$
$$y = mx + b$$
$$-2 = 2(3) + b$$
$$-2 = 6 + b$$
$$-8 = b$$
$$y = 2x - 8$$

53. CREATE Create a real-world situation that fits the graph at the right. Define the two quantities and describe the functional relationship between them. Write an equation to represent this relationship, and describe what the slope and y-intercept mean.

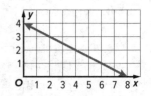

Writing Equations in Standard and Point-Slope Forms

Explore Forms of Linear Equations

 Online Activity Use graphing technology to complete the Explore.

> ⊗ **INQUIRY** How are the point-slope and slope-intercept forms of a linear equation related?

Learn Creating Linear Equations in Point-Slope Form

When the slope and the coordinates of one point of a line are known, an equation for the line can be written in point-slope form.

Key Concept • Point-Slope Form

Words	The linear equation $y - y_1 = m(x - x_1)$ is written in point-slope form, where (x_1, y_1) is a given point on a nonvertical line and m is the slope of the line.
Symbols	$y - y_1 = m(x - x_1)$
Example	(graph showing a line through point (x_1, y_1) with x- and y-axes and origin O)

If you are given two points on the line or a point on the line and its slope, you can write an equation for the line in point-slope form.

Key Concept • Writing Equations of Lines in Point-Slope Form

Given the Slope and One Point		Given Two Points	
Step 1	Let the x and y coordinates be (x_1, y_1).	Step 1	Find the slope.
Step 2	Substitute the values of m, x_1, and y_1 into the equation of a line in point-slope form.	Step 2	Choose one of the two points to use.
		Step 3	Follow the steps for writing an equation given the slope and one point.

Today's Goals
• Write equations of lines in point-slope form.
• Create and identify equations of parallel or perpendicular lines.

Today's Vocabulary
parallel lines
perpendicular lines

💬 **Talk About It!**
Why must a line be nonvertical in order to be written in point-slope form? Explain.

🔗 **Go Online**
You may want to complete the Concept Check to check your understanding.

Example 1 Equation in Point-Slope Form Given Slope and a Point

Write an equation in point-slope form for the line that passes through (−2, 7) with a slope of $-\frac{3}{2}$. Then graph the equation.

$$y - y_1 = m(x - x_1)$$ Point-slope form

$$y - 7 = -\frac{3}{2}[x - (-2)]$$ $(x_1, y_1) = ($ ___ , ___ $)$ and $m =$ ___

_____   Simplify.

Step 1 Plot the given point (−2, 7).

Step 2 Use the slope, $-\frac{3}{2}$, to plot another point on the line.

Step 3 Draw a line through the points.

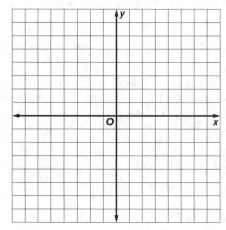

Check

Determine the equation in point-slope form for the line that passes through (7, 5) with a slope of −3. Then graph the equation.

$$y \underline{\quad} = \underline{\quad}(x \underline{\quad})$$

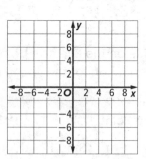

🔊 **Go Online** You can complete an Extra Example online.

Example 2 Equation in Point-Slope Form Given Two Points

Write an equation in point-slope form for the line that passes through the given points.

(2, −7) and (6, −3)

Step 1 Find the slope.

$$m = \frac{y_2 - y_1}{x_2 - x_1}$$ Slope Formula

$$m = \frac{-3 - (-7)}{6 - 2} = \frac{4}{4} \text{ or } \underline{\quad}$$ $(x_1, y_1) = (2, -7)$ and $(x_2, y_2) = (6, -3)$

Step 2 Write an equation.

You can select either point for (x_1, y_1) in point-slope form.

$$y - y_1 = m(x - x_1)$$ Point-slope form

$$y - \underline{\quad} = 1(x - \underline{\quad})$$ $(x_1, y_1) = (6, -3)$ and $m = 1$

$$\underline{\hspace{5cm}}$$ Simplify.

Check

Select an equation in point-slope form for the line that passes through (−16, 18) and (−11, −2). _____

A. $y + 2 = -4(x + 11)$

B. $y + 2 = -\frac{1}{4}(x + 11)$

C. $y - 2 = -4(x + 11)$

D. $y - 2 = -\frac{1}{4}(x - 11)$

E. None of these

Example 3 Change to Slope-Intercept Form

Write $y + 4 = -2(x - 6)$ in slope-intercept form.

$$y + 4 = -2(x - 6)$$ Original Equation

$$y + 4 = -2x + 12$$ $\underline{\hspace{4cm}}$

$$y = -2x + 8$$ Subtract _____ from each side.

Check

Write $y + 3 = -\frac{1}{2}(x - 8)$ in slope-intercept form.

$$y = \underline{\quad} x + \underline{\quad}$$

Go Online You can complete an Extra Example online.

Think About It!

Write another equation in point-slope form for the line with the points given.

Why are there multiple correct answers with the same given information?

Study Tip

Checking Your Work
To check your work, you can substitute the point from the original point-slope form of the equation, in this case (6, −4), into the slope-intercept form of the equation. If it is a true statement, the equation is correct.

$$y = -2x + 8$$
$$-4 = -2(6) + 8$$
$$-4 = -12 + 8$$
$$-4 = -4 \checkmark$$

Example 4 Apply Point-Slope Form

READING Nadia's book club is ordering new novels. She knows that the total cost of 5 books is $61.25, and 15 books cost $159.75. Write an equation in point-slope form to represent the total cost y of ordering x books.

Step 1 Find the slope.

$$m = \frac{y_2 - y_1}{x_2 - x_1}$$ Slope Formula

$$m = \frac{159.75 - 61.25}{15 - 5} = \frac{98.5}{10} \text{ or } 9.85$$ $(x_1, y_1) = (5, 61.25)$ and
$(x_2, y_2) = (15, 159.75)$

Step 2 Write an equation.

$$y - y_1 = m(x - x_1)$$ Point-slope form

$$y - \underline{\quad} = 9.85(x - \underline{\quad})$$ $(x_1, y_1) = (5, 61.25)$ and $m = 9.85$

Think About It!
Use the equation to find the cost of purchasing 12 books.

Check

TAXIS The total cost of a taxi fare is given in the table. Determine the equation(s) in point-slope form that model(s) this situation if x represents the distance in miles and y represents the cost in dollars. _____

Distance (miles)	1.5	4	7.5	12.25
Cost (dollars)	6.90	13.40	22.50	34.85

A. $y - 13.4 = 2.6(x - 4)$ B. $y - 22.5 = 2.6(x - 7.5)$

C. $y = 2.6x + 3$ D. $y - 6.9 = \frac{5}{13}(x - 1.5)$

E. $y + 34.85 = 2.6(x + 12.25)$

Example 5 Change to Standard Form

Write $y - 1 = -\frac{2}{5}(x + 3)$ in standard form.

$$y - 1 = -\frac{2}{5}(x + 3)$$ Original equation

$$\underline{\quad}(y - 1) = \underline{\quad}(x + 3)$$ Multiply each side by 5 to eliminate the fraction.

$$\underline{\hspace{4cm}}$$ Distributive Property

$$5y = -2x - 1$$ Add 5 to each side.

$$\underline{\hspace{4cm}}$$ Add 2x to each side.

Check

Write $y = -\frac{7}{2}x + 5$ in standard form.

$$\underline{\hspace{4cm}}$$

 Go Online You can complete an Extra Example online.

Study Tip
Fractional Slopes
When working with an equation with a fractional slope, it is often simpler to first multiply each side of the equation by the denominator. This will eliminate distributing a fraction later in the equation.

Example 6 Standard Form Given Two Points

Write an equation in standard form for the line that passes through (8, −4) and (−6, −11).

Step 1 Find the slope.

$$m = \frac{y_2 - y_1}{x_2 - x_1} \qquad \text{Slope Formula}$$

$$m = \frac{-11 - (-4)}{-6 - 8} = \frac{-7}{-14} \text{ or } \underline{\hspace{1cm}} \qquad (x_1, y_1) = (8, -4) \text{ and } (x_2, y_2) = (-6, -11)$$

Go Online

An alternate method is available for this example.

Step 2 Write an equation in slope-intercept form.

$y = mx + b$	Slope-intercept form
$-4 = \frac{1}{2}(8) + b$	$(x, y) = (\underline{\hspace{0.5cm}}, \underline{\hspace{0.5cm}})$ and $m = \underline{\hspace{0.5cm}}$
$-4 = \underline{\hspace{0.5cm}} + b$	Simplify.
$-8 = b$	Subtract 4 from each side.
$y = \underline{\hspace{1cm}}$	Replace m with $\frac{1}{2}$ and b with -8.

Think About It!

In Step 2, why is it possible to write an equation in either slope-intercept form or point-slope form and still get the same equation in standard form?

Step 3 Write the equation in standard form.

$2y = 2\left(\frac{1}{2}x - 8\right)$	Multiply each side by 2.
$2y = \underline{\hspace{1.5cm}}$	Distributive Property
$-x + 2y = -16$	Subtract x from each side.
$\underline{\hspace{2cm}}$	Multiply each side by −1.

Check

Select the equation in standard form for the line that passes through (−9, 8) and (1, −12). _____

A. $x + 2y = -20$

B. $2x + y = -13$

C. $2x + y = 7$

D. $2x + y = -10$

E. $x + 2y = 20$

Go Online You can complete an Extra Example online.

Learn Equations of Parallel and Perpendicular Lines

Nonvertical lines in the same plane that have the same slope are called **parallel lines**. Nonvertical lines in the same plane for which the product of the slopes is −1 are called **perpendicular lines**.

Key Concept • Slopes of Parallel and Perpendicular Lines

Parallel Lines	Perpendicular Lines
If two nonvertical lines are parallel, their slopes are the same.	If two nonvertical lines are perpendicular, the product of their slopes is −1.
Since both lines have a slope of $\frac{1}{2}$, $\overleftrightarrow{AB} \parallel \overleftrightarrow{CD}$.	Since $\frac{1}{2}(-2) = -1$, $\overleftrightarrow{AB} \perp \overleftrightarrow{EF}$.

You can write an equation of a line parallel or perpendicular to a given line if you know a point on the line and an equation of the given line.

Key Concept • Writing Equations of Lines Parallel or Perpendicular to a Given Line

Parallel Lines		Perpendicular Lines	
Step 1	Identify the slope m of the given line.	Step 1	Identify the slope m of the given line. The slope of the line perpendicular to the original line is $-\frac{1}{m}$.
Step 2	Use the point-slope form with slope m and the coordinates of the given point.	Step 2	Use the point-slope form with slope $-\frac{1}{m}$ and the coordinates of the given point.
Step 3	Rewrite the equation in the needed form.	Step 3	Rewrite the equation in the needed form.

Example 7 Parallel Line Through a Given Point

Write an equation in slope-intercept form for the line that passes through (−4, 2) and is parallel to the graph of $y = 3x - 5$.

Step 1 Identify the slope of the given line.

The slope of the line with equation $y = 3x - 5$ is _____. The line parallel to that line has the same slope, _____.

📱 **Go Online** You can complete an Extra Example online.

💭 **Think About It!**

If the given line is vertical, what is the slope of any line parallel to the given line? perpendicular to the given line?

📱 **Go Online**
You may want to complete the Concept Check to check your understanding.

Steps 2, 3 Write the equation of the parallel line.

Use the point-slope form to rewrite the equation in slope-intercept form.

$$y - y_1 = m(x - x_1)$$ Point-slope form

$$y - \underline{\quad} = \underline{\quad}[x - (-4)]$$ $(x_1, y_1) = (-4, 2)$ and $m = 3$

$$y - \underline{\quad} = \underline{\qquad}$$ Simplify.

$$y - 2 = \underline{\qquad}$$ Distributive Property

$$y = 3x \underline{\quad}$$ Add 2 to each side.

Study Tip

Checking Your Work
To check that your equation represents the correct line, graph both lines. Verify that the lines appear to be parallel and that your line passes through the given point.

Check

Write an equation for the line that passes through (8, 2) and is parallel to the graph of $y = \frac{3}{4}x + 2$.

_____

Example 8 Perpendicular Line Through a Given Point

Write an equation in slope-intercept form for the line that passes through (1, −2) and is perpendicular to the graph of $3x + 2y = 12$.

Step 1 Identify the slope of the given line.

Write the equation in slope-intercept form.

$$3x + 2y = 12$$ Original equation

$$3x \underline{\quad} + 2y = 12 \underline{\quad}$$ Subtract 3x from each side.

$$2y = \underline{\quad} + 12$$ Simplify.

$$\frac{2y}{2} = \frac{-3x}{2} + \frac{12}{2}$$ Divide each side by 2.

$$y = \underline{\qquad}$$ Simplify.

The slope of the line with equation $3x + 2y = 12$ is $-\frac{3}{2}$. The slope of the line perpendicular to that line is the opposite reciprocal, _____.

Steps 2, 3 Write the equation of the perpendicular line.

Use the point-slope form to rewrite the equation in slope-intercept form.

$$y - y_1 = m(x - x_1)$$ Point-slope form

$$y - \underline{\quad} = \underline{\quad}(x - 1)$$ $(x_1, y_1) = (1, -2)$ and $m = \frac{2}{3}$

$$y + \underline{\quad} = \frac{2}{3}(x \underline{\quad})$$ Simplify.

$$y + 2 = \underline{\qquad}$$ Distributive Property

$$y = \frac{2}{3}x \underline{\quad}$$ Subtract 2 from each side.

 Go Online You can complete an Extra Example online.

Check

Select the equation in slope-intercept form for the line that passes through (5, 0) and is perpendicular to the graph of $x - 6y = 1$. _____

A. $y = -6x + 30$

B. $6x + y = 30$

C. $y = -\frac{1}{6}x + 2$

D. $x - 6y = 5$

Example 9 Determine Line Relationships

Determine whether $\overleftrightarrow{AB}$ and $\overleftrightarrow{EF}$ are *parallel*, *perpendicular*, or *neither* for A(6, 8), B(2, 5), E(−6, −3), and F(0, 5).

Step 1 Find the slope of each line.

slope of $\overleftrightarrow{AB} = \frac{8-5}{6-2} =$ _____   slope of $\overleftrightarrow{EF} = \frac{-3-5}{-6-0} = \frac{-8}{-6}$ or _____

Step 2 Determine the relationship.

parallel   To determine whether the lines are parallel, compare their slopes. The two lines _____ have the same slope, so they _____ parallel.

perpendicular   To determine whether the lines are perpendicular, find the product of their slopes. $\frac{3}{4} \cdot \frac{4}{3} =$ _____

Since the product of the slopes _____ −1, $\overleftrightarrow{AB}$ and $\overleftrightarrow{EF}$ _____ perpendicular.

Check

Determine whether $\overleftrightarrow{CD}$ and $\overleftrightarrow{KL}$ are *parallel*, *perpendicular*, or *neither* for C(4, 10), D(−1, 12), K(6, −5), and L(1, −3).

$\overleftrightarrow{CD}$ and $\overleftrightarrow{KL}$ are _____.

Complete each sentence given $y = ax - 5$ and $y = bx + 3$.

When $a = 4$ and $b = 4$, the graphs are _____.

When $a = -3$ and $b = 5$, the graphs are _____.

When $a = -2$ and $b = \frac{1}{2}$, the graphs are _____.

Go Online to practice what you've learned about writing linear equations in the Put It All Together over Lessons 5-1 and 5-2.

 Go Online You can complete an Extra Example online.

Practice

Go Online You can complete your homework online.

Example 1

Write an equation in point-slope form for the line that passes through each point with the given slope. Then graph the equation.

1. $(-6, -3)$, $m = -1$

2. $(-7, 6)$, $m = 0$

3. $(-2, 11)$, $m = \frac{4}{3}$

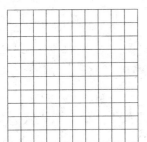
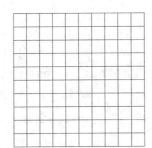

Example 2

Write an equation in point-slope form for the line that passes through the given points.

4. $(-4, 6)$, $(-2, 22)$

5. $(1, -3)$, $(4, -15)$

6. $(4, -6)$, $(6, -4)$

7. $(3, 3)$, $(6, 7)$

Example 3

Write each equation in slope-intercept form.

8. $y - 1 = \frac{4}{5}(x + 5)$

9. $y + 5 = -6(x + 7)$

10. $y + 6 = -\frac{3}{4}(x + 8)$

11. $y + 2 = \frac{1}{6}(x - 4)$

Example 4

12. NATURE The frequency of a male cricket's chirp is related to the outdoor temperature. The relationship is expressed by the graph, where y is the temperature in degrees Fahrenheit and x is the number of chirps the cricket makes in 14 seconds. Write an equation for the line in point-slope form.

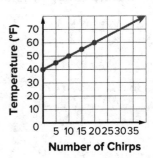

Number of Chirps

13. CANOEING Geoff paddles his canoe at an average speed of 3.5 miles per hour. After 5 hours of canoeing, Geoff has traveled 18 miles. Write an equation in point-slope form to find the total distance y Geoff travels after x hours.

14. GEOMETRY The perimeter of a square is four times the length of one side. If the side length of a square is 1 centimeter, then the perimeter of the square is 4 centimeters. Write an equation in point-slope form to find the perimeter y of a square with side length x.

Example 5

Write each equation in standard form.

15. $y - 10 = 2(x - 8)$

16. $y + 7 = -\frac{3}{2}(x + 1)$

17. $2y + 3 = -\frac{1}{3}(x - 2)$

18. $4y - 5x = 3(4x - 2y + 1)$

19. $y = x + 1$

20. $y = \frac{1}{3}x - 10$

Example 6

Write an equation in standard form for the line that passes through the given points.

21. $(-2, -3), (4, -7)$

22. $(2, 7)$ and $(-5, 2)$

23. $(-4, 9), (2, -9)$

24. $(-1, 19)$ and $(3, 35)$

Examples 7 and 8

Write an equation in slope-intercept form for the line that passes through the given point and is parallel to the graph of the equation. Then write an equation for the line that passes through the given point and is perpendicular to the graph of the equation.

25. $(3, -2); y = x + 4$

26. $(4, -3); y = 3x - 5$

27. $(0, 2); y = -5x + 8$

28. $(-4, 2); y = -\frac{1}{2}x + 6$

29. $(-2, 3); y = -\frac{3}{4}x + 4$

30. $(9, 12); y = 13x - 4$

Example 9

Determine whether the graphs of each pair of equations are *parallel*, *perpendicular*, or *neither*.

31. $y = 4x + 3$
$4x + y = 3$

32. $y = -2x$
$2x + y = 3$

33. $3x + 5y = 10$
$5x - 3y = -6$

34. $-3x + 4y = 8$
$-4x + 3y = -6$

35. $2x + 5y = 15$
$3x + 5y = 15$

36. $2x + 7y = -35$
$4x + 14y = -42$

Mixed Exercises

37. Write an equation in standard form with an *x*-intercept of 4 and a *y*-intercept of 5.

Write each equation in slope-intercept and standard forms.

38. $y + 3 = -\frac{1}{3}(2x + 6)$

39. $y + 4 = 3(3x + 3)$

40. $y - 6 = -3(x + 2)$

41. $y - 9 = -6(x + 9)$

42. $y + 4 = \frac{2}{3}(x + 7)$

43. $y + 7 = \frac{9}{10}(x + 3)$

44. Consider the graphs of the following equations.

$y = -2x \qquad 2y = x \qquad 4y = 2x + 4$

a. Which equations are parallel? Explain your reasoning.

b. Which equations are perpendicular? Explain your reasoning.

45. INSPECTIONS Mrs. Sanchez is inspecting a shed to determine if it is safe to use for storing football equipment. Mrs. Sanchez mapped the top view of the ceiling walls of the shed on a coordinate plane. If one of the walls lies from (−6, 11.5) to (2, 9.5) and the second wall lies from (−1, −2.5) to (2, 9.5), are the walls perpendicular? Explain your reasoning.

46. Nya mapped a quadrilateral on a coordinate plane. If she plots one segment from (−3, −3) to (3, 9) and another segment from (−5, 12) to (1, 0), are the segments parallel? Justify your reasoning.

47. STRUCTURE Immediately after take-off, a jet plane consistently climbs 20 feet for every 40 feet it moves horizontally. The graph shows the trajectory of the jet.

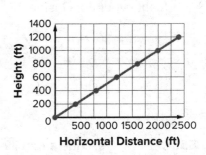

a. Write an equation in point-slope form for the line representing the jet's trajectory.

b. Write the equation from **part a** in slope-intercept form.

c. Write the equation in standard form.

48. CONSTRUCT ARGUMENTS Consider three points, (3, 7), (−6, 1), and (9, p) on the same line. Find the value of p. Justify your argument.

49. WRITE What information is needed to write the equation of a line? Explain.

50. ANALYZE Levy claims that the line through (−6, −2) and (2, 10) is perpendicular to the graph of $3x − 2y = 10$. Do you agree? Justify your argument.

51. ANALYZE Jeremiah says the line through (7, −10) and (3, −2) is parallel to $2x − y = −5$. Do you agree? Justify your argument.

52. FIND THE ERROR Alonae says that the line through (1, −4) and (5, −6) is parallel to the line through (2, −7) and (5, −6). How can you tell she is mistaken without determining the slope? Explain your reasoning.

53. PERSEVERE Write an equation in point-slope form for the line that passes through the points (f, g) and (h, j).

54. WHICH ONE DOESN'T BELONG? Identify the equation that does not belong. Justify your conclusion.

$y − 5 = 3(x − 1)$	$y + 1 = 3(x + 1)$	$y + 4 = 3(x + 1)$	$y − 8 = 3(x − 2)$

55. CREATE Describe a real-life scenario that has a constant rate of change and a value of y for a particular value of x. Represent this situation using an equation in point-slope form and an equation in slope-intercept form.

Scatter Plots and Lines of Fit

Learn Scatter Plots

Bivariate data consists of pairs of values. A **scatter plot** is a graph of bivariate data that consists of ordered pairs on a coordinate plane. Using a scatter plot can help you see the **trend,** or general pattern, in the data. Trends can represent linear or nonlinear associations in the data. Trends can be described as positive or negative correlations.

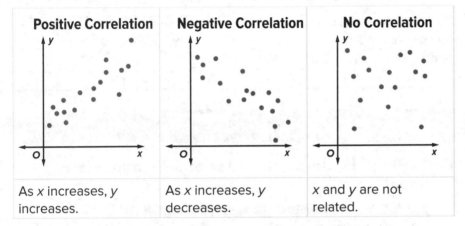

Positive Correlation	**Negative Correlation**	**No Correlation**
As *x* increases, *y* increases.	As *x* increases, *y* decreases.	*x* and *y* are not related.

Notice that in the graphs for positive and negative correlations, many of the points form **clusters** of points that slope upward or downward. Points outside of clusters are **outliers.**

🌐 Example 1 Evaluate Correlation

FOOTBALL The scatter plot displays the height and weight of New Orleans Saints football players. Determine whether the scatter plot shows a *positive, negative,* or *no* correlation. If the correlation is positive or negative, describe its meaning in the situation.

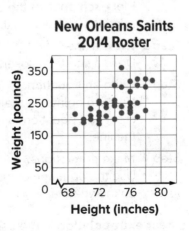

New Orleans Saints 2014 Roster

The scatter plot shows a positive correlation. As the height of the football player increases, weight usually _____.

Copyright © McGraw-Hill Education

Today's Goals
- Categorize the correlation of a set of data in a scatter plot.
- Make and evaluate predictions by fitting linear functions to sets of data.

Today's Vocabulary
bivariate data
scatter plot
trend
positive correlation
negative correlation
no correlation
line of fit
linear extrapolation
linear interpolation

Study Tip

Labeling Axes
Because scatter plots display bivariate data, it is critical to label axes with their corresponding units. Otherwise, the graph may not make sense.

💬 Think About It!

What type of correlation would you expect between a player's jersey number and his birth month?

 Go Online You can complete an Extra Example online.

Study Tip

Determining Correlation
Similar to slope, when data points are generally increasing from left to right, there is a positive correlation. Negative correlation occurs when the data points generally decrease from left to right. If you are unable to tell if the data are increasing or decreasing, there is probably no correlation.

Think About It!
How can you ensure that your data predictions that are outside the range of data are as accurate as possible?

Check

TELEPHONES The scatter plot displays the number of landline telephones in the United States, in 100 millions, since 2000.

Determine whether the scatter plot shows a *positive, negative,* or *no correlation*. Describe the correlation's meaning in the situation.

The scatter plot shows _____ correlation.

As time increases, the number of landlines generally _____.

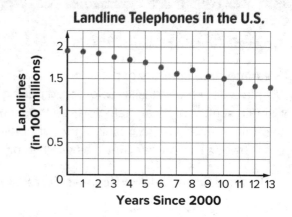

Landline Telephones in the U.S.

Landlines (in 100 millions) vs. Years Since 2000

Explore Make Predictions by Using a Scatter Plot

Online Activity Use a real-world situation to complete the Explore.

INQUIRY How can you use a scatter plot to estimate unknown data?

Learn Lines of Fit

A **line of fit** is used to describe the trend of the data in a scatter plot.

Key Concept • Using a Linear Function to Model Data
Step 1 Make a scatter plot. Plot each point of the data and determine whether any relationship exists in the data.
Step 2 Draw a line. Draw a line that closely follows the trend in the data.
Step 3 Write an equation. Use two points on the line of fit to find the slope of the line and create an equation for the line using the slope and a point on the line.
Step 4 Make predictions. Use the equation of the line of fit to make predictions about unknown data.

Linear extrapolation is the use of a linear equation to predict values that are outside of the range of data. **Linear interpolation** is the use of a linear equation to predict values that are inside of the data range.

 Go Online You can complete an Extra Example online.

🌐 Example 2 Write an Equation for a Line of Fit

BOATS The table shows the average cost of a jet boat in the years after 2000. Write an equation to represent the data. Then, use the equation to predict the cost of a jet boat in 2005 and 2025.

Years Since 2000	Cost ($)
0	17,663
1	19,144
2	21,176
3	20,584
4	23,280
6	24,443
7	27,784

Years Since 2000	Cost ($)
8	28,088
9	29,774
10	32,752
11	34,082
12	35,589
13	37,618

Watch Out!

Variations Equations for scatter plots generally do not have an exact correct solution. Equations will vary depending on how the line of fit was drawn and which points were selected when writing the equation. So, your solutions may not be exactly the same as another student's solutions or the sample answers given.

Step 1 Make a scatter plot.

The independent variable is the number of years since 2000 and the dependent variable is the cost of the jet boats. As the years increase, the cost of the jet boats also increases. This scatter plot shows positive correlation.

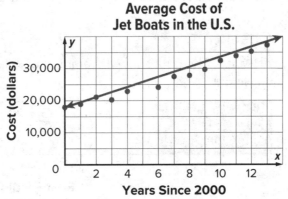

Average Cost of Jet Boats in the U.S.

Step 2 Draw a line of fit.

A line is drawn that follows the trend of the data points and passes close to most of the points.

Step 3 Write an equation.

The line of fit passes close to the data points (5, 25,108) and (10, 32,752).

Find the slope.

$$m = \frac{y_2 - y_1}{x_2 - x_1} \qquad \text{Slope Formula}$$

$$= \frac{32,752 - 25,108}{10 - 5} \qquad (x_1, y_1) = (5, 25,108) \text{ and } (x_2, y_2) = (10, 32,752)$$

$$= \frac{7644}{5} \text{ or } 1528.8 \qquad \text{Simplify.}$$

Use $m = 1528.8$ and a point to write an equation.

$$y - y_1 = m(x - x_1) \qquad \text{Point-slope form}$$

$$y - 25,108 = 1528.8(x - 5) \qquad (x_1, y_1) = (5, 25,108)$$

$$y - 25,108 = 1528.8x - 7644 \qquad \text{Distribute.}$$

$$y = 1528.8x + 17,464 \qquad \text{Simplify.}$$

(continued on the next page)

🧭 **Go Online** You can complete an Extra Example online.

💭 Think About It!

What do the slope and *y*-intercept mean in the context of this example?

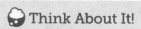
Step 4 Predict the cost in 2005.

Use evaluation to predict the cost of a jet boat in 2005.

Since the independent variable represents the number of years after 2000, the value of x is 2005 − 2000 or 5.

$y = 1528.8\underline{\hspace{1cm}} + 17{,}464$ Equation of the line of fit

$y = 1528.8\underline{\hspace{1cm}} + 17{,}464$ $x = 5$

$y = 7644 + 17{,}464 \text{ or } 25{,}108$ Simplify.

We can predict that the cost of a jet boat in 2005 was about $25,108.

Step 5 Predict the cost in 2025.

Extrapolate the data to determine the cost of a jet boat in 2025.

$y = 1528.8\underline{\hspace{1cm}} + 17{,}464$ Equation of the line of fit

$y = 1528.8\underline{\hspace{1cm}} + 17{,}464$ $x = 25$

$y = 38{,}220 + 17{,}464 \text{ or } \underline{\hspace{2cm}}$ Simplify.

We can predict that the cost of a jet boat in 2025 will be about $55,684.

Check

ANIMALS The data show the amount of milk that a baby goat needs by its weight.

Weight (pounds)	5	7	10	15	20	25	30	40	50
Milk (ounces)	12	16	20	28	32	40	48	64	80

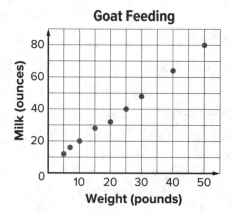

Goat Feeding

Part A Use the data points (10, 20) and (50, 80), which are contained in the line of fit, to write an equation of the line in slope-intercept form. _____

Part B Use the equation from Part A to predict the amount of milk needed for a 17-pound goat and a 55-pound goat.

17-pound goat: _____ ounces

55-pound goat: _____ ounces

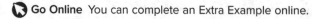 **Go Online** You can complete an Extra Example online.

Practice

Go Online You can complete your homework online.

Example 1

Determine whether each scatter plot shows a *positive, negative,* or *no* correlation. If the correlation is positive or negative, describe its meaning in the situation.

1. Calories Burned During Exercise

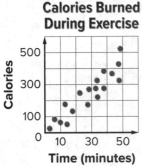

2. Library Fines

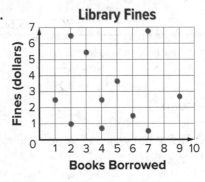

3. Weight-Lifting

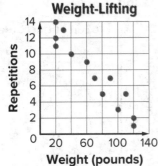

4. Car Dealership Revenue

Example 2

5. MUSIC The scatter plot shows the number of CDs in millions that were sold from 2011 to 2016.

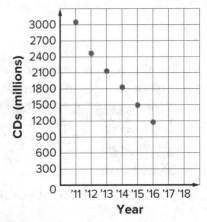

a. Use the points (2, 2485.6) and (6, 1172.5) to write an equation of the line of fit in slope-intercept form. Let x be the years since 2010.

b. If the trend continued, about how many CDs were sold in 2019?

6. HOUSING The data show the median price of an existing home from 2010 to 2015.

Year	2010	2011	2012	2013	2014	2015
Price	222,900	226,900	238,400	258,400	275,200	296,500

a. Use the points (1, 226.9) and (4, 275.2) to write an equation for the line of fit in slope-intercept form, where x is the number of years since 2010 and y is the median price in thousands of dollars.

b. If the trend continues, what will be the approximate median price of an existing home in 2025?

Mixed Exercises

7. FAMILY The table shows the predicted annual cost for a middle-income family to raise a child from birth until adulthood.

Cost of Raising a Child Born in 2013					
Child's Age	2	5	8	11	14
Annual Cost ($)	12,940	12,970	12,800	13,600	14,420

a. Make a scatter plot and describe what relationship exists within the data.

b. Use the points (8, 12,800) and (14, 14,420) to write the equation of the line of fit in slope-intercept form.

c. If the trend continues, what will be the approximate annual cost of raising a child born in 2013 at age 17?

Determine whether each scatter plot shows a *positive, negative,* or *no* correlation. If the correlation is positive or negative, describe its meaning in the situation.

8.

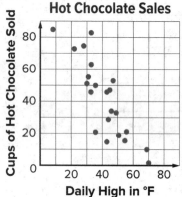

9.

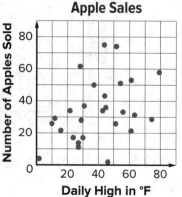

10.

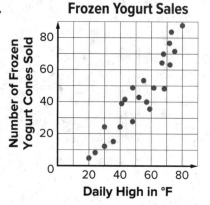

11. BASEBALL The table shows the average length in minutes of professional baseball games in selected years.

Average Length of Major League Baseball Games							
Year	2005	2007	2009	2011	2013	2015	2017
Time (min)	169	175	175	176	184	180	189

a. Make a scatter plot and draw a line of fit.

b. Write the equation of the line of fit in slope-intercept form where *x* is the number of years since 2005. Explain your process.

c. If the trend continues, what will be the approximate length of a major league baseball game in 2021?

d. How reliable is the predicted length of a major league baseball game in 2021? Justify your argument.

12. INCOME The table shows the average median income for selected ages.

Age (years)	26	27	28	29	30
Median Income ($1000)	16.8	19.1	23.3	25.8	33.9

a. Make a scatter plot relating age to median income. Then draw a line of fit for the scatter plot.

b. Determine whether the graph shows a *positive, negative,* or *no correlation.* If the correlation is positive or negative, describe its meaning in the situation.

c. Use the table to write an equation of the line of fit.

d. Use the line of fit to predict the median income for 32-year-olds.

13. FOOTBALL The scatter plot shows the average price of a National Football League ticket from 2007 to 2016.

a. Determine what relationship, if any, exists in the data. Explain.

b. Use the points (2007, 67.11) and (2016, 92.98) to write the slope-intercept form of an equation for the line of fit shown, where *x* is the number of years since 2006. Round to the nearest hundredth.

c. Predict the price of a ticket in 2030.

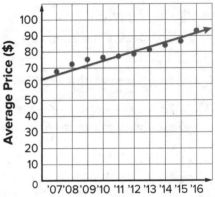

14. STRUCTURE Refer to the scatter plot at the right.

a. Describe the trend in the data shown in the scatter plot and the relationship between *x* and *y.*

b. Describe a real-life situation that could be modeled by the given scatter plot. Explain your reasoning.

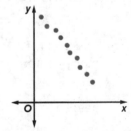

15. USE A MODEL The table gives the life expectancy of a child born in the United States in a given year.

a. Make a scatter plot of the data. Then draw a line of fit.

b. Use the data to predict the life expectancy of a baby born in 2023. Round to the nearest tenth. Does your answer follow the trend of the data? Explain.

c. Explain any assumptions you made when using the line of fit to extrapolate and find the life expectancy of a baby born in 2023.

Years of Life Expected at Birth	
Year of Birth	Life Expectancy (years)
1930	59.7
1940	62.9
1950	68.2
1960	69.7
1970	70.8
1980	73.7
1990	75.4
2000	77.0
2010	78.7

16. USE TOOLS Several groups volunteered to clean up litter along a mile of the highway near their town. The table shows how many people were in each group and how long it took each group to finish the job.

Workers	9	16	18	8	15	11	9	17	9	15	11	12
Minutes	80	40	35	90	60	60	70	30	70	50	80	70

a. Graph the data on a scatter plot.

b. Draw a line of fit to show the trend of the data.

c. Choose two points on the line of fit. Then find the equation of the line in slope-intercept form.

d. Another group wants to get done in 45 minutes. About how many workers should they have? Explain your reasoning.

e. Find the *y*-intercept of the line of fit. Does the *y*-intercept make sense in the context of the situation? Justify your argument.

🧠 Higher-Order Thinking Skills

17. CREATE Describe a real-life situation that can be modeled using a scatter plot. Describe whether there is *positive, negative,* or *no* correlation.

18. WHICH ONE DOESN'T BELONG? Analyze the following situations and determine which one does not belong. Justify your conclusion.

hours worked and amount of money earned	height of an athlete and favorite color
seedlings that grow an average of 2 centimeters each week	number of photos stored on a camera and capacity of camera

19. ANALYZE Determine which line of fit shown is a better fit for the data in the scatter plot. Justify your argument.

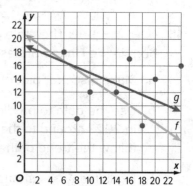

20. WRITE Does an accurate line of fit always predict what will happen in the future? Explain your reasoning.

21. CREATE Make a scatter plot that shows the height of a person and age. Explain how you could use the scatter plot to predict the age of a person given his or her height. How can the information from a scatter plot be used to identify trends and make decisions?

Correlation and Causation

Copyright © McGraw-Hill Education

Explore Collecting Data to Determine Correlation and Causation

🔍 **Online Activity** Use a real-wolrd situation to complete an Explore.

> ⦿ **INQUIRY** What is the difference between correlation and causation? ✕

Learn Correlation and Causation

Causation occurs when a change in one variable produces a change in another variable. It is the relationship between cause and effect. Correlation, however, can be observed between many variables.

Key Concept • Correlation and Causation	
Step 1	Graph ordered pairs to create a scatter plot.
Step 2	Determine whether the scatter plot shows a positive or negative correlation.
Step 3	Determine whether the two sets of data are related. Does one variable *cause* the other? Could other factors be influencing the data results?
Step 4	Decide if the data illustrate correlation or causation.

🌐 Example 1 Correlation and Causation by Graphing

ANALYSIS The data show the per capita consumption of mozzarella cheese and the number of civil engineering doctoral degrees awarded in the United States. Determine whether the data plotted on the graph illustrate a *correlation* or *causation*.

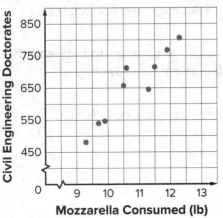

(continued on the next page)

🔍 **Go Online** You can complete an Extra Example online.

Today's Goal
• Determine correlation or causation.

Today's Vocabulary
causation

💭 Think About It!
Why does correlation not prove causation?

Copyright © McGraw-Hill Education

💬 Talk About It!

Describe an experiment that could be conducted to show causation between the number of civil engineers who were awarded a doctoral degree and another factor.

Step 1 Determine the correlation.

As the amount of mozzarella consumed increases, the number of civil engineering doctorates also increases. The scatter plot shows a _____ correlation.

Step 2 Determine causation.

Consumption of mozzarella _____ cause anyone to obtain a doctoral degree in civil engineering. These two sets of data _____ related. Many factors may affect the increase in these two areas. As the demand for more roadways, airports, and water and sewage treatment plants grows, the demand for more civil engineers also increases. An increase in the per capita consumption of mozzarella may be related to increased pizza sales or dairy production. Both variables are affected by a general increase in population.

Step 3 Determine whether the data illustrate a *correlation* or *causation*.

The data exhibit a _____, but there is no _____.

Check

ANALYSIS Determine whether the data illustrate a *correlation* or *causation*.

Month	March	April	May	June	July	August
Sunscreen Sold	14	37	84	117	135	98
Sunglasses Sold	6	11	28	36	40	39

The data show a _____ correlation. As the number of bottles of sunscreen sold increases, the number of sunglasses sold _____. These data illustrate a _____.

Example 2 Correlation and Causation by Situation

Determine whether the situation illustrates a *correlation* or *causation*. Explain your reasoning, including other factors that might be involved.

A university experiment showed a negative correlation between the average weekly time spent exercising and the probability of developing heart disease.

This situation models _____. Exercise and heart disease _____ related, and lack of exercise _____ a cause of heart disease. Other factors that might have led to heart disease are _____ _____.

 Go Online You can complete an Extra Example online.

Practice

🔗 **Go Online** You can complete your homework online.

Example 1

1. FROZEN DESSERTS The table shows the number of pounds of frozen yogurt and the number of pounds of sherbet consumed per capita in the United States from 2009 to 2016.

Year	2009	2010	2011	2012	2013	2014	2015	2016
Pounds of Frozen Yogurt	0.9	1	1.2	1.1	1.4	1.3	1.4	1.2
Pounds of Sherbet	1	1	0.9	0.8	0.9	0.9	0.8	0.8

   a. Graph the ordered pairs (pounds of frozen yogurt, pounds of sherbet) to create a scatter plot.

   b. Does the scatter plot show a *positive, negative,* or *no* correlation? Explain.

   c. Determine whether the data illustrate a *correlation* or *causation.* What other factors may influence the data?

2. LEISURE ACTIVITIES The table shows the average number of minutes a person reads per weekday and the average number of minutes a person watches television per weekday.

Age	15	25	35	45	55	65
Minutes Reading	7	9	12	17	30	50
Minutes Watching Television	117	115	113	127	155	236

   a. Graph the ordered pairs as a scatter plot (minutes reading, minutes watching television).

   b. Does the scatter plot show a *positive, negative,* or *no* correlation? Explain.

   c. Determine whether the data illustrate a *correlation* or *causation.* What other factors may influence the data?

Example 2

Determine whether each situation illustrates a *correlation* or *causation.* Explain your reasoning.

3. A class experiment shows a negative correlation between the width of a person's palm and the amount of time they spend watching television each day.

4. The larger a person's shoe size, the higher a person's reading level.

5. At a grocery store, there is a negative correlation between the price of cereal and number of boxes of cereal sold.

6. Hae notices that the lower the daily temperature is, the less time she spends outside.

Mixed Exercises

7. GARDENING Jalen weighs each type of fruit his garden produces each week.

Week	1	2	3	4	5
Strawberries (lb)	6.5	8	12	13.5	20
Blueberries (lb)	6	5	4.5	3	2.5

a. Graph the ordered pairs (pounds of strawberries, pounds of blueberries) to create a scatter plot.

b. Does the scatter plot show a *positive, negative,* or *no* correlation? Explain.

c. Determine whether the data illustrate a *correlation* or *causation.* What other factors may influence the data?

8. SHOES The table shows the number of pairs of sandals and snow boots sold at a certain store during various months of the year.

Month	January	April	July	December
Sandals	12	153	215	27
Snow Boots	268	34	6	272

a. Graph the ordered pairs (sandals sold, snow boots sold) to create a scatter plot.

b. Does the scatter plot show a *positive, negative,* or *no* correlation? Explain.

c. Determine whether the data illustrate a *correlation* or *causation.* What other factors may influence the data?

Determine whether each situation illustrates a *correlation* or *causation.* If there is a correlation, describe the trend. Explain your reasoning.

9. PIZZA The more pizzas a restaurant sells, the more cheese it uses.

10. BOOKS Sam notices that as the number of words in a book increases, the number of pages in the book increases.

🌧 **Higher-Order Thinking Skills**

11. CONSTRUCT ARGUMENTS What is meant by this statement: *Correlation does not imply causation*? Justify your argument.

A study compared the average monthly amount spent on swimsuits with the average monthly amount spent on air conditioning for several months in Sunnyside. The data is shown in the scatter plot. Use this information for Exercises 12 and 13.

12. ANALYZE Explain what the scatter plot shows and describe any correlation.

13. WRITE Explain whether this statement is accurate: "There is a strong positive correlation between spending money on swimsuits and spending money on air conditioning. Therefore, to cut down the amount of electricity used in Sunnyside, people should buy fewer swimsuits."

Linear Regression

Learn Linear Regression and Best-Fit Lines

A calculator can find the line that most closely approximates data in a scatter plot, called the **best-fit line**. **Linear regression** is one algorithm used to find a precise line of fit for a set of data.

Calculators may also compute a number *r* called the **correlation coefficient**. This measure shows how well data are modeled by a linear equation. It will tell you if a correlation is positive or negative and how closely the equation is modeling the data. The closer the correlation coefficient is to 1 or –1, the more closely the equation models the data.

Weak Correlation	Moderate Correlation	Strong Correlation
$r = 0.02$	$r = 0.72$	$r = -0.97$

Example 1 Find a Best-Fit Line

BASEBALL **The table shows Jackie Robinson's total hits during each season of his major league career. Use a graphing calculator to write an equation for the best-fit line for the data. Then find and interpret the correlation coefficient.**

Year	1947	1948	1949	1950	1951	1952	1953	1954	1955	1956
Total Hits	175	170	203	140	185	157	159	120	81	98

Step 1 Enter the data.

Before you begin, make sure that your Diagnostic setting is on. You can find this under the **CATALOG** menu. Press **D** and then scroll down and click **DiagnosticOn**. Then press .

(continued on the next page)

Today's Goals
- Write equations of best-fit lines using linear regressions.
- Determine how well functions fit sets of data.

Today's Vocabulary
best-fit line
linear regression
correlation coefficient
residual

Think About It!
Write the following correlation coefficients in order from weakest to strongest.

0.85   0.3   1   –0.78

0.54   –0.06   –0.9

Study Tip
Correlation Coefficient The table shows a rule of thumb for determining how well the equation models the data based on the correlation coefficient.

Correlation Coefficient	Strength of Correlation
$\lvert r \rvert \geq 0.8$	Strong
$0.5 \leq \lvert r \rvert < 0.8$	Moderate
$\lvert r \rvert < 0.5$	Weak

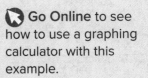 **Go Online** to see how to use a graphing calculator with this example.

Your Notes

Math History Minute

One of the areas of interest of British statistician **Florence Nightingale David (1909–1993)**, who was named after family friend Florence Nightingale, was the distribution of correlation coefficients. In 1938, she released a book entitled *Tables of the Correlation Coefficient*, for which all of the calculations were done on a hand-cranked mechanical calculator.

Use a Source

Choose another baseball player and research the total number of hits they have he has by season. Use a graphing calculator to write an equation for the best-fit line, and decide whether the equation models the data well.

Enter the data by pressing **stat** and selecting the **Edit** option. Let the year 1947 be represented by year ____. Enter the years since 1947 into List 1 (**L1**). These will represent the ____-values. Enter the total hits into List 2 (**L2**). These will represent the ____-values.

Step 2 Perform the regression.

Perform the regression by pressing **stat** and selecting the **CALC** option. Scroll down to **LinReg (ax+b)** and press **enter**. Make sure **L1** is the **Xlist** and **L2** is the **Ylist**. Then select **Calculate**.

Step 3 Interpret the results.

Write the equation of the regression line by rounding the *a* and *b* values on the screen. The form that we chose for the regression was $ax + b$, so the equation is $y =$ _____ + _____. The correlation coefficient is about _____, which means that the equation models the data _____. Its negative value means that as the years since 1947 increase, the total number of Jackie Robinson's hits _____.

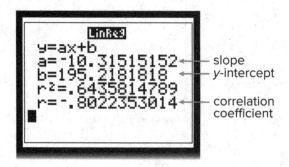

Check

TEMPERATURE The table shows the average annual temperature for the top 10 most populous states in 2014.

Rank	1	2	3	4	5	6	7	8	9	10
Temperature (°F)	59.4	64.8	70.7	45.4	51.8	48.8	50.7	63.5	59	44.4

Part A Use a graphing calculator to write an equation for the best-fit line for the data. Round to the nearest hundredth.

$y =$ _____ $x +$ _____

Part B Find the correlation coefficient *r*. Round to the nearest hundredth.

$r =$ _____

Part C Based on your answer to **part b**, does the equation model the data well? Yes or No?

_____

🔁 **Go Online** You can complete an Extra Example online.

Best-fit lines can be used to estimate values that are not in the data. Recall that when we estimate values that are between known values, this is called linear interpolation. When we estimate a number outside the range of data, it is called linear extrapolation.

🌐 Example 2 Use a Best-Fit Line

SHOPPING The table shows U.S. desktop online sales on Cyber Monday since 2009. Estimate the Cyber Monday sales in 2025.

Year	2009	2010	2011	2012	2013	2014	2015	2016
Sales (millions of dollars)	887	1028	1251	1465	1735	2038	2280	2671

Step 1 Graph the data.

Enter the data from the table into the lists. Let _____ be represented by 0. Then the years since 2009 are the x-values. Let the _____ be the y-values. Graph the scatter plot. Turn on **Plot1** under the **STAT PLOT** menu and choose 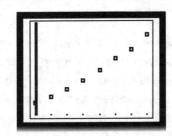. Use **L1** for the **Xlist** and **L2** for the **Ylist**.

[−0.7, 7.7] scl: 1 by [583.72, 2974.28] scl: 1

Change the viewing window so that all data are visible by pressing [zoom] and then selecting **ZoomStat**.

Step 2 Perform the regression.

Perform the regression using the data in the lists. The equation is about $y =$ _____ $x +$ _____. The correlation coefficient is _____, which means that the equation models the data _____.

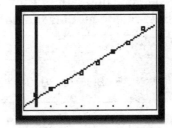

[−0.7, 7.7] scl: 1 by [583.72, 2974.28] scl: 1

Step 3 Graph the best-fit line.

Graph the best-fit line. Press [y =] [vars] and choose **Statistics**.

From the **EQ** menu, choose **RegEQ**. Press [graph].

Step 4 Extrapolate.

Use the graph to predict the _____ Cyber Monday sales. Change the viewing window to include the x-value to be evaluated, _____. Also increase **Ymax** to accommodate the increasing y-values. Press [2nd] **CALC** [enter] 16 [enter] to find that when $x = 16, y \approx$ _____.

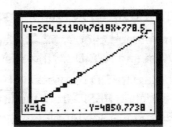

[−0.7, 17] scl: 1 by [583.72, 5500] scl: 1

We can estimate that in 2025, Cyber Monday sales will be about $_____.

🔵 **Go Online** You can complete an Extra Example online.

🌱 **Think About It!**

Why is it helpful to define x as *years since 2009* instead of *years*?

Study Tip

Assumptions Using a best-fit line to make predictions requires you to assume that the trend continues at a constant rate and that more people choose to shop on Cyber Monday each year.

Check

SOCIAL MEDIA The table shows the number of daily users on a social media site in various years.

Year	2011	2012	2013	2014	2015
Daily Users (millions)	372	526	665	802	936

Use linear regression to estimate the number of daily users in millions on the site in 2030. _____

A. 3187.4 users

B. 4591.4 users

C. 285,391.4 users

D. 3047 users

Learn Residuals

When finding a best-fit line, not all data will lie on the line. The difference between an observed *y*-value and its predicted *y*-value on a regression line is called a **residual**. When residuals are plotted on a scatter plot, they can help assess how well the best-fit line describes the data. If there is no pattern in the residual plot, then the best-fit line is a good fit.

🌐 **Example 3** Graph and Analyze a Residual Plot

THANKSGIVING **The table shows the average price of a 10-person Thanksgiving dinner from 2004 to 2014. Determine whether the best-fit line models the data well by graphing a residual plot.**

Year	2004	2005	2006	2007	2008	2009	2010	2011	2012	2013	2014
Price ($)	35.68	36.78	38.10	42.26	44.61	42.91	43.47	49.20	49.48	49.04	49.41

Step 1 Find the best-fit line.

Enter the data from the table into the lists. Let 2004 be represented by 0. Then the years since _____ are the *x*-values. Let the _____ be the *y*-values. Perform the linear regression using the data in the lists.

```
LinReg
y=ax+b
a=1.496090909
b=36.24136364
r²=.9054617629
r=.9515575458
```

Step 2 Graph the residual plot.

Turn on **PLOT2** under the **STAT PLOT** menu and choose 📊. Use **L1** for the **Xlist** and **RESID** for the **Ylist**. You can obtain **RESID** by pressing **2nd** [LIST] and selecting **RESID** from the list of names. Graph the scatter plot of the residuals by pressing **zoom** and choosing **ZoomStat**. The residuals appear to be randomly scattered and centered about the line $y = 0$. Thus, the best-fit line seems to model the data _____.

[−1, 1] scl: 1 by [−2.52, 3.21] scl: 1

🌐 **Go Online** You can complete an Extra Example online.

💬 Talk About It!

Why would a residual plot where the residuals are almost on the line $y = 0$ indicate a very good fit? Explain your reasoning.

☁ Think About It!

Use a calculator to find the correlation coefficient of the best-fit line. Does the correlation coefficient also suggest a good-fit? Justify your argument.

🖱 Go Online

to see how to use a graphing calculator with this example.

Practice

⬧ **Go Online** You can complete your homework online.

Example 1

1. **SOCCER** The table shows the number of goals a soccer team scored each season since 2010. Let x be the number of years since 2010.

Year	2010	2011	2012	2013	2014	2015
Goals Scored	48	52	50	46	48	42

a. Write the equation for the best-fit line for the data.

b. Find and interpret the correlation coefficient.

2. **REVENUE** The table shows the estimated revenue earned for ringtone and ringback purchases, in millions of dollars, each year since 2010.

Year	2010	2011	2012	2013	2014	2015	2016
Revenue ($)	448	276.2	166.9	97.9	66.3	54.5	40.1

a. Write the equation for the best-fit line for the data.

b. Find and interpret the correlation coefficient.

3. **SALES** The table shows the sales of a health and beauty supply company, in millions of dollars, for several years. Let x be the number of years since 2010.

Year	2011	2012	2013	2014	2015
Sales	12.2	19.1	29.4	37.3	45.7

a. Write the equation for the best-fit line for the data.

b. Find and interpret the correlation coefficient.

Example 2

4. **PURCHASING** A supermarket chain closely monitors how many bottles of sunscreen it sells each year so that it can reasonably predict how many bottles to stock in the following year. Let x be the number of years since 2010.

Year	2013	2014	2015	2016	2017
Bottles of Sunscreen	60,200	65,000	66,300	65,200	70,600

a. Find the equation for the best-fit line for the data.

b. How many bottles of sunscreen should the supermarket expect to sell in 2025?

5. GOLD Ounces of gold are traded by large investment banks in commodity exchanges much the same way that shares of stock are traded. The table below shows the cost of a single ounce of gold on the last day of trading in given years. Let x be the number of years since 2000.

Year	2002	2004	2006	2008	2010	2012	2014
Price	$342.75	$435.60	$635.70	$869.75	$1420.25	$1664.00	$1199.25

a. Find the equation for the best-fit line for the data.

b. According to the equation, what would be the price of an ounce of gold on the last day of trading in 2030?

6. GOLF SCORES Emmanuel is practicing golf as part of his school's golf team. Each week he plays a full round of golf and records his total score. His scores for the first five weeks are shown.

Week	1	2	3	4	5
Golf Score	112	107	108	104	98

a. Find the equation for the best-fit line for the data.

b. What score can Emmanuel expect to get after 10 weeks?

Example 3

7. MODELING For a science project, Noah measured the effect of light on plant growth. At the end of 3 weeks, he recorded the height of each plant and how many hours of light it received each day.

Hours of Sunlight Per Day (x)	0	3	6	10	4	9	7	8	12	11	5
Height in Inches (y)	1	3	4	8	4	6	7	8	9	6	5

a. Find the equation for the best-fit line for the data.

b. Graph and analyze the residual plot.

8. STRUCTURE For his project, Darius measured the effect of fertilizer on plant growth. At the end of 3 weeks, he recorded the height of each plant and how many drops of fertilizer it received each day.

Drops of Fertilizer (x)	5	15	20	25	18	22	21	30	10	13	16
Height in Inches (y)	5	8	9	0	8	0	9	0	7	6	9

a. Find the equation of the best-fit line for the data.

b. Graph and analyze the residual plot.

Mixed Exercises

9. PHYSICAL FITNESS The table shows the percentage of students in public school who have met all six of California's physical fitness standards each year since the 2011–2012 school year.

Year	2011–2012	2012–2013	2013–2014	2014–2015
Percentage	20.5%	22.1%	22.6%	21.2%

a. Write the equation for the best-fit line for the data.

b. Find and interpret the correlation coefficient.

c. What constraints are there in the situation? Explain.

10. FARMING Some crops, such as barley, are very sensitive to the acidity of the soil. Barley grows best in soil with a pH range of 6 to 7.5. To determine the ideal level of acidity, a farmer measures how many bushels of barley he harvests in different fields with varying acidity levels.

Soil Acidity (pH)	5.7	6.2	6.6	6.8	7.1
Bushels Harvested	3	20	48	61	73

a. Find the equation for the best-fit line for the data and the correlation coefficient.

b. Use the equation of the best-fit line to estimate how many bushels the farmer would harvest if the soil had a pH of 10.

c. Could the equation of the best-fit line be used to extrapolate the data for extremely high levels of soil acidity? Explain.

11. FOOTBALL A college running back ran for 1732 total yards in the regular season. The table shows his cumulative total number of yards gained after select games.

Game Number	1	3	6	9	12
Cumulative Yards	184	431	818	1257	1732

a. Find the equation for the best-fit line for the total yards *y* gained after *x* games.

b. Find and interpret the correlation coefficient.

c. Use the trend of the data and the table to estimate when the running back will have run for 950 yards. Explain your reasoning.

d. During which game would you expect the running back to reach a total of 1000 yards?

12. REGULARITY Consider the linear regression equation that models a set of data very well to be $y = 1.43x - 4.2$. Would there be any restrictions on what the correlation coefficient value could be? Justify your reasoning.

13. STRUCTURE The table shows the number of student athletes participating in college athletics since the 2010-2011 school year.

Year	2010-2011	2011-2012	2012-2013	2013-2014	2014-2015
Student Athletes	444,077	453,347	463,202	472,625	482,533

    a. Find the equation for the best-fit line for the data.

    b. Find and interpret the correlation coefficient.

    c. Graph and analyze the residual plot. Does this support your conclusion from **part b?**

    d. Predict the number of college athletes in 2035.

🧠 Higher-Order Thinking Skills

14. WRITE How are lines of fit and linear regression similar? different?

15. CREATE For a class project, the scores that 10 randomly selected students earned on the first 8 tests of the school year are given. Explain how to find a line of best fit. Could it be used to predict the scores of the other students? Explain your reasoning.

16. ANALYZE Determine whether the following statement is *sometimes, always,* or *never* true: *If the correlation coefficient in a given situation is 0.946, the change in the independent variable* **causes** *change in the dependent variable.* Justify your argument.

17. PERSEVERE The table shows the number of participants in high school athletics.

Years Since 1980	0	10	20	25	30
Number of Athletes	5,356,913	5,298,671	6,705,223	7,159,904	7,667,955

    a. Find an equation for the regression line.

    b. According to the equation, how many participated in 2008?

Inverses of Linear Functions

Copyright © McGraw-Hill Education

Learn Inverses of Relations

Two relations are **inverse relations** if and only if one relation contains points of the form (a, b) when the other relation contains points of the form (b, a). So, the x-coordinates are exchanged with the y-coordinates for each ordered pair in the relation.

Key Concept • Inverse Relations

Words	If one relation contains the element (a, b), then the inverse relation will contain the element (b, a).
Symbols	$(a, b) \rightarrow (b, a)$
Example	A and B are inverse relations.
	$\quad A \qquad\quad B$
	$(-8, 12) \;\rightarrow (12, -8)$
	$(-2, -5) \rightarrow (-5, -2)$
	$\;\;(0, 4) \quad \rightarrow (4, 0)$
	$\;\;(7, 16) \quad \rightarrow (16, 7)$
Graph	The graph of an inverse is the graph of the original relation reflected over the line $y = x$. For every point (a, b) on the graph of the original relation, the graph of the inverse will include (b, a).

Example 1 Inverse Relations

Determine the inverse of {(−8, 3), (0, 14), (11, 52), (12, −6)}.

Write the coordinates in the ordered pairs to complete the inverse relation.

$(-8, 3) \rightarrow (3, -8)$ $(11, 52) \rightarrow (\underline{\quad}, \underline{\quad})$

$(0, 14) \rightarrow (14, \underline{\quad})$ $(12, -6) \rightarrow (\underline{\quad}, \underline{\quad})$

The inverse relation is {(3, −8), (14, 0), (52, 11), (−6, 12)}.

Check

Determine the inverse of the relation.

{(−1.4, 5), (1.3, 6.5), (3, −8), (3.05, 9)}

Today's Goals
- Construct the inverses of relations.
- Find inverses of linear functions.

Today's Vocabulary
inverse relations

inverse functions

😮 **Think About It!**

Describe the relationship between the domains and ranges of inverse relations.

🧭 **Go Online** You can complete an Extra Example online.

Example 2 Find Inverse Relations from a Table

Find the inverse of the relation shown in the table.

x	−11	0.3	−3	3.5
y	9	−2	−8	2

Write the coordinates in the ordered pairs to complete the inverse relation.

$(-11, 9) \rightarrow (9, -11)$ $(-3, -8) \rightarrow (\underline{\quad}, \underline{\quad})$

$(0.3, -2) \rightarrow (-2, \underline{\quad})$ $(3.5, 2) \rightarrow (\underline{\quad}, \underline{\quad})$

The inverse relation is _____.

Example 3 Graph Inverse Relations

Graph the inverse of the relation.

The graph of the relation passes through the points $(-1, -5)$, $(0, -2)$, $(1, 1)$, and $(2, 4)$.

Exchange the x-coordinates and y-coordinates to find points of the inverse relation.

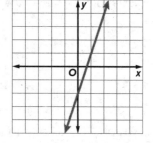

$(-1, -5) \rightarrow (-5, -1)$ $(1, 1) \rightarrow (\underline{\quad}, \underline{\quad})$

$(0, -2) \rightarrow (-2, \underline{\quad})$ $(2, 4) \rightarrow (\underline{\quad}, \underline{\quad})$

Plot the points of the inverse relation and draw a line passing through them.

Study Tip

Plotting Points While it only takes two points to graph a line, use several points when graphing inverse relations to create more accurate graphs.

💭 Think About It!

Describe the graph of the inverse of the horizontal line $y = 3$.

Explore Comparing a Function and Its Inverse

🔎 **Online Activity** Use graphing technology to complete the Explore.

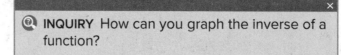

@ **INQUIRY** How can you graph the inverse of a function?

Learn Inverses of Linear Functions

A linear relation that is described by a function may have an **inverse function** that can generate ordered pairs of the inverse relation. The inverse of the linear function $f(x)$ is written as $f^{-1}(x)$ and is read *f inverse of x* or *the inverse of f of x*.

🌐 **Go Online** You can complete an Extra Example online.

Key Concept • Finding Inverse Functions

To find the inverse function $f^{-1}(x)$ of the linear function $f(x)$, complete the following steps.

Step 1	Replace $f(x)$ with y in the equation for $f(x)$.
Step 2	Interchange y and x in the equation.
Step 3	Solve the equation for y.
Step 4	Replace y with $f^{-1}(x)$ in the new equation.

Watch Out!

In $f^{-1}(x)$, the −1 is not an exponent. It is a way to indicate that $f^{-1}(x)$ is an inverse of another function called $f(x)$.

Example 4 Find an Inverse Linear Function

Find the inverse of $f(x) = 5x + 10$.

Step 1 $f(x) = 5x + 10$ Original equation

          _____ $= 5x + 10$ Replace $f(x)$ with y.

Step 2 $x = 5y + 10$ Interchange y and x.

Step 3   _____ $= 5y$ Subtract 10 from each side.

          $\dfrac{x - 10}{5} = y$ Divide each side by 5.

Step 4 $\dfrac{x - 10}{5} =$ _____   Replace y with $f^{-1}(x)$.

The inverse of $f(x) = 5x + 10$ is $f^{-1}(x) = \dfrac{x - 10}{5}$ or $f^{-1}(x) = \dfrac{1}{5}x - 2$.

Example 5 Find Inverses of Linear Functions

Find the inverse of $f(x) = -\dfrac{2}{3}x - 8$.

Step 1 $f(x) = -\dfrac{2}{3}x - 8$ Original equation

          $y = -\dfrac{2}{3}x - 8$ Replace $f(x)$ with y.

Step 2 $x = -\dfrac{2}{3}y - 8$ Interchange x and y.

Step 3 $x + 8 = -\dfrac{2}{3}y$ Add 8 to each side.

          $-\dfrac{3}{2}(x + 8) = y$ Multiply each side by $-\dfrac{3}{2}$.

          $-\dfrac{3}{2}x - 12 = y$ Simplify.

Step 4 $-\dfrac{3}{2}x - 12 = f^{-1}(x)$ Replace y with $f^{-1}(x)$.

The inverse of $f(x) = -\dfrac{2}{3}x - 8$ is $f^{-1}(x) = -\dfrac{3}{2}x - 12$.

Talk About It!

What is the inverse of $f(x) = -x$? How could you check your solution?

Go Online to see Example 5.

🌐 Example 6 Apply Inverse Linear Functions

BOATING Skyler and Carmen rent a paddle boat at a state park for $15 plus $4 for each hour it is used. The function $C(x) = 4x + 15$ represents the total cost $C(x)$ for x hours.

Part A Determine the inverse function.

Step 1	$C(x) = 4x + 15$	Original equation
	$y = 4x + 15$	Replace $C(x)$ with y.
Step 2	$\underline{\quad} = 4\underline{\quad} + 15$	Interchange y and x.
Step 3	$x - 15 = 4y$	Subtract 15 from each side.
	$\dfrac{x - 15}{4} = y$	Divide each side by 4.
Step 4	$\dfrac{x - 15}{4} = \underline{\quad\quad}$	Replace y with $C^{-1}(x)$.

Study Tip

Function Notation
Function notation is a way to give an equation a name, such as $f(x)$, $g(x)$, or $C(x)$. In Step 1 of finding the inverse function, replace the function notion with y regardless of the name of the function.

Part B Interpret the inverse function.

x is the _____ of renting the paddle boat, and $C^{-1}(x)$ is the number of _____ that Skyler and Carmen use the paddle boat.

Part C Evaluate using the inverse function.

Skyler and Carmen have $35 to rent the paddle boat. How long can they rent it?

To find the length of time that they can rent the boat, find $C^{-1}(\underline{\quad})$.

$C^{-1}(x) = \dfrac{x - 15}{4}$	Original equation
$C^{-1}(35) = \dfrac{35 - 15}{4}$	Substitute 35 for x.
$= \dfrac{20}{4}$ or $\underline{\quad}$	Simplify.

Check

CANDLES Javi is making candles to sell at an upcoming festival. He has already made 38 candles, and he makes 24 candles each day. The function $C(x) = 24x + 38$ represents the total number of candles $C(x)$ he has in inventory, where x is the number of days since he began making more candles.

Part A Select the inverse of the function $C(x)$. _____

A. $C^{-1}(x) = \dfrac{1}{24}x - \dfrac{19}{12}$ B. $C^{-1}(x) = \dfrac{1}{24}x - 38$

C. $C^{-1}(x) = \dfrac{1}{38}x - \dfrac{19}{12}$ D. $C^{-1}(x) = \dfrac{12}{19}x - 24$

Part B Estimate the amount of time it would take Javi to make 350 candles. It would take Javi between ____ and ____ days to make 350 candles.

 Go Online You can complete an Extra Example online.

Practice

🔾 **Go Online** You can complete your homework online.

Examples 1 and 2
Find the inverse of each relation.

1.

x	y
−9	−1
−7	−4
−5	−7
−3	−10
−1	−13

2.

x	y
1	8
2	6
3	4
4	2
5	0

3.

x	y
−4	−2
−2	−1
0	1
2	0
4	2

4. {(−3, 2), (−1, 8), (1, 14), (3, 20)}

5. {(5, −3), (2, −9), (−1, −15), (−4, −21)}

6. {(4, 6), (3, 1), (2, −4), (1, −9)}

7. {(−1, 16), (−2, 12), (−3, 8), (−4, 4)}

8. {(−5, 13), (6, 10.8), (3, 11.4), (−10, 14)}

9. {(−4, −49), (8, 35), (−1, −28), (4, 7)}

Example 3
Graph the inverse of each function.

10.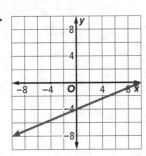

11.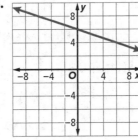

12.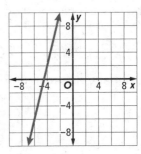

13.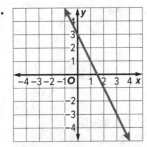

14.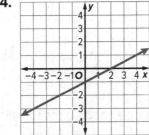

15.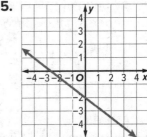

Examples 4 and 5

Find the inverse of each function.

16. $f(x) = 8x - 5$

17. $f(x) = 6(x + 7)$

18. $f(x) = \frac{3}{4}x + 9$

19. $f(x) = -16 + \frac{2}{5}x$

20. $f(x) = \frac{3x + 5}{4}$

21. $f(x) = \frac{-4x + 1}{5}$

Example 6

22. LEMONADE Bernardo spent $15 on supplies for his lemonade stand. He charges $1.25 per glass. The function $P(x) = 1.25x - 15$ represents his profit, where x is the number of glasses of lemonade sold.

   a. Find the inverse function, $P^{-1}(x)$.

   b. What do x and $P^{-1}(x)$ represent in the context of the inverse function?

   c. How many glasses must Bernardo sell in order to make $10 in profit?

23. BUSINESS Alisha started a baking business. She spent $36 initially on supplies and can make 5 dozen brownies for $12. She charges her customers $10 per dozen brownies. The function $P(x) = 7.6x - 36$ represents her profit, where x is the number of dozens of brownies sold.

   a. Find the inverse function, $P^{-1}(x)$.

   b. What do x and $P^{-1}(x)$ represent in the context of the inverse function?

   c. How many dozens of brownies does Alisha need to sell in order to make a profit?

24. SEASON PASS A season pass to an amusement park costs $70 per family member plus an additional $50 fee for parking. The function $C(x) = 70x + 50$ represents the total cost of the season pass for a family, where x is the number of family members on the season pass.

   a. Find the inverse function, $C^{-1}(x)$.

   b. What do x and $C^{-1}(x)$ represent in the context of the inverse function?

   c. How many family members purchased a season pass to the amusement park if the total charge was $470?

25. GARDENING Kara is building raised garden beds for her backyard. The total cost $C(x)$ in dollars is given by $C(x) = 125 + 16x$, where x is the number of pieces of wood required for the boxes.

   a. Find the inverse function $C^{-1}(x)$.

   b. If the total cost was $269 and each piece of wood was 12 feet long, how many total feet of wood were used?

26. GEOMETRY The area of the base of a cylindrical water tank is 12π square feet. The volume of water in the tank is dependent on the height of the water h and is represented by the function $V(h) = 12\pi h$.

   a. Find $V^{-1}(h)$.

   b. What will the height of the water be when the volume reaches 420π cubic feet?

Mixed Exercises

Write the inverse of each function in $f^{-1}(x)$ notation.

27. $3y - 12x = -72$ **28.** $x + 3y = 10$

29. $-42 + 6y = x$ **30.** $3y + 24 = 2x$

31. $-7y + 2x = -28$ **32.** $12y - x = 7$

Write an equation for the inverse function $f^{-1}(x)$ that satisfies the given conditions.

33. slope of $f(x)$ is 7; graph of $f^{-1}(x)$ contains the point (13, 1)

34. graph of $f(x)$ contains the points $(-3, 6)$ and $(6, 12)$

35. graph of $f(x)$ contains the point (10, 16); graph of $f^{-1}(x)$ contains the point $(3, -16)$

36. slope of $f(x)$ is 4; $f^{-1}(5) = 2$

Match each function with the graph of its inverse.

37. $f(x) = \frac{1}{2}x + 2$ **38.** $f(x) = \frac{1}{2}x - 2$ **39.** $f(x)\ x + 2$

A.   **B.**   **C.**

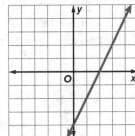 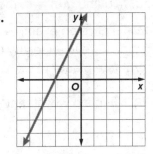

40. STRUCTURE Write the inverse of each function.

 a. $f(x) = \frac{x - 10}{3}$

 b. $g(x) = \frac{3}{4}x + 6$

 c. $h(x) = -5x - 7$

 d. Graph $h(x)$ and $h^{-1}(x)$ on the same coordinate plane to check your answer.

41. REASONING Suppose the inverse of a relation is $\{(b, -k), (-g, p), (-w, -m), (r, q)\}$. What is the relation?

🤔 Higher-Order Thinking Skills

42. ANALYZE How can you use ordered pairs to check if the inverse of a function is correct?

43. WRITE What is the relationship between the slopes of two lines that are inverse functions of one another? Give an example.

44. WRITE What is the relationship between the x- and y-intercepts of two lines that are inverse functions of one another? Give an example.

45. FIND THE ERROR A student claims that there is a simple method to find the inverse of the function $f(x)$. To find the inverse, $f^{-1}(x)$, we need only remember that raising something to the power of -1 is the same as taking its reciprocal. Is this claim correct? Include an example or counterexample.

46. PERSEVERE If $f(x) = 5x + a$ and $f^{-1}(10) = -1$, find a.

47. PERSEVERE If $f(x) = \frac{1}{a}x + 7$ and $f^{-1}(x) = 2x - b$, find a and b.

ANALYZE Determine whether the following statements are *sometimes*, *always*, or *never* true. Explain your reasoning.

48. If $f(x)$ and $g(x)$ are inverse functions, then $f(a) = b$ and $g(b) = a$.

49. If $f(a) = b$ and $g(b) = a$, then $f(x)$ and $g(x)$ are inverse functions.

50. CREATE Give an example of a function and its inverse. Verify that the two functions are inverses by graphing the functions and the line $y = x$ on the same coordinate plane.

51. WRITE Explain why it may be helpful to find the inverse of a function.

Essential Question

What can a function tell you about the relationship that it represents?

Module Summary

Lessons 5-1 and 5-2

Writing Equations

- Slope-intercept form is $y = mx + b$, where m is the slope of the line and b is the y-intercept.

- Point-slope form is $y - y_1 = m(x - x_1)$, where (x_1, y_1) is a given point on a nonvertical line and m is the slope of the line.

- Standard form is $Ax + By = C$, where A, B, and C are integers, $A > 0$, A and B are both not equal to 0, and the GCF of A, B, and C is 1.

- To write a linear equation given two points on a line, first find the slope. Then use either point to write the equation in point-slope form or find the y-intercept to write the equation in slope-intercept form.

Lessons 5-3 through 5-5

Scatter Plots

- A scatter plot shows the relationship between a set of bivariate data, graphed as ordered pairs on a coordinate plane.

- A positive correlation exists when, as x increases, y increases. A negative correlation exists when, as x increases, y decreases. No correlation exists when x and y are not related.

- A line of fit is used to describe the trend of the data in a scatter plot.

- To determine causation, determine whether one variable influences the other variable.

- The correlation coefficient tells you how well the equation for the best-fit line models the data.

- A correlation coefficient close to 1 has a strong positive correlation. A correlation coefficient close to −1 has a strong negative correlation.

- Residuals measure how much the data deviate from the regression line.

Lesson 5-6

Inverses of Linear Functions

- Two relations are inverse relations if and only if one relation contains the element (a, b) when the other relation contains the element (b, a).

- In inverse relations, the x-coordinates are exchanged with the y-coordinates for each ordered pair in the relation.

- To find the inverse of $f(x)$, replace $f(x)$ with y in the equation for $f(x)$. Interchange y and x in the equation. Solve the equation for y. Replace y with $f^{-1}(x)$ in the new equation.

Study Organizer

 Foldables

Use your Foldable to review the module. Working with a partner can be helpful. Ask for clarification of concepts as needed .

Creating
Linear
Equations

Test Practice

1. **MULTIPLE CHOICE** What is the equation of the line that passes through the points $(-2, 1)$ and $(6, 3)$? (Lesson 5-1)

   Ⓐ $y = \frac{1}{4}x + \frac{3}{2}$

   Ⓑ $y = \frac{3}{2}x + \frac{1}{4}$

   Ⓒ $y = \frac{2}{3}x + \frac{1}{4}$

   Ⓓ $y = \frac{3}{2}x + \frac{1}{4}$

2. **MULTIPLE CHOICE** Select the equation of a line with a slope of 5 that passes through the point $(2, -3)$. (Lesson 5-1)

   Ⓐ $y = 5x + 2$

   Ⓑ $y = 5x - 3$

   Ⓒ $y = 5x + 7$

   Ⓓ $y = 5x - 13$

Use the table for exercises 3 and 4. A movie streaming service charges a set fee for membership each month, plus an additional fee for the number of movies streamed each month. This table shows the total charge for different numbers of movies.

Number of movies streamed (x)	2	4	6
Total cost (y)	$14	$17	$20

3. **OPEN RESPONSE** Write the slope-intercept form of the equation that models the linear relationship in the table. (Lesson 5-1)

4. **OPEN RESPONSE** Explain the meaning of the slope and y-intercept in the context of the situation. (Lesson 5-1)

5. **MULTIPLE CHOICE** Which equation represents a line that passes through the point $(3, -4)$ with a slope of 7? (Lesson 5-2)

   Ⓐ $y + 4 = 7(x - 3)$

   Ⓑ $y + 4 = 7(x + 3)$

   Ⓒ $y - 4 = 7(x - 3)$

   Ⓓ $y - 4 = 7(x + 3)$

6. **MULTI-SELECT** Select all of the equations that represent the line. (Lesson 5-2)

   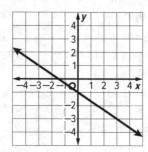

   Ⓐ $2x - 3y = -1$

   Ⓑ $2x + 3y = -3$

   Ⓒ $3x - 2y = 2$

   Ⓓ $y + 3 = -\frac{2}{3}(x - 3)$

   Ⓔ $y - 1 = -\frac{2}{3}(x + 3)$

   Ⓕ $y + 1 = -\frac{3}{2}(x + 3)$

7. **OPEN RESPONSE** A city parking garage charges $4 to park for up to two hours. After that, an additional charge of $2.50 per hour applies. Write an equation in point-slope form that models the total cost y for parking x hours, where $x > 2$. (Lesson 5-2)

8. **MULTIPLE CHOICE** Which scatter plot shows the best line of fit? (Lesson 5-3)

Ⓐ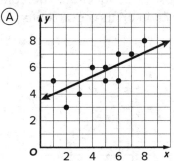

Ⓑ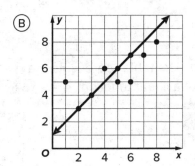

Ⓒ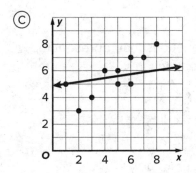

Ⓓ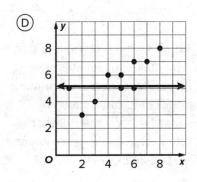

9. **MULTIPLE CHOICE** Which equation represents the best line of fit for the scatter plot? (Lesson 5-3)

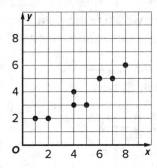

Ⓐ $y = 0.6x + 1$

Ⓑ $y = 0.5x + 2$

Ⓒ $y = x - 2$

Ⓓ $y = 0.75x$

10. **OPEN RESPONSE** Adriana keeps the statistics for her favorite basketball team and creates the scatter plot shown. (Lesson 5-3)

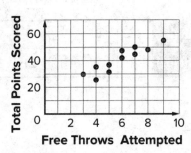

Adriana then draws a line of fit for her scatter plot. What does the slope of the line represent?

11. **OPEN RESPONSE** A researcher found that students who spent more time exercising each week also had higher average test scores. Describe the correlation, if any, between time spent exercising and test scores. (Lesson 5-4)

12. OPEN RESPONSE Lalita tracked the amount of time she studied each week and her score on a weekly chemistry quiz for eight weeks. She made this scatter plot from the data. Determine whether the data illustrate a *correlation* or *causation*. Explain. (Lesson 5-4)

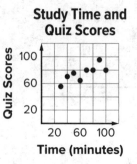

Study Time and Quiz Scores

13. MULTIPLE CHOICE Use linear regression to estimate the weight, in ounces, of a bluegill that has a length of 9.5 inches. Round your answer to the nearest tenth of an ounce. (Lesson 5-5)

Length (in.)	7	8	9	11	12	13
Weight (oz)	4	7	11	21	26	32

(A) 12.9

(B) 13.4

(C) 14.5

(D) 15.1

14. OPEN RESPONSE Use a graphing calculator and linear regression to write the equation of a best-fit line for the data in slope-intercept form. Round to the nearest tenth. (Lesson 5-5)

x	2.4	2.8	3.4	4.3	5.1	7.6	8.4	9.1
y	6.2	9.6	8.4	6.5	7.2	2.5	1.8	4.2

15. OPEN RESPONSE The graph of a function passes through $(-3, 2)$, $(-1, 1)$, $(1, 0)$, and $(3, -1)$. Find the inverse function. Then graph the inverse function on the coordinate plane. (Lesson 5-6)

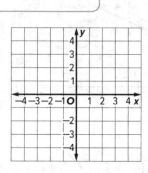

16. MULTI-SELECT The table represents the coordinates of a linear function. (Lesson 5-6)

x	y
−6	5
4	1
2	−3

Select the equations that represent the inverse of the function.

(A) $f^{-1}(x) = \frac{5}{2}x + \frac{5}{2}$

(B) $f^{-1}(x) = -\frac{5}{2}x + \frac{13}{2}$

(C) $f(x) = 2.5x - 6.5$

(D) $f^{-1}(x) = -1.25x + \frac{13}{2}$

(E) $f^{-1}(x) = -2.5x + 6.5$

17. MULTIPLE CHOICE Shakir is running in a long-distance race. If he maintains an average speed of 8 miles per hour, then the distance in miles that he has left to run is given by $D(x) = -\frac{2}{15}x + 10$, where x is the number of minutes since Shakir started the race. Which function is the inverse of $D(x)$? (Lesson 5-6)

(A) $D^{-1}(x) = -\frac{15}{2}x + 75$

(B) $D^{-1}(x) = \frac{2}{15}x - 10$

(C) $D^{-1}(x) = -\frac{15}{2}x + \frac{1}{10}$

(D) $D^{-1}(x) = -\frac{2}{15}x - \frac{4}{3}$

Linear Inequalities

e Essential Question

How can writing and solving inequalities help you solve problems in the real world?

What Will You Learn?

Place a check mark (✓) in each row that corresponds with how much you already know about each topic **before** starting this module.

KEY	Before			After		
👎 — I don't know. 👈 — I've heard of it. 👍 — I know it!	👎	👈	👍	👎	👈	👍
graph linear inequalities						
solve one-step linear inequalities using addition and subtraction						
solve one-step linear inequalities using multiplication and division						
solve multi-step linear inequalities						
solve compound linear inequalities						
solve absolute value linear inequalities						
graph inequalities in two-variables						

📙 Foldables Make this Foldable to help you organize your notes about linear inequalities. Begin with one sheet of 11″ × 17″ paper.

1. **Fold** each side so the edges meet in the center.

2. **Fold** in half.

3. **Unfold** and cut from each end until you reach the vertical line.

4. **Label** the front of each flap.

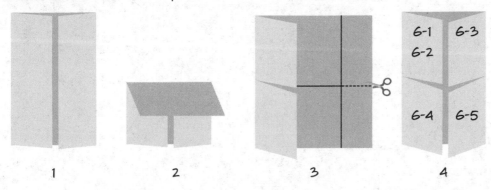

What Vocabulary Will You Learn?

Check the box next to each vocabulary term that you may already know.

- ☐ boundary
- ☐ closed half-plane
- ☐ compound inequality
- ☐ half-plane
- ☐ inequality

- ☐ intersection
- ☐ open half-plane
- ☐ set-builder notation
- ☐ union

Are you ready?

Complete the Quick Review to see if you are ready to start this module.
Then complete the Quick Check.

Quick Review

Example 1

Solve −2(x − 4) = 7x − 19.

$-2(x - 4) = 7x - 19$	Original equation
$-2x + 8 = 7x - 19$	Distributive Property
$-2x + 8 + 2x = 7x - 19 + 2x$	Add 2x to each side.
$8 = 9x - 19$	Simplify.
$8 + 19 = 9x - 19 + 19$	Add 19 to each side.
$27 = 9x$	Simplify.
$3 = x$	Divide each side by 3.

Example 2

Solve |x − 4| = 9.

if $|x - 4| = 9$, then $x - 4 = 9$ or $x - 4 = -9$.

$$x - 4 = 9 \qquad \text{or} \qquad x - 4 = -9$$
$$x - 4 + 4 = 9 + 4 \qquad\qquad x - 4 + 4 = -9 + 4$$
$$x = 13 \qquad\qquad\qquad x = -5$$

So, the solution set is {−5, 13}.

Quick Check

Solve each equation.

1. $2x + 1 = 9$

2. $4x - 5 = 15$

3. $9x + 2 = 3x - 10$

4. $3(x - 2) = -2(x + 13)$

Solve each equation.

5. $|x + 11| = 18$

6. $|3x - 2| = 16$

7. $|x - 7| = 8$

8. $|2x| = -9$

How did you do?

Which exercises did you answer correctly in the Quick Check? Shade those exercise numbers below.

① ② ③ ④ ⑤ ⑥ ⑦ ⑧

Solving One-Step Inequalities

Explore Graphing Inequalities

Online Activity Use graphing technology to complete the Explore.

> ✕
>
> ⓠ **INQUIRY** How can you graph the solution set of an inequality of the form $x < a$ or $x > a$ for some number a?

Learn Graphing Inequalities

An **inequality** is a mathematical sentence that contains the symbol $<$, $>$, $\leq$, $\geq$ or $\neq$. An inequality compares the value of two numbers or expressions using these symbols.

Example 1 Graph Inequalities

Graph the solution set of $y \leq 4$.

The dot at 4 shows that 4 _____ a solution. The heavy arrow pointing to the _____ shows that the solution includes all numbers _____ 4.

Check

Graph the solution set of $y > \frac{1}{2}$.

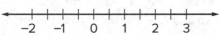

Example 2 Write Inequalities from a Graph

Write an inequality that represents the graph.

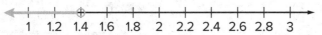

The endpoint is shown with a circle at 1.4, so 1.4 is _____ in the solution. The inequality must be $<$ or $>$.

The arrow points to values _____ than 1.4.
The graph represents the solution of _____ .

Check

Write an inequality that represents the graph.

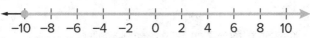

⬀ **Go Online** You can complete an Extra Example online.

Copyright © McGraw-Hill Education

Today's Goals
- Graph the solutions of an inequality.
- Solve linear inequalities by using addition.
- Solve linear inequalities by using subtraction.

Today's Vocabulary
inequality
set-builder notation

🍥 **Think About It!**
Compare and contrast the graphs of $x \geq 6$ and $x > 6$.

🍥 **Think About It!**
How do you know which inequality symbol to use by looking at a graph?

Explore Properties of Inequalities

► Online Activity Use graphing technology to complete the Explore.

> ⊗
>
> **@ INQUIRY** Do the properties of equality hold true for inequalities?

Learn Solving Inequalities by Using Addition and Subtraction

Addition and subtraction can be used to solve inequalities.

Key Concept • Addition Property of Inequalities	
Words	If the same number is added to each side of a true inequality, the resulting inequality is also true.
Symbols	For any real numbers a, b, and c, the following are true. If $a > b$, then $a + c > b + c$. If $a < b$ then $a + c < b + c$.

Key Concept • Subtraction Property of Inequalities	
Words	If the same number is subtracted from each side of a true inequality, the resulting inequality is also true.
Symbols	For any real numbers a, b, and c, the following are true. If $a > b$, then $a - c > b - c$. If $a < b$ then $a - c < b - c$.

When solving inequalities, you can write the solution set in a more concise way using **set-builder notation**. For example, $\{x \mid x \geq -4\}$ represents the set of all numbers x such that x is greater than or equal to -4.

Copyright © McGraw-Hill Education

Example 3 Solve Inequalities by Adding

Solve $x - 10 < 15$.

$x - 10 < 15$	Original inequality
$x - 10 + 10 < 15 + 10$	Add 10 to each side to isolate x.
$x < \underline{\quad}$	Simplify.

The solution set is { _____ }.

Check

Solve $-9 + b \leq 16$. _____

► Go Online You can complete an Extra Example online.

💬 Talk About It!

How many solutions of the inequality are there? Justify your argument.

Study Tip

Set-Builder Notation
$\{x \mid x < 25\}$ is read *the set of all numbers x such that x is less than 25.*

Example 4 Solve Inequalities by Subtracting

Solve $x + 24 \geq 61$.

$$x + 24 \geq 61 \qquad \text{Original inequality}$$

$$x + 24 - 24 \geq 61 - 24 \qquad \text{Subtract 24 from each side.}$$

$$x \geq \underline{\hspace{1cm}} \qquad \text{Simplify.}$$

The solution set is { _____ }.

Check

Write the solution set for $88 < x + 13$. _____

Think About It!
How can you check the solution of the inequality?

Example 5 Add or Subtract to Solve Inequalities with Variables on Each Side

Solve $9y + 3 \geq 10y$.

$$9y + 3 \geq 10y \qquad \text{Original inequality}$$

$$9y - 9y + 3 \geq 10y - 9y \qquad \text{Subtract } 9y \text{ from each side.}$$

$$\underline{\hspace{1cm}} \geq y \qquad \text{Simplify.}$$

Since $3 \geq y$ is the same as $y \leq 3$, { _____ }.

Check

Write the solution set for $7x + 6 < 8x$. _____

Study Tip

Writing Inequalities Simplifying the inequality so that the variable is on the left side, as in $y \leq 3$, prepares you to write the solution set in set-builder notation and graph the inequality on a number line.

🌐 Example 6 Use an Inequality to Solve a Problem

DATA USAGE Hassan's wireless contract allows him to use at most 5 gigabytes (GB) of data per month. At this point, Hassan has used 3.7 GB of data. How many gigabytes of data can Hassan use during the rest of the month without exceeding the maximum allowance?

Complete the table to write an inequality to represent how many gigabytes of data Hassan can use. Then solve the inequality.

Words	Hassan can use	_____	5 GB of data.
Variables	Let g = the _____ that Hassan has left to use.		
Inequality	_____	_____	_____

(continued on the next page)

Use a Source

Research data plans for wireless carriers in your area. Write and solve your own inequality to represent the amount of data remaining if you have already used 5.2 GB.

Study Tip

Inequalities Verbal problems containing phrases like *greater than* and *less than* can be solved by using inequalities. Some other phrases that include inequalities are:

< less than; fewer than
> greater than; more than
≤ less than or equal to; at most; no more than
≥ greater than or equal to; at least; no less than

 Think About It!

If a, b, and c are positive real numbers, what must be true if ac is greater than or equal to bc?

What must happen to an inequality symbol when you divide each side by a negative number if the inequality is to remain true?

$$3.7 + g \leq 5 \qquad \text{Original inequality}$$

$$3.7 - 3.7 + g \leq 5 - 3.7 \qquad \text{Subtract 3.7 from each side.}$$

$$g \leq \underline{\hspace{1cm}} \qquad \text{Simplify.}$$

The solution set is { _____ }.

Hassan can use up to _____ GB of data without exceeding his maximum allowance. Notice that negative numbers are solutions to the inequality, but they are _____ solutions to the problem because Hassan cannot use a negative amount of data.

Learn Solving Inequalities by Using Multiplication and Division

If you multiply or divide each side of an inequality by a positive number, then the inequality remains true.

If you multiply or divide each side of an inequality by a negative number, the inequality symbol changes direction.

Key Concept • Multiplication Property of Inequalities		
Words	If each side of a true inequality is multiplied by a positive number, the resulting inequality is also true.	If each side of a true inequality is multiplied by a negative number, the direction of the inequality sign must be reversed to make the resulting inequality also true.
Symbols	For any real numbers a and b and any positive real number c: If $a > b$, then $ac > bc$. If $a < b$, then $ac < bc$.	For any real numbers a and b and any negative real number c: If $a > b$, then $ac < bc$. If $a < b$, then $ac > bc$.

Key Concept • Division Property of Inequalities		
Words	If each side of a true inequality is divided by a positive number, the resulting inequality is also true.	If each side of a true inequality is divided by a negative number, the direction of the inequality sign must be reversed to make the resulting inequality also true.
Symbols	For any real numbers a and b and any positive real number c: If $a > b$, then $\frac{a}{c} > \frac{b}{c}$. If $a < b$, then $\frac{a}{c} < \frac{b}{c}$.	For any real numbers a and b and any negative real number c: If $a > b$, then $\frac{a}{c} < \frac{b}{c}$. If $a < b$, then $\frac{a}{c} > \frac{b}{c}$.

These properties also hold true for inequalities involving ≤ and ≥.

🌐 Apply Example 7 Write and Solve an Inequality

BOOKS Alisa has read approximately $\frac{1}{4}$ of a novel. If she has read at least 112 pages, how many pages are there in the novel?

1. What is the task?
Describe the task in your own words. Then list any questions that you may have. How can you find answers to your questions?

_____

_____

_____

2. How will you approach the task? What have you learned that you can use to help you complete the task?

_____

_____

3. What is your solution?
Estimate the number of pages in the novel. _____

Write an inequality to represent this situation. Let n = the number of pages in the novel.

_____

_____

_____

There are at least _____ pages in the novel.

4. How can you know that your solution is reasonable?
✏️ **Write About It!** Write an argument that can be used to defend your solution.

_____

_____

_____

Check

ELECTRIC CAR For every hour x that Eva's electric car charges, she can drive the car 7.5 miles. Eva needs to drive at least 60 miles tomorrow.

Part A What inequality represents the situation in terms of x hours?
_____

Part B What is the least amount of time that Eva will need to charge her car? _____ hours

Math History Minute

German mathematician **Emmy Noether (1882–1935)** has been described as one of the greatest mathematicians of the twentieth century. She devised theorems for several concepts later found in Einstein's theory of relativity and was one of the founders of abstract algebra. One person wrote, "The development of abstract algebra, which is one of the most distinctive innovations of twentieth century mathematics, is largely due to her."

🔄 **Go Online** You can complete an Extra Example online.

Multiplicative Inverses
The multiplicative inverse, or reciprocal, of a number can be used to undo multiplication. Multiplying $-\frac{2}{5}x$ by the reciprocal $-\frac{5}{2}$ in the example at the right is the same as dividing by $-\frac{2}{5}$, but is easier to compute mentally.

 Think About It!

Why was the inequality symbol reversed?

 Think About It!

Why is the solution of $-13z \geq 117$ shaded to the left when the original inequality symbol is greater than or equal to?

Study Tip

Negatives A negative sign in an inequality does not necessarily mean that the direction of the inequality symbol should change. For example, when solving $\frac{x}{3} \geq -9$, do not change the direction of the inequality symbol.

Example 8 Solve an Inequality by Multiplying

Solve $-\frac{2}{5}x \leq 11$. Graph the solution set on a number line.

$$-\frac{2}{5}x \leq 11 \qquad \text{Original inequality}$$

$$\underline{\qquad} -\frac{2}{5}x \leq 11 \underline{\qquad} \qquad \text{Multiply each side by } -\frac{2}{5}. \text{ Reverse the inequality symbol.}$$

$$x \underline{\qquad} \qquad \text{Simplify.}$$

The solution set is { _____ }.

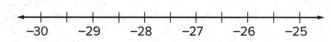

Example 9 Solve an Inequality by Dividing

Solve $20x < 4$. Graph the solution set on a number line.

$$20x < 4 \qquad \text{Original inequality}$$

$$\frac{20x}{20} < \frac{4}{20} \qquad \text{Divide each side by 20.}$$

$$x < \underline{\qquad} \qquad \text{Simplify.}$$

The solution set is { _____ }.

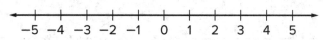

Check

Solve $7x > -161$. _____

Example 10 Solve an Inequality with a Negative Coefficient

Solve $-13z \geq 117$. Graph the solution set on a number line.

$$-13z \geq 117 \qquad \text{Original inequality}$$

$$-\frac{13z}{-13} \geq \frac{117}{-13} \qquad \text{Divide each side by } -13.$$

$$z \leq \underline{\qquad} \qquad \text{Simplify.}$$

The solution set is { _____ }.

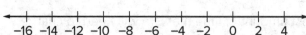

Check

Select the solution set for $-13x > -169$. ____

A. $\{x \mid x > 13\}$ **B.** $\{x \mid x < 13\}$ **C.** $\{x \mid x > -13\}$ **D.** $\{x \mid x < -13\}$

 Go Online You can complete an Extra Example online.

Practice

Go Online You can complete your homework online.

Example 1

Graph the solution set of each inequality.

1. $x \leq -5$

2. $y \geq -2$

3. $g > 5$

4. $h < -6$

5. $a < 7$

6. $b \leq 6$

Example 2

Write an inequality that represents each graph.

7.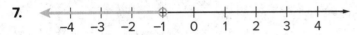

8.

9.

10.

11.

12.

Examples 3–5

Solve each inequality.

13. $m - 4 < 3$

14. $p - 6 \geq 3$

15. $r - 8 \leq 7$

16. $t - 3 > -8$

17. $b + 2 \geq 4$

18. $13 > 18 + r$

19. $5 + c \leq 1$

20. $-23 \geq q - 30$

21. $11 + m \geq 15$

22. $h - 26 < 4$

23. $8 \leq r - 14$

24. $-7 > 20 + c$

25. $2a \leq -4 + a$

26. $z + 4 \geq 2z$

27. $w - 5 \leq 2w$

28. $3y \leq 2y - 6$

29. $6x + 5 \geq 7x$

30. $-9 + 2a < 3a$

Example 6

31. PIZZA Tara and friends order a pizza. Tara eats 3 of the 10 slices and pays $4.50 for her share. Assuming that Tara has paid at least her fair share, write and solve an inequality to represent the cost of the pizza.

32. WEATHER Theodore Fujita of the University of Chicago developed a classification of tornadoes according to wind speed and damage. The table shows the classification system.

Level	Name	Wind Speed Range (mph)
F0	Gale	40–72
F1	Moderate	73–112
F2	Significant	113–157
F3	Severe	158–206
F4	Devastating	207–260
F5	Incredible	261–318
F6	Inconceivable	319–379

Source: National Weather Service

a. Suppose an F3 tornado has winds that are 162 miles per hour. Write and solve an inequality to determine how much the winds would have to increase before the F3 tornado becomes an F4 tornado.

b. A tornado has wind speeds that are at least 158 miles per hour. Write and solve an inequality that describes how much greater these wind speeds are than the slowest tornado.

Example 7

33. GARBAGE The amount of garbage that the average American adds to a landfill each day is 4.6 pounds. If at least 2.5 pounds of a person's daily garbage could be recycled, how much would still go into a landfill?

34. SUPREME COURT The first Chief Justice of the U.S. Supreme Court, John Jay, served 2079 days as Chief Justice. He served 10,463 days fewer than John Marshall, who served as Supreme Court Chief Justice for the longest period of time. How many days must the current Supreme Court Chief Justice John Roberts serve to surpass John Marshall's record of service?

35. AIRLINES On average, at least 25,000 pieces of luggage are lost or misdirected each day by United States airlines. Of these, 98% are located by the airlines within 5 days. From a given day's lost luggage, at least how many pieces of luggage are still lost after 5 days?

36. SCHOOL Gilberto earned these scores on the first three tests in biology this term: 86, 88, and 78. What is the lowest score that Gilberto can earn on the fourth and final test of the term if he wants to have an average of at least 83?

Examples 8–10

Solve each inequality. Graph the solution on a number line.

37. $\frac{1}{4}m \le -17$

38. $\frac{1}{2}a < 20$

39. $-11 > -\frac{c}{11}$

40. $-2 \ge -\frac{d}{34}$

41. $-10 \le \frac{x}{-2}$

42. $-72 < \frac{f}{-6}$

43. $\frac{2}{3}h > 14$

44. $-\frac{3}{4}j \ge 12$

45. $-\frac{1}{6}n \le -18$

46. $6p \le 96$

47. $4r < 64$

48. $32 > -2y$

49. $-26 < 26t$

50. $-6v > -72$

51. $-33 \ge -3z$

52. $4b \le -3$

53. $-2d < 5$

54. $-7f > 5$

Mixed Exercises

Match each inequality with its corresponding statement.

55. $3n < 9$

    a. Three times a number is at most nine.

56. $\frac{1}{3}n \ge 9$

    b. One third of a number is no more than nine.

57. $3n \le 9$

    c. Negative three times a number is more than nine.

58. $-3n > 9$

    d. Three times a number is less than nine.

59. $\frac{1}{3}n \le 9$

    e. Negative three times a number is at least nine.

60. $-3n \ge 9$

    f. One third of a number is greater than or equal to nine.

Define a variable, write an inequality, and solve each problem. Check your solution.

61. Seven more than a number is less than or equal to –18.

62. Twenty less than a number is at least 15.

63. A number plus 2 is at most 1.

64. One eighth of a number is less than or equal to 3.

65. Negative twelve times a number is no more than 84.

66. Eight times a number is at least 16.

STRUCTURE Solve each inequality. Check your solution, and then graph it on a number line.

67. $14c > 56$

68. $20b \geq -120$

69. $\frac{x}{4} < 9$

70. $\frac{x}{2.5} \leq 8$

71. $m + 3.7 < 9.1$

72. $n - \frac{1}{5} > \frac{4}{5}$

73. $c + (-1.4) \geq 2.3$

74. $k + \frac{3}{4} > \frac{1}{3}$

75. EVENT PLANNING The Community Center does not charge a rental fee as long as a rentee orders a minimum of $5000 worth of food. Antonio is planning a banquet. If he is expecting 225 people to attend, what is the minimum he will have to spend on food per person to avoid paying a rental fee?

76. VITAMINS The minimum daily requirement of vitamin C for 14-year-olds is at least 50 milligrams per day. An average-sized apple contains 6 milligrams of vitamin C. How many apples would a person have to eat each day to satisfy this requirement? Define a variable and write and solve an inequality to represent this situation.

77. USE A SOURCE The loudest insect is the African cicada. It produces sounds as loud as 105 decibels. The blue whale is the loudest mammal. The call of the blue whale can reach levels up to 83 decibels louder than the African cicada. Write and solve an inequality to represent the situation. How loud are the calls of the blue whale? Use a source to verify your answer.

78. USE A MODEL In a mathematics exam with a maximum score of 100, Machelle loses less than 27 points. The table shows the grade that matches the exam score. Compare points to grades and identify which grade Machelle can get.

Grade	Points
A	92–100
B	83–91
C	74–82
D	65–73
F	64 and below

   a. Define a variable. Then write and solve an inequality to represent the number of points Machelle received on her exam.

   b. Interpret the solution to your inequality. What do you know about Machelle's grade on the exam?

79. WHICH ONE DOESN'T BELONG Which inequality does *not* have the solution $\{x \mid x < -2\}$?

   A $-3x > 6$ **B** $-\frac{x}{2} < 1$ **C** $7x < -14$ **D** $\frac{4}{3}x < -\frac{8}{3}$

80. FIND THE ERROR Marty and Heath solved the same exercise in different ways. Is either correct? Explain your reasoning.

Marty	Heath
$3m \geq -21$	$3m \geq -21$
$\frac{3m}{3} \geq \frac{-21}{3}$	$\frac{3m}{3} \geq \frac{21}{-3}$
$m \leq -7$	$m \geq -7$

81. Solve each inequality in terms of x. Assume that a does not equal 0.

   a. $ax < 7$

   b. $ax \geq 12$

   c. $ax > 3a$

   d. $ax \geq \frac{a}{4}$

82. ANALYZE Determine whether the statement is *sometimes*, *always*, or *never* true. If $a > b$, then $\frac{1}{a} > \frac{1}{b}$. Justify your argument.

Solving Multi-Step Inequalities

Explore Modeling Multi-Step Inequalities

Online Activity Use algebra tiles to complete the Explore.

> **INQUIRY** How can you model and solve a
> multi-step inequality? ✕

Learn Solving Inequalities Involving More Than One Step

Step 1 Isolate the variable terms on one side of the inequality using addition or subtraction.

Step 2 Multiply or divide to isolate the variable.

🌐 Example 1 Apply Multi-Step Inequalities

PUBLISHING Suzy wants to self-publish her comic book. One printing company offers to publish the book for a $220 flat rate plus $3 per copy of the book. Her maximum budget is $400.

Part A Write an inequality.

Words	$ _____ flat rate	Plus	_____	is at most	_____
Inequality	_____	_____	_____	_____	_____

Part B Solve the inequality.

_____	Original inequality
_____	Subtract 220 from each side.
_____	Divide each side by 3.

Suzy can have up to _____ books printed while not exceeding her budget.

Check

TICKETS Jamal has $40 to buy tickets to a performance for himself and his friends. If he buys a $10 membership, he can buy tickets for $5 each. How many tickets can he buy while remaining within his budget?

If x represents the number of tickets Jamal purchases, write an inequality that represents the situation. _____

Solve the inequality. _____

🌐 **Go Online** You can complete an Extra Example online.

Today's Goals
• Solve multi-step linear inequalities.

Study Tip

Negative Numbers
When multiplying or dividing by a negative number, the direction of the inequality symbol changes. This holds true for multi-step as well as one-step inequalities.

🔵 **Think About It!**
Can x be any real number less than or equal to 60? Explain your reasoning.

Example 2 Write and Solve a Multi-Step Inequality

Consider the inequality *The opposite of a number divided by two minus seventeen is less than seven.*

Part A Translate the sentence into an inequality.

_____

Part B Solve the inequality.

_____   Original inequality

_____   Add 17 to each side.

_____   Multiply each side by 2.

_____   Divide each side by −1,
                           reversing the inequality symbol.

Part C Graph the solution on a number line.

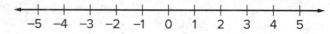

−50 −48 −46 −44 −42 −40 −38 −36 −34 −32 −30

Example 3 Solve an Inequality with the Distributive Property

Solve the inequality $4(2x − 11) \le −12 + 2(x − 4)$. Then graph the solution on a number line.

$4(2x − 11) \le −12 + 2(x − 4)$ Original inequality

___x − ___ $\le −12 +$ ___x − ___   Distributive Property

___x − ___ $\le$ ___ $+$ ___x Simplify.

___$x \le$ ___ $+$ ___x Add 44 to each side.

___$x \le$ ___   Subtract 2x from each side.

$x \le$ ___   Divide each side by 6.

Graph the solution of $4(2x − 11) \le −12 + 2(x − 4)$ on a number line.

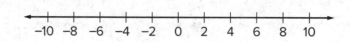

−5 −4 −3 −2 −1 0 1 2 3 4 5

Check

Solve $88 \ge −33 + 11(x + 8)$. Then graph the inequality on a number line.

_____

−10 −8 −6 −4 −2 0 2 4 6 8 10

Study Tip

Empty Set or All Real Numbers If an inequality simplifies to a false statement, then there is no solution to the inequality. The solution set is the empty set, ∅. However, if all values of a variable make the inequality true, then the solution set is all real numbers.

 Go Online You can complete an Extra Example online.

Practice

Go Online You can complete your homework online.

Example 1

1. **BEACHCOMBING** Jay wants to rent a metal detector. A rental company charges a one-time rental fee of $15 plus $2 per hour to rent a metal detector. Jay has only $35 to spend.

   a. Write an inequality to represent this situation, where h is the number of hours Jay will rent a metal detector.

   b. Solve the inequality. What is the maximum amount of time he can rent the metal detector?

2. **AGES** Pedro, Sebastian, and Manuel Martinez are each one year apart in age. The sum of their ages is greater than the age of their father, who is 60.

   a. Write an inequality to represent this situation, where x is the age of the youngest brother.

   b. Solve the inequality.

   c. How old can the oldest brother be? Explain your reasoning.

3. **RIDE SHARE** Demetri lives in the city and sometimes uses a ride share service. A ride costs $1.50 for the first $\frac{1}{5}$ mile and $0.25 for each additional $\frac{1}{5}$ mile. Demetri does not want to spend more than $3.75 on a ride.

   a. Write an inequality to represent this situation, where x is the number of miles.

   b. Solve the inequality. What is the maximum distance he can travel if he does not tip the driver?

   c. Generalize a method for writing an inequality for this situation if the service charges $1.50 for the first $\frac{1}{a}$ mile and $0.25 for each additional $\frac{1}{a}$ mile.

4. **POST OFFICE** Keshila goes to the post office to mail a package and a few letters. Stamps cost 49 cents each. It will cost $7.65 to mail the package. Keshila has $10.00.

   a. Write an inequality to represent this situation, where x is the number of stamps.

   b. Solve the inequality. What is the maximum number of stamps Keshila can purchase to mail letters?

5. **BANQUET** A charity is hosting a benefit dinner. They are asking $100 per table plus $40 per person. Nathaniel is purchasing tickets for his friends and does not want to spend more than $250.

   a. Write an inequality to represent this situation, where x is the number of people.

   b. Solve the inequality. What is the maximum number of people Nathaniel can invite to the dinner?

Example 2

Translate each sentence into an inequality. Then solve the inequality and graph the solution on a number line.

6. Five times a number minus one is greater than or equal to negative eleven.

7. Twenty-one is greater than the sum of fifteen and two times a number.

8. Negative nine is greater than or equal to the sum of two-fifths times a number and seven.

9. A number divided by eight minus thirteen is greater than negative six.

10. The sum of the opposite of a number and six is less than or equal to five.

11. Thirty-seven is less than the difference of seven and ten times a number.

12. Eight minus a number divided by three is greater than or equal to eleven.

13. Negative five-fourths times a number plus six is less than twelve.

14. The difference of three times a number and six is greater than or equal to the sum of fifteen and twenty-four times a number.

15. The sum of fifteen times a number and thirty is less than the difference of ten times a number and forty-five.

Example 3

Solve each inequality. Then graph the solution on a number line.

16. $-3(7n + 3) < 6n$

17. $21 \geq 3(a - 7) + 9$

18. $2y + 4 > 2(3 + y)$

19. $3(2 - b) < 10 - 3(b - 6)$

20. $7 + t \leq 2(t + 3) + 2$

21. $8a + 2(1 - 5a) \leq 20$

Mixed Exercises

Solve each inequality. Check your solution.

22. $2(x - 4) \leq 2 + 3(x - 6)$

23. $\frac{2x - 4}{6} \geq -5x + 2$

24. $5.6z + 1.5 < 2.5z - 4.7$

25. $0.7(2m - 5) \geq 21.7$

26. $2(-3m - 5) \geq -28$

27. $-6(w + 1) < 2(w + 5)$

USE TOOLS Use a graphing calculator to solve each inequality.

28. $3x + 7 > 4x + 9$

29. $13x - 11 \leq 7x + 37$

30. $2(x - 3) < 3(2x + 2)$

31. $\frac{1}{2}x - 9 < 2x$

32. $2x - \frac{2}{3} \geq x - 22$

33. $\frac{1}{3}(4x + 3) \geq \frac{2}{3}x + 2$

STRUCTURE Solve each inequality. Then graph it on a number line.

34. $9.1g + 4.5 < 10.1g$

35. $\frac{3}{2}p - \frac{2}{3} \le \frac{4}{9} + \frac{1}{2}p$

36. $3.3r - 8.3 \ge 5.3r - 12.9$

37. TREEHOUSE DESIGN Devontae is building a treehouse in his backyard. He researches city restrictions on building codes. The height of the treehouse cannot exceed 13 feet. He wants to build a tree house with 2 levels of equal height that is 4 feet off the ground.

a. Write and solve an inequality.

b. What is the maximum height of one level?

c. Devontae decides to build one level higher off the ground. If the level is 8 feet tall, how high can the tree house be off the ground?

38. MEDICINE Clark's Rule is a formula used to determine pediatric dosages of over-the-counter medicines: $\frac{\text{weight of child (lb)}}{150} \times$ adult dose = child dose.

a. If an adult dose of acetaminophen is 1000 milligrams and a child weighs no more than 90 pounds, what is the recommended child's dose?

b. The label below appears on a child's cold medicine. What is the adult minimum dosage in milliliters?

Weight (lb)	Age (yr)	Dose
under 48	under 6	call a doctor
48-95	6-11	2 tsp or 10 mL

c. What is the maximum adult dosage in milliliters?

39. CONSTRUCT ARGUMENTS Eric says that 15 more than 6 times the number of pencils he has is less than 20. What can you conclude about the number of pencils Eric has? Justify your argument.

40. REASONING The perimeter of a rectangular playground can be no greater than 120 meters. The width of the playground cannot exceed 22 meters. What are the possible lengths of the playground?

41. STRUCTURE Solve $10n - 7(n + 2) > 5n - 12$. Explain each step in your solution.

42. WRITE What is the solution set of the inequality $2(2x + 4) < 4(x + 1)$? Why? How is the solution set related to the solution set of $2(2x + 4) \geq 4(x + 1)$? Explain.

43. PERSEVERE Mei got scores of 76, 80, and 78 on her last three history exams. Write and solve an inequality to determine the score she needs on the next exam so that her average is at least 82.

44. ANALYZE A triangular carpet has sides of length a feet, b feet, and c feet. The maximum perimeter is 20 feet.

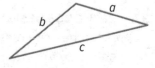

a. Side b is 2 feet longer than a and c is 2 feet longer than b. Which side is the shortest? Explain.

b. What are the possible lengths of the shortest side of the carpet? Explain.

45. CREATE Write an inequality that has the solution set graphed at the right. Solving the inequality should require the Distributive Property, Addition Property of Inequalities, and the Division Property of Inequalities.

```
◄──┼──┼──┼──┼──┼──┼──┼──⊕──┼──┼──►
  −5 −4 −3 −2 −1  0  1  2  3  4  5
```

46. PERSEVERE Let $b > 2$. Describe how you would determine if $ab > 2a$.

47. CREATE Four times the number of baseball cards in Ted's collection is more than five times that number minus 15. Define a variable and write an inequality to represent Ted's baseball cards. Solve the inequality and interpret the results.

48. WRITE Explain how you could solve $-3p + 7 \geq -2$ without multiplying or dividing each side by a negative number.

49. PERSEVERE If $ax + b < ax + c$ is true for all real values x, what will be the solution of $ax + b > ax + c$? Explain your reasoning.

50. WHICH ONE DOESN'T BELONG? Name the inequality that does not belong. Justify your conclusion.

$4y + 9 > -3$	$3y - 4 > 5$	$-2y + 1 < -5$	$-5y + 2 < -13$

51. WRITE Explain when the solution set of an inequality will be the empty set or the set of all real numbers. Show an example of each.

52. PERSEVERE Solve each inequality in terms of a. Assume that a does not equal 0.

a. $ax + 5 < 11$

b. $ax - 4 \geq 12$

c. $ax - 5 > 3a$

d. $ax + 1 \geq \frac{a}{2}$

Solving Compound Inequalities

Explore Guess the Range

⬇ **Online Activity** Use a real-world situation to complete the Explore.

> ⊘ **INQUIRY** How can you tell if a value will satisfy a compound inequality that includes the word *and*? ✕

Learn Solving Compound Inequalities Using the Word *and*

A **compound inequality** is two or more inequalities that are connected by the word *and* or *or*. A compound inequality containing the word *and* is only true when both of the inequalities are true. So, its graph is where the graphs of the inequalities overlap. This overlapping section that represents the compound inequality is called the **intersection**. To determine where the graphs intersect, graph each inequality and identify where they overlap.

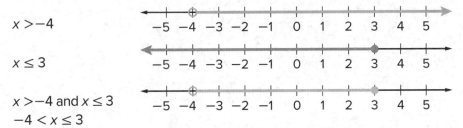

$x > -4$

$x \leq 3$

$x > -4$ and $x \leq 3$
$-4 < x \leq 3$

The compound inequality $-4 < x \leq 3$ can be read in two ways. It can be read as *x is greater than −4 and less than or equal to 3* or *x is between −4 and 3 including 3*.

Example 1 Solve and Graph an Intersection

Solve $-8 \leq h - 2 < 1$. Then graph the solution set.

Express the compound inequality as two inequalities joined by the word *and*.

_____   and   _____
                    Write the inequality
                        using *and*.

_____   Add 2 to each side.   _____

_____   Simplify.   _____

Today's Goals

* Solve and graph linear inequalities containing the word *and*.

* Solve and graph linear inequalities containing the word *or*.

Today's Vocabulary

compound inequality

intersection

union

Study Tip

Inequality Solutions
For inequalities using the word *and*, a number has to be true for both inequalities in order to be a solution for the compound inequality.

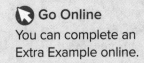 **Go Online**
You can complete an Extra Example online.

The solution set is $\{h \mid -6 \leq h < 3\}$.

Graph the solution set on a number line.

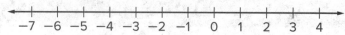

Check

Solve $-7 \leq 3x + 2 \leq 5$. Then graph the solution set.

_____

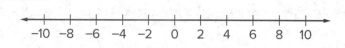

🌐 **Example 2** Apply Compound Inequalities

MANUFACTURING **A cereal manufacturer distributes cases of cereal to grocery stores. Each case contains 12 boxes of cereal. In order to pass the manufacturer's quality assurance test, the case must weigh between 336.4 ounces and 331.6 ounces, which includes 34 ounces for the weight of the case's cardboard box.**

Part A
Write an inequality that describes the weight of a box of cereal.

_____, where b is the weight of each box.

Part B Solve the inequality.

_____   and   _____

_____   Subtract 34 from each side.   _____

_____   Divide each side by 12.   _____

Part C Graph the solution set on a number line.

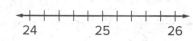

A cereal box must weigh between 24.8 and 25.2 ounces in order to pass the manufacturer's quality assurance test. The compound inequality is {_____}.

Check

CARS Keshawn has been saving to buy his first car. He wants the total cost of the car and fees to be more than $5000 but at most $7000. The fees for buying a used car, such as title, registration, and dealership fees, will be $700.

Graph the list price of the cars Keshawn could buy.

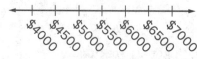

🌀 **Go Online** You can complete an Extra Example online.

Learn Solving Compound Inequalities Using the Word *or*

A compound inequality containing the word *or* is true if at least one of the inequalities is true. A **union** is the graph of a compound inequality containing *or*; the solution is a solution of either inequality, not necessarily both.

$x \geq 1$

$x \leq -3$

$x \geq 1 \text{ or } x \leq -3$

Example 3 Solve and Graph a Union

Solve $4n + 8 \leq 16$ or $-3n + 7 < -11$. Then graph the solution set.

Express the compound inequality as two inequalities joined by the word *or*.

	or	
_____		_____
_____	Subtract.	_____
_____	Simplify.	_____
_____	Divide.	_____
_____	Simplify.	_____

Graph the solution set on a number line.

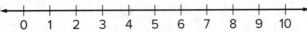

The union contains all points with coordinates less than or equal to 2 and all points with coordinates greater than 6. So, the solution set is {_____}.

Go Online You can complete an Extra Example online.

Go Online You can watch a video to see how to solve inequalities involving *and* and *or*.

Study Tip

Inequality Solutions
For inequalities using the word *or*, a number has to be true for at least one of the inequalities in order for it to be a solution for the compound inequality. The solution must work for the first inequality or the second inequality.

Check

Solve $5x + 1 < 11$ or $-3x + 10 \leq -11$. Then graph the solution set.

Part A

Select the solution set for $5x + 1 < 11$ or $-3x + 10 \leq -11$. _____

A. $\{x \mid x < 2 \text{ or } x \geq 7\}$

B. $\{x \mid 2 < x \leq 7\}$

C. $\{x \mid x < 2\}$

D. $\{x \mid x \leq -7 \text{ or } x > 2\}$

Part B

Graph the solution set for $5x + 1 < 11$ or $-3x + 10 \leq -11$.

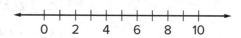

Talk About It!

Why does the union $4k + 12 < 2$ or $4 - 2k > 4$ include $-2\frac{1}{2}$ even though one of the solutions, $k < -2\frac{1}{2}$, does not?

Example 4 Overlapping Intervals

Solve $4k + 12 < 2$ or $4 - 2k > 4$. Then graph the solution set.

Express the compound inequality as two inequalities joined by the word *or*.

	or	
_____		_____
_____	Subtract.	_____
_____	Simplify.	_____
_____	Divide.	_____
_____	Simplify.	_____

Graph the solution set on a number line.

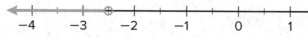

The graph of $k < -2\frac{1}{2}$ contains all points with coordinates less than $-2\frac{1}{2}$.

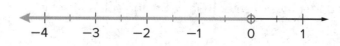

The graph of $k < 0$ contains all points with coordinates less than 0.

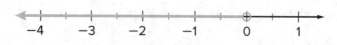

The union contains all points with coordinates less than 0.

Because $k < -2\frac{1}{2}$ is contained within $k < 0$, the solution set is $\{$_____$\}$.

Go Online You can complete an Extra Example online

Check

Solve $4m + 7 \geq 19$ or $-m + 5 \leq 0$. Then graph the solution set.

Part A

Select the solution set for $4m + 7 \geq 19$ or $-m + 5 \leq 0$. ____

A. $\{m|m \geq 3\}$

B. $\{m|3 \leq m \leq 5\}$

C. $\{m|m \leq 3 \text{ or } m \geq 5\}$

D. $\{m|m \geq 5\}$

Part B

Graph the solution set for $4m + 7 \geq 19$ or $-m + 5 \leq 0$.

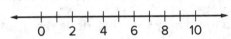

Example 5 Write a Compound Inequality for an Intersection

Write a compound inequality that describes the graph.

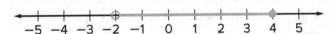

The graph shows an interval between two numbers. Because a compound inequality with the word *and* represents the intersection of two inequalities, its graph shows the overlap as an interval.

Step 1

Analyze the leftmost endpoint of the interval. The endpoint is shown with a circle at −2, so −2 is _____ in the solution. Points to the _____ of the endpoint are shaded, so the graph represents solutions of _____.

Step 2

Analyze the rightmost endpoint of the interval. The endpoint is shown with a dot at 4, so 4 is _____ in the solution. Points to the _____ of the endpoint are shaded, so the graph represents solutions of _____.

Step 3

The shaded interval represents the intersection of the solutions of _____ and _____, so the compound inequality _____ describes the graph.

🔵 **Go Online** You can complete an Extra Example online.

Check

Which compound inequality describes the graph? ____

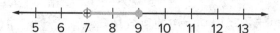

A. $7 \leq x < 9$

B. $x > 7$ or $x \leq 9$

C. $9 \leq x < 7$

D. $7 < x \leq 9$

Example 6 Write a Compound Inequality for a Union

Write a compound inequality that describes the graph.

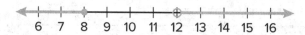

The graph shows the union of two inequalities. Because a compound inequality with the word *or* represents the union of two inequalities, its graph includes the graphs of both inequalities.

The leftmost endpoint is shown with a ____ at 8, so 8 ____ included in the solution. Points to the ____ of the endpoint are shaded, so the graph represents solutions of x ____ 8.

The rightmost endpoint is shown with a ____ at 12, so 12 ____ included in the solution. Points to the ____ of the endpoint are shaded, so the graph represents solutions of x ____ 12.

The solutions represented on the graph represent the union of the solutions of ____ and ____, so the compound inequality ____ or ____ describes the graph.

Check

Which compound inequality describes the graph? ____

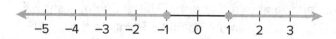

A. $x \leq -1$ or $x \geq 1$

B. $-1 \leq x \geq 1$

C. $x \leq -1$ and $x \geq 1$

D. $x < -1$ or $x > 1$

🡒 **Go Online** to practice what you've learned about solving linear inequalities in the Put It All Together over Lessons 6-1 through 6-3.

 Go Online You can complete an Extra Example online

Practice

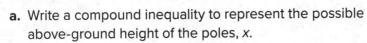

Go Online You can complete your homework online.

Examples 1, 3, and 4

Solve each compound inequality. Then graph the solution set.

1. $f - 6 < 5$ and $f - 4 \geq 2$

2. $n + 2 \leq -5$ and $n + 6 \geq -6$

3. $y - 1 \geq 7$ or $y + 3 < -1$

4. $t + 14 \geq 15$ or $t - 9 < -10$

5. $-5 < 3p + 7 \leq 22$

6. $-3 \leq 7c + 4 < 18$

7. $5h - 4 \geq 6$ and $7h + 11 < 32$

8. $22 \geq 4m - 2$ or $5 - 3m \leq -13$

9. $-y + 5 \geq 9$ or $3y + 4 < -5$

10. $-4a + 13 \geq 29$ and $10 < 6a - 14$

11. $3b + 2 < 5b - 6 \leq 2b + 9$

12. $-2a + 3 \geq 6a - 1 > 3a - 10$

13. $10m - 7 < 17m$ or $-6m > 36$

14. $5n - 1 < -16$ or $-3n - 1 < 8$

15. $m + 3 \geq 5$ and $m + 3 < 7$

16. $y - 5 < -4$ or $y - 5 \geq 1$

Example 2

17. STORE SIGNS In Randy's town, all stand-alone signs must be exactly 8 feet high. When mounted atop a pole, the combined height of the sign and pole must be less than 20 feet or greater than 35 feet so that they do not interfere with the power and phone lines.

    a. Write a compound inequality to represent the possible above-ground height of the poles, x.

    b. Solve the inequality. Explain any restrictions.

    c. Graph the inequality.

18. HEALTH The human heart circulates from 770,000 to 1,600,000 gallons of blood through a person's body every year.

    a. Write a compound inequality to represent the number of gallons of blood that the heart circulates through the body in one day, x.

    b. Solve the inequality. Round to the nearest whole gallon.

    c. Graph the inequality.

Write a compound inequality that describes each graph.

19.
```
←——+——⊕——+——+——+——+——●——+——→
  -4  -3  -2  -1   0   1   2   3   4
```

20.
```
←——+——+——+——◆——+——+——◆——+——+——→
    -2  -1   0   1   2   3   4   5   6
```

21.
```
←——+——+——⊕——+——+——◆——+——+——→
  -4  -3  -2  -1   0   1   2   3   4
```

22.
```
←——+——+——⊕——+——+——+——⊕——+——+——→
    -4  -3  -2  -1   0   1   2   3   4
```

23.
```
←——+——+——+——+——◆——+——+——⊕——+——→
  -4  -3  -2  -1   0   1   2   3   4
```

24.
```
←——+——+——+——◆——+——+——+——◆——+——→
    -4  -3  -2  -1   0   1   2   3   4
```

25.
```
←——+——+——+——⊕——+——◆——+——+——→
  -4  -3  -2  -1   0   1   2   3   4
```

26.
```
←——+——+——◆——+——+——+——⊕——+——+——→
    -3  -2  -1   0   1   2   3   4   5
```

Solve each compound inequality. Then graph the solution set.

27. $4 < f + 6$ and $f + 6 < 5$

28. $w + 3 \leq 0$ or $w + 7 \geq 9$

29. $-6 < b - 4 < 2$

30. $p - 2 \leq -2$ or $p - 2 > 1$

31. $-5 \leq 2a - 1 < 9$

32. $-1 < 2x - 1 \leq 5$

Define a variable, write an inequality, and solve each problem. Check your solution.

33. A number decreased by two is at most four or at least nine.

34. The sum of a number and three is no more than eight or is more than twelve.

35. **WEATHER** Kenya saw this graph in the local weather forecast. It shows the predicted temperature range for the following day. Write an inequality to represent the number line.

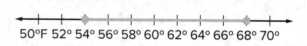

```
←——+——+——◆——+——+——+——+——+——◆——+——+——→
  50°F 52° 54° 56° 58° 60° 62° 64° 66° 68° 70°
```

36. **REASONING** The pH of a person's eyes is 7.2. Therefore, the ideal pH for the water in a swimming pool is between 7.0 and 7.6. Write a compound inequality to represent pH levels that could cause physical discomfort to a person's eyes.

37. **FIELD TRIP** It costs $1000 to rent a bus that holds 100 students. A school is planning to rent one of these buses for a field trip to an aquarium. The trip will also have a cost of $15 per student for the tickets to the aquarium. Given that the total expense for the trip must be between $2000 and $3000, find the minimum and maximum number of students who can go on the trip. Explain.

38. HEALTH Body mass index (BMI) is a measure of weight status. The BMI of a person over 20 years old is calculated using the following formula.

$$BMI = 703 \times \frac{\text{weight in pounds}}{(\text{height in inches})^2}$$

a. The recommended BMI for a person over 20 years old is 18.5–24.9. Write a compound inequality to represent the recommended BMI range.

b. Write a compound inequality to represent the weight of an adult who is 6 feet tall that is within the recommended BMI range. Round to the nearest tenth if necessary.

39. STATE YOUR ASSUMPTION The Triangle Inequality states that in any triangle, the sum of the lengths of any two sides is greater than the length of the third side. In the figure, this means $a + b > c$, $a + c > b$, and $b + c > a$.

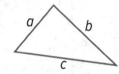

a. Suppose a triangle has a side that is 5 meters long and a perimeter of 14 meters. Let one of the unknown sides be x. Write a compound inequality that you can use to determine the value of x. Explain.

b. What assumption about the two unknown side lengths of the triangle can you make? Explain.

40. CONSTRUCT ARGUMENTS Bianca said that if k is a real number, then the solution set of the compound inequality $x < k$ or $x > k$ is all real numbers. Do you agree? Justify your argument.

41. STRUCTURE Write the solution set of the following compound inequality. Then graph the solution set: $-x + 1 < 8$ and $-x + 1 < 3$ and $-x + 1 > -4$?

42. PHYSICS The table shows common chemical solutions and their densities. The density, in grams per milliliter, of a substance determines whether it will float or sink in a liquid. Any object with a greater density will sink and any object with a lesser density will float. Density is given by the formula $d = \frac{m}{v}$, where m is mass and v is volume.

Solution	Density (g/mL)
concentrated calcium chloride	1.40
70% isopropyl alcohol	0.92

Source: American Chemistry Council

Plastics vary in density when they are manufactured; therefore, their volumes are variable for a given mass. A tablet of polystyrene (a manufactured plastic) sinks in 70% isopropyl alcohol solution and floats in calcium chloride solution. The tablet has a mass of 0.4 gram. Write an inequality to represent the range of values for v, the volume of the tablet.

43. COMPUTER SALE Marietta is shopping during a computer store's sale. She is considering buying computers that range in cost from $500 to $1000.

a. How much are the computers after the 20% discount?

b. If sales tax is 7%, how much should Marietta expect to pay?

44. CREATE The figure shows the solution set of a compound inequality. Write a compound inequality that has the given solution set. Solving the inequality should require the Distributive Property, Addition Property of Inequalities, and the Division Property of Inequalities.

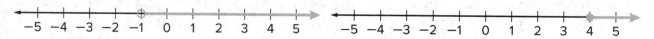

45. ANALYZE Which value of x is not a solution to $3x - 1 < 5$ or $7 - x \le 3$?

A 0   **B** 2   **C** 4   **D** 5

46. FIND THE ERROR Sierra solved the compound inequality $-3x + 7x - 1 < -5$ or $-3x + 7x - 1 > 11$ as shown. What error did she make in solving the inequality?

Step	Property	Step	Property
$-3x + 7x - 1 < -5$	Original inequality	$-3x + 7x - 1 > 11$	Original inequality
$4x - 1 < -5$	Combine like terms.	$-4x - 1 > 11$	Combine like terms.
$4x < -4$	Addition Property	$-4x > 12$	Addition Property
$x < -1$	Division Property	$x < -3$	Division Property

Refer to the graphs for Exercises 47–49.

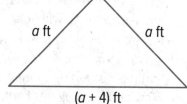

47. WRITE Write a compound inequality whose solution is the union of the two graphs. Then explain how the compound inequality can be expressed as a single inequality.

48. WRITE Write a compound inequality whose solution is the intersection of the two graphs. Then explain how the compound inequality can be expressed as a single inequality.

49. ANALYZE How are the graphs of the solution sets for the inequalities in Exercise 47 and Exercise 48 related to the given graphs?

50. PERSEVERE Jocelyn is planning to place a fence around the triangular flower bed shown. The fence costs \$1.50 per foot. Assuming that Jocelyn spends between \$60 and \$75 for the fence, what is the shortest possible length for a side of the flower bed? Use a compound inequality to explain your answer.

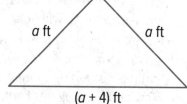

51. PERSEVERE Solve each inequality for x. Assume a is constant and $a > 0$

a. $-3 < ax + 1 \le 5$ **b.** $-\frac{1}{a}x + 6 < 1$ or $2 - ax > 8$

52. CREATE Create an example of a compound inequality containing *or* that has infinitely many solutions.

53. ANALYZE Determine whether the following statement is *always*, *sometimes*, or *never* true. Justify your argument. *The graph of a compound inequality that involves an* or *statement is bounded on the left and right by two values of x.*

54. WRITE Give an example of a compound inequality you might encounter at an amusement park. Does the example represent an intersection or a union?

Solving Absolute Value Inequalities

Explore Solving Absolute Value Inequalities

 Online Activity Use a graph to complete the Explore.

> ☒
> ⊘ **INQUIRY** How is solving an absolute value inequality similar to solving an absolute value equation?

Learn Solving Inequalities Involving < and Absolute Value

For a real number a, the inequality $|x| < a$ means that the distance between x and 0 is less than a.

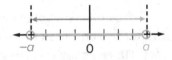

When solving absolute value inequalities, there are two cases to consider.

Case 1 The expression inside the absolute value symbols is nonnegative. If x is nonnegative, then $|x| = x$.

Case 2 The expression inside the absolute value symbols is negative. If x is negative, then $|x| = -x$.

Example 1 Solve Absolute Value Inequalities (<)

Solve $|m + 5| < 3$. Then graph the solution set.

Part A Rewrite $|m + 5| < 3$ for Case 1 and Case 2.

Case 1 If $m + 5$ is nonnegative, _____.

$\quad m + 5 < 3$ \qquad Case 1

$\quad$ _____ \qquad Subtract 5 from each side.

Case 2 If $m + 5$ is negative, _____.

$\quad -(m + 5) < 3$ \qquad Case 2

$\quad$ _____ \qquad Distributive Property

$\quad$ _____ \qquad Add 5 to each side.

$\quad$ _____ \qquad Divide each side by −1. Reverse the inequality symbol.

So, _____ and _____. The solution set is

_____.

(continued on the next page)

Today's Goals
- Solve absolute value inequalities (<).
- Solve absolute value inequalities (>).

Study Tip
Absolute Value Inequalities The inequality $|x|$ can be rewritten as $|x - 0| < a$, which is why it is read as *the distance between x and 0 is less than a.*

Watch Out
Absolute Value Cases Assigning the correct inequality symbol in Case 2 of an absolute value inequality can be confusing. Think of an inequality like $|x - 72| < 1.8$ as *the distance from x to 72 is less than 1.8 units.* Visualizing the graph of the inequality as an interval of 1.8 units on each side of the graph of 72 can help you ensure you have the correct symbol.

1.8 units 72 1.8 units

Part B Graph the solution set on a number line.

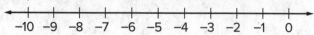

−10 −9 −8 −7 −6 −5 −4 −3 −2 −1 0

Check

Solve $|6m + 12| < 12$. _____

Graph the solution set.

−10   −5   0   5

Example 2 Absolute Value Inequalities (<) with No Solutions

Solve $|n - 1| < -5$. Then graph the solution set.

Because $|n - 1|$ is an absolute value expression, it cannot be negative. So it is not possible for $|n - 1|$ to be less than −5. Therefore, there is no solution, and the solution set is the empty set, ∅.

Study Tip

Absolute Value as Distance The solution set for $|n - 1| < -5$ might be easier to understand if you think of the inequality in terms of distance. The inequality can be read as *the distance between a number and 1 is less than negative five.* However, that would make the distance a negative number. Because distance cannot be negative, the solution set is ∅.

Example 3 Use Absolute Value Inequalities

SURVEY Jonas is a software developer who wants to determine whether the changes he made to his program are popular with users. He releases a survey to get some feedback and finds that 72% of users like the changes. The margin of error is within 1.8%, which means that with a reasonable level of certainty, the actual percentage can be said to fall within 1.8% of 72%.

Part A

Complete the table to write an inequality that represents the percent of users who like the changes.

Words			
Variable	Let x be the actual percent of users who like the changes.		
Inequality			

Talk About It!

Is the solution set of $|n - 1| - 6 < -5$ also the empty set? Explain your reasoning.

Part B

Solve each case of the inequality.

Case 1 $x - 72$ is nonnegative.

$x - 72 \leq 1.8$ Case 1

_____   Add 72 to each side.

Case 2 $x - 72$ is negative.

$$-(x - 72) \le 1.8 \qquad \text{Case 2}$$

_____ Distributive Property

_____ Subtract 72 from each side.

_____ Divide each side by −1. Reverse the inequality symbol.

The percent of users who favor the changes Jonas made to his software is between 70.2% and 73.8%, so the solution set is {_____}. This solution set is a small interval of possible values close to the percent that Jonas found, so the solution set seems reasonable for the situation.

Learn Solving Inequalities Involving > and Absolute Value

For a real number a, the inequality $|x| > a$ means that the distance between x and 0 is greater than a.

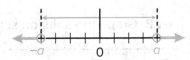

When solving absolute value inequalities, there are two cases to consider.

Case 1 The expression inside the absolute value symbols is nonnegative. If x is nonnegative, $|x| = x$.

$$x > a \qquad \text{Case 1}$$

Case 2 The expression inside the absolute value symbols is negative. If x is negative, $|x| = -x$.

$$-x > a \qquad \text{Case 2}$$

$$\frac{-x}{-1} < \frac{a}{-1} \qquad \text{Divide each side by } -1. \text{ Reverse the inequality.}$$

$$x < -a \qquad \text{Simplify.}$$

The solution set is the union of the solutions to these two cases. So, $x > a$ or $x < -a$. The solution set is $\{x \mid x < -a \text{ or } x > a\}$.

Example 4 Solve Absolute Value Inequalities (>)

Solve $|2m - 9| > 13$. Then graph the solution set.

Part A Rewrite $|2m - 9| > 13$ for Case 1 and Case 2.

Case 1 If $2m - 9$ is nonnegative, _____.

$$2m - 9 > 13 \qquad \text{Case 1}$$

_____ Add 9 to each side.

_____ Divide each side by 2.

(continued on the next page)

Go Online You can complete an Extra Example online.

Study Tip

> and < If an absolute value inequality involves > or ≥, the solution set uses the word _or_. If an absolute value inequality involves < or ≤, the solution set uses the word _and_.

Case 2 If $2m - 9$ is negative, $|2m - 9| = -(2m - 9)$.

$$-(2m - 9) > 13 \qquad \text{Case 2}$$

$$-2m + 9 > 13 \qquad \text{Distributive Property}$$

_____ Subtract 9 from each side.

_____ Divide each side by -2. Reverse the inequality.

So, _____ or _____. The solution set is _____.

Part B Graph the solution set on a number line.

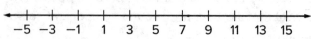

$$-5 \quad -3 \quad -1 \quad 1 \quad 3 \quad 5 \quad 7 \quad 9 \quad 11 \quad 13 \quad 15$$

Check

Part A Solve $|4m - 20| \geq 12.$ _____

Part B Graph the solution set.

$$-5 \qquad 0 \qquad 5 \qquad 10 \qquad 15$$

Example 5 Absolute Value Inequalities (>) with Overlapping Case Solutions

Solve $|n - 6| \geq -5$. Then graph the solution set.

Part A Rewrite $|n - 6| \geq -5$ for Case 1 and Case 2.

Case 1 $n - 6$ is nonnegative.

_____ Case 1

_____ Add 6 to each side.

Case 2 $n - 6$ is negative.

_____ Case 2

$$-n + 6 \geq -5 \qquad \text{Distributive Property}$$

_____ Subtract 6 from each side.

_____ Divide each side by -1. Reverse the inequality.

So, _____ or _____. The solution set is _____, which is equivalent to _____.

Part B Graph the solution set on a number line.

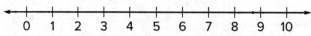

$$0 \quad 1 \quad 2 \quad 3 \quad 4 \quad 5 \quad 6 \quad 7 \quad 8 \quad 9 \quad 10$$

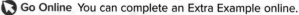

Go Online You can complete an Extra Example online.

🗨 Think About It!

Why is the empty set the solution of $|n - 6| \leq -5$, but not the solution of $|n - 6| \geq -5$?

Practice

Go Online You can complete your homework online.

Examples 1, 2, 4, 5

Solve each inequality. Then graph the solution set.

1. $|x + 8| < 16$

2. $|r + 1| \leq 2$

3. $|2c - 1| \leq 7$

4. $|3h - 3| < 12$

5. $|m + 4| < -2$

6. $|w + 5| < -8$

7. $|r + 2| > 6$

8. $|k - 4| > 3$

9. $|2h - 3| \geq 9$

10. $|4p + 2| \geq 10$

11. $|5v + 3| > -9$

12. $|-2c - 3| > -4$

13. $|4n + 3| \geq 18$

14. $|5t - 2| \leq 6$

15. $\left|\dfrac{3h + 1}{2}\right| < 8$

16. $\left|\dfrac{2p - 8}{4}\right| \geq 9$

17. $\left|\dfrac{7c + 3}{2}\right| \leq -5$

18. $\left|\dfrac{2g + 3}{2}\right| > -7$

19. $|-6r - 4| < 8$

20. $|-3p - 7| > 5$

Example 3

21. SPEEDOMETERS The government requires speedometers on cars sold in the United States to be accurate within ±2.5% of the actual speed of the car. If your speedometer meets this requirement, find the range of possible actual speeds at which your car could be traveling when your speedometer reads 60 miles per hour.

22. BAKING Pablo is making muffins for a bake sale. Before he starts baking, he goes online to research different muffin recipes. The recipes that he finds all specify baking temperatures between 350°F and 400°F, inclusive. Write an absolute value inequality to represent the possible temperatures t called for in the muffin recipes Pablo is researching.

23. PAINT A manufacturer claims that their cans of paint contain exactly 130 fluid ounces of paint. The amount of paint in each can of paint must be accurate within ±3.05 fluid ounces of the actual amount of paint.

    a. Write an absolute value inequality to represent the possible amount of paint, in fluid ounces, p for which the manufacturer's claim is correct.

    b. Graph the solution set of the inequality you wrote in part **a**.

24. CATS During a recent visit to the veterinarian's office, Mrs. Vasquez was informed that a healthy weight for her cat is approximately 10 pounds, plus or minus one pound. Write an absolute value inequality that represents unhealthy weights w for her cat.

25. STATISTICS In a recent year, the mean score on the mathematics section of the SAT test was 515 and the standard deviation was 114. This means that people within one deviation of the mean have SAT math scores that are no more than 114 points higher or 114 points lower than the mean.

   a. Write an absolute value inequality to find the range of SAT mathematics test scores within one standard deviation of the mean.

   b. What is the range of SAT mathematics test scores ±2 standard deviation from the mean?

Mixed Exercises

REGULARITY Write an open sentence involving absolute value for each graph.

26.

      −3 −2 −1 0 1 2 3 4 5 6 7

27.

      −6 −5 −4 −3 −2 −1 0 1 2 3 4

28.

      −10 −9 −8 −7 −6 −5 −4 −3 −2 −1 0

29.

      −4 −3 −2 −1 0 1 2 3 4 5 6

30.

      0 1 2 3 4 5 6 7 8 9 10

31.

      −4 −3 −2 −1 0 1 2 3 4

REASONING Match each open sentence with the graph of its solution set.

32. $|x| > 2$

   a.   −5 −4 −3 −2 −1 0 1 2 3 4 5

33. $|x - 2| \leq 3$

   b.   −5 −4 −3 −2 −1 0 1 2 3 4 5

34. $|x + 1| < 4$

   c.   −8 −7 −6 −5 −4 −3 −2 −1 0 1 2 3 4 5 6

35. $|-x + 1.5| < 3$

   d.   −4 −3 −2 −1 0 1 2 3 4 5 6

USE A MODEL Express each statement using an inequality involving absolute value. Then solve and graph the absolute value inequality.

36. The meteorologist predicted that the temperature would be within 3° of 52°F.

37. Serena will make the B team if she scores within 8 points of the team average of 92.

38. The dance committee expects attendance to number within 25 of last year's 87 students.

Solve each inequality. Then graph the solution set.

39. $\left|\dfrac{x-1}{2}\right| \le 1$

40. $|2x-1| \ge 3$

41. $\left|\dfrac{x+3}{3}\right| \le 2$

42. $|x+7| \ge 4.5$

43. $\left|\dfrac{2x-1}{7}\right| > 5$

44. $|-4x-2| < 10$

45. CONSTRUCT ARGUMENTS Is the solution to this inequality $|x-2| > -1$ all real numbers? Justify your argument.

46. USE TOOLS Forensic scientists use the equation $h = 2.4f + 46.2$ to estimate the height h of a woman given the length in centimeters f of her femur bone. Suppose the equation has a margin of error of 3 centimeters. Could a female femur bone measuring 47 centimeters be that of a woman who was 170 centimeters tall?

47. REGULARITY A box of cereal should weigh 516 grams. The quality control inspector randomly selects boxes to weigh. The inspector sends back any box that is not within 4 grams of the ideal weight.

    a. Explain how to write an absolute value inequality to represent this situation.

    b. Explain the steps to solve this inequality. What do the solutions represent?

48. REASONING Write a compound inequality in which the solution set is the given set.

    a. $\{x \mid 4 \le x\}$

    b. $\{4\}$

49. ANALYZE Determine if the open sentence $|x-2| > 4$ and the compound inequality $-2x < 4$ or $x > 6$ have the same solution set.

50. ARCHITECTURE An architect is designing a house for the Frazier family. In the design, she must consider the desires of the family and the local building codes. The rectangular lot on which the house will be built is 158 feet long and 90 feet wide.

    a. The building codes state that one can build no closer than 20 feet to the lot line. Write an inequality to represent the possible widths of the house along the 90-foot dimension. Solve the inequality.

    b. The Fraziers requested that the rectangular house contain no less than 2800 square feet and no more than 3200 square feet of floor space. If the house has only one floor, use the maximum value for the width of the house from part a, and explain how to use an inequality to find the possible lengths.

    c. The Fraziers have asked that the cost of the house be about $175,000 and are willing to deviate from this price no more than $20,000. Write an open sentence involving an absolute value and solve. Explain the meaning of the answer.

51. **FIND THE ERROR** Jordan and Chloe are solving $|x + 3| > 10$.

Jordan	
$\|x + 3\| > 10$	
$x + 3 > 10$	$-(x + 3) > 10$
$x > 7$	$-x - 3 > 10$
	$-x > 13$
	$x < -13$

Chloe	
$\|x + 3\| > 10$	
$x + 3 > 10$	$-(x + 3) > 10$
$x > 7$	$-x + 3 > 10$
	$-x > 7$
	$x < -7$

Is either correct? Explain your reasoning.

52. **CREATE** Write an absolute value inequality using the numbers 3, 2, and –7. Then solve the inequality.

53. **PERSEVERE** Solve $2 < |n + 1| \leq 7$. Explain your reasoning and graph the solution set.

54. **ANALYZE** Which of the following inequalities could be represented by the graph?

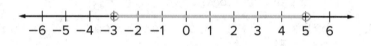

I. $|m - 1| < 4$ **II.** $3x < 15$ or $-x < 1$ **III.** $1 < 2k + 7 < 17$

A. I only   **B.** I and II   **C.** I and III   **D.** II and III

55. **FIND THE ERROR** Lucita sketched a graph of her solution to $|2a - 3| > 1$. Is she correct? Explain your reasoning.

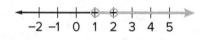

56. **ANALYZE** The graph of an absolute value inequality is *sometimes, always,* or *never* the union of two graphs. Explain.

57. **ANALYZE** Determine why the solution of $|t| > 0$ is not all real numbers. Explain your reasoning.

58. **WRITE** How are symbols used to represent mathematical ideas? Use an example to justify your reasoning.

59. **WRITE** Explain how to determine whether an absolute value inequality uses a compound inequality with *and* or a compound equality with *or*. Then summarize how to solve absolute value inequalities.

60. **WHICH ONE DOESN'T BELONG?** Which inequality does not belong? Justify your conclusion.

$$|x + 4| - 7 \geq 3 \qquad |-6x - 1| \leq \frac{1}{2} \qquad -2|10x + 4| < 6 \qquad |3x + 5| < -\frac{3}{5}$$

Graphing Inequalities in Two Variables

Explore Graphing Linear Inequalities on the Coordinate Plane

 Online Activity Use graphing technology to complete the Explore.

> ⊘ **INQUIRY** How is graphing a linear inequality on the coordinate plane similar to and different from graphing on the number line?   ×

Learn Graphing Linear Inequalities in Two Variables

The graph of a linear inequality represents the set of all points that are solutions of the inequality.

The edge of the graph is a **boundary**. Depending on the inequality, the boundary will or will not be included in the solution set.

The boundary divides the coordinate plane into regions called **half-planes**.

When the boundary is included, the solution of the linear inequality is a **closed half-plane**.

When the boundary is not included, it is an **open half-plane**.

Key Concept • Graphing Linear Inequalities	
Step 1	Graph the boundary. Use a solid boundary when the inequality contains ≤ or ≥. Use a dashed boundary when the inequality contains < or >.
Step 2	Use a test point to determine which half-plane should be shaded.
Step 3	Shade the half-plane that contains the solution.

Example 1 Graph an Inequality with an Open Half-Plane

Graph $3x - 2y < 8$.

Step 1 Graph the boundary.

$3x - 2y < 8$ Original inequality

_____   Subtract $3x$ from each side.

_____   Divide each side by -2.

(continued on the next page)

Today's Goals
- Graph the solutions of linear inequalities in two variables.

Today's Vocabulary
boundary

half-plane

closed half-plane

open half-plane

 Go Online
You can watch a video to see how to graph inequalities in two variables.

Watch Out!

Selecting a Test Point
When selecting a test point to use, make sure that the point does not lie on the boundary. While using the point (0, 0) will make calculations easier, you cannot use that point if the boundary passes through the origin. Instead, try using (1, 1) or (0, 1).

🧁 **Think About It!**

What does the shaded area of this graph represent?

Step 2 Use a test point. Select (0, 0) as a test point.

$$3x - 2y < 8 \qquad \text{Original inequality}$$

$$3 \underline{\quad} - 2 \underline{\quad} < 8 \qquad x = 0 \text{ and } y = 0$$

$$\underline{\quad} < 8 \qquad \text{Simplify.}$$

Step 3 Shade the half-plane.
Because the test point is a solution of the inequality, shade the half-plane containing the test point.

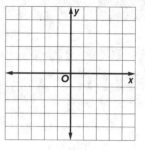

Example 2 Graph an Inequality with a Closed Half-Plane

Graph $3x + 4y \leq 0$.

Step 1 Graph the boundary.

$$3x + 4y \leq 0 \qquad\qquad \text{Original inequality}$$

$$\underline{\hspace{3cm}} \qquad\qquad \text{Subtract } 3x \text{ from each side.}$$

$$\underline{\hspace{3cm}} \qquad\qquad \text{Divide each side by 4.}$$

Step 2 Use a test point. Use (1, 1) as a test point.

$$3x + 4y \leq 0 \qquad\qquad \text{Original inequality}$$

$$3 \underline{\quad} + 4 \underline{\quad} \leq 0 \qquad\qquad x = 1 \text{ and } y = 1$$

$$\underline{\quad} \text{ is not less than or equal to 0} \qquad \text{Simplify.}$$

Step 3 Shade the half-plane.
Because the test point *is not* a solution of the inequality, shade the half-plane that does not contain the test point.

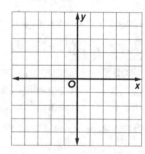

🖱 **Go Online** You can watch a video to to see how to use a graphing calculator with this example.

Check

Graph the inequalities on the coordinate planes provided.

a. $2x + y < -4$

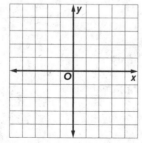

b. $x - 2y > -4$

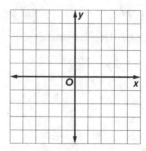

🖱 **Go Online** You can complete an Extra Example online.

🌐 Example 3 Apply Graphing Inequalities in Two Variables

REFRESHMENTS **Dominique can spend up to $20 to provide the dance squad with drinks after their practice. A bottle of water costs $0.80, and a sports drink costs $1.25. How many bottles of water and sports drinks can Dominique buy for the dance squad?**

Step 1. Write an inequality.

Words	$0.80 times the number of bottles of water	plus	$1.25 times the number of sports drinks	is less than or equal to	$20
Variables	Let x = the number of bottles of water and y = the number of sports drinks that Dominique can buy.				
Inequality					

Step 2 Solve the inequality for y.

$$0.8x + 1.25y \leq 20$$ Original inequality

_____   Subtract 0.8x from each side.

_____   Divide each side by 1.25.

Step 3 Graph the inequality.
Because Dominique cannot buy a negative number of drinks, negative values of x and y are nonviable options. So the domain and range must be nonnegative numbers. Graph the boundary.

x	y
0	
10	
15	
25	

Sports Drinks vs. Bottles of Water

The test point (0, 0) _____ a solution of the inequality. Shade the closed half-plane that _____ (0, 0).

Step 4 Interpret the solution in the context of the situation.
Notice that there are _____ solutions of the inequality. Because buying fractional bottles of water or sports drinks is not reasonable, only the solutions in which both x and y are _____ _____ are viable. One viable solution is 10 bottles of water and 8 sports drinks.

Problem-Solving Tip
Use a Graph You can use a graph to visualize data, analyze trends, and make predictions.

Study Tip
Specifying Units
Because we assigned the variables for the different types of drinks, it is critical to label the axes.

💬 Talk About It!
Are there any viable solutions in which Dominique spends a total of $20? If so, where do those solutions appear on the graph?

Example 4 Solve Linear Inequalities

Graph $-4x + 7 \geq 11$.

Step 1 Graph the boundary.

$$-4x + 7 \geq 11 \qquad \text{Original inequality}$$

$$\underline{\hspace{2cm}} \qquad \text{Subtract 7 from each side.}$$

$$\underline{\hspace{2cm}} \qquad \text{Divide each side by } -4. \text{ Reverse the inequality.}$$

Step 2 Use a test point. Use $(0, 0)$ as a test point.

$$-4x + 7 \geq 11 \qquad \text{Original inequality}$$

$$-4(0) + 7 \geq 11 \qquad x = 0 \text{ and } y = 0$$

$$\underline{\hspace{1cm}} \ngeq 11 \qquad \text{Simplify.}$$

Step 3 Shade the half-plane.

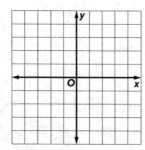

Since the test point *is not* a solution of the inequality, shade the half-plane that does not contain the test point.

Check

The graph of $y = 3x - 4$ is shown.

Consider the solutions of $y > 3x - 4$. Write each point in the appropriate column.

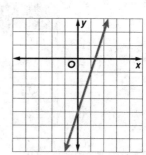

$(-5, -3)$ $(-3, 4)$

$(0, -4)$ $(0, 0)$

$(1, -7)$ $(1, 1)$

$(2, 2)$ $(4, 2)$

In Solution	Not in Solution

 Go Online You can complete an Extra Example online.

Copyright © McGraw-Hill Education

Practice

Go Online You can complete your homework online.

Examples 1 and 2

Graph each inequality.

1. $y < x - 3$

2. $y > x + 12$

3. $y \geq 3x - 1$

4. $y \leq -4x + 12$

5. $6x + 3y > 12$

6. $2x + 2y < 18$

7. $5x + y > 10$

8. $2x + y < -3$

9. $-2x + y \geq -4$

10. $8x + y \leq 6$

11. $10x + 2y \leq 14$

12. $-24x + 8y \geq -48$

Example 3

13. INCOME In 2006 the median yearly family income was about $48,200 per year. Suppose the average annual rate of change since then is $1240 per year.

   a. Write and graph an inequality for the annual family incomes y that are less than the median for x years after 2006.

   b. Determine whether each of the following points is part of the solution set.

   (2, 51,000)   (8, 69,200)   (5, 50,000)   (10, 61,000)

14. FUNDRAISING Troop 200 sold cider and donuts to raise money for charity. They sold small boxes of donut holes for $1.25 and cider for $2.50 a gallon. In order to cover their expenses, they needed to raise at least $100. Write and graph an inequality that represents this situation.

Example 4

Graph each inequality.

15. $2y + 6 \geq 0$

16. $\frac{1}{2}x + 1 < 3$

17. $\frac{2}{3}x - \frac{10}{3} > -4$

Mixed Exercises

Graph each inequality.

18. $y < -1$

19. $y \geq x - 5$

20. $y > 3x$

21. $y \leq 2x + 4$

22. $y + x > 3$

23. $y - x \geq 1$

24. Kumiko has a $50 gift card for a Web site that sells apps and games. Games cost $2.50 each, and apps cost $1.25 each.

   a. Write an inequality that represents the number of apps a and the number of games g that Kumiko can buy and describe any constraints.

   b. Graph the solution of the inequality on a coordinate plane.

   c. Use your graph to find three different combinations of apps and games that Kumiko can buy.

   d. Kumiko decides to buy the same number of apps and games, and she decides to spend as much of the $50 as possible. How many apps and games does she buy? How much money does she have left on her gift card?

25. USE A MODEL A café sells peach smoothies and berry smoothies. The café makes a profit of $2.25 for each peach smoothie that is sold and a profit of $2 for each berry smoothie that is sold. The owner of the café wants to make a total profit of more than $90 per day from the sales of smoothies.

a. Write an inequality that represents the number of peach smoothies p and berry smoothies b that the café needs to sell. Describe the constraints on the variables.

b. Graph the solution of the inequality on a coordinate plane.

c. On Monday, the café sold 20 peach smoothies and made the daily profit goal. What can you say about the number of berry smoothies that were sold on Monday?

d. On Tuesday, the café made the daily profit goal by selling the minimum number of smoothies. How many smoothies did they sell? Explain.

26. REASONING Oleg is training for a triathlon. One day, he jogged for 2 hours at x miles per hour. Then he bicycled for 2 hours at y miles per hour. Finally, he swam a distance of 2 miles. The total number of miles did not exceed 30 miles.

a. Write an inequality to represent the distance that he traveled that day. Describe the constraints on the variables.

b. Graph the solution of the inequality on the coordinate plane shown. Label the axes with a description of the quantity that each axis represents. Include the unit of measure.

c. What is the greatest possible speed that Oleg could have bicycled that day? How do you know?

27. CONSTRUCT ARGUMENTS The solution of the inequality $ax + by < c$ is a half-plane that includes the point (0, 0). What conclusion can you make about the value of c? Justify your argument.

Higher-Order Thinking Skills

28. FIND THE ERROR Reiko and Kristin are solving $4y \leq \frac{8}{3}x$ by graphing. Is either of them correct? Explain your reasoning.

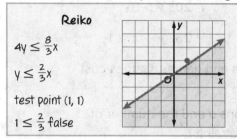
Reiko
$$4y \leq \frac{8}{3}x$$
$$y \leq \frac{2}{3}x$$
test point (1, 1)
$1 \leq \frac{2}{3}$ false

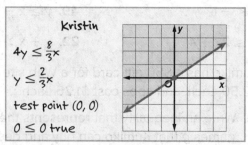
Kristin
$$4y \leq \frac{8}{3}x$$
$$y \leq \frac{2}{3}x$$
test point (0, 0)
$0 \leq 0$ true

29. CREATE Write a linear inequality for which (–1, 2), (0, 1), and (3, –2) are solutions but (1, 1) is not.

30. ANALYZE Explain why a point on the boundary should not be used as a test point.

31. CREATE Write a two-variable inequality with a restricted domain and range to represent a real-world situation. Give the domain and range, and explain why they are restricted.

32. WRITE Summarize the steps to graph an inequality in two variables.

@ **Essential Question**

How can writing and solving inequalities help you solve problems in the real world?

Module Summary

Lessons 6-1, 6-2

Solving One-Step and Multi-Step Inequalities

- A solution set can be graphed on a number line. If the endpoint is not included in the solution, use a circle; if the endpoint is included, use a dot.

- If a number is added to or subtracted from each side of a true inequality, the resulting inequality is also true.

- If each side of a true inequality is multiplied or divided by the same positive number, the resulting inequality is also true.

- If each side of a true inequality is multiplied or divided by the same negative number, the direction of the inequality symbol must be changed to make the resulting inequality true.

- Multi-step inequalities can be solved by undoing the operations in the same way you would solve a multi-step equation.

Lesson 6-3

Solving Compound Inequalities

- To determine the solution set of a compound inequality, graph each inequality and identify where they overlap.

- If a compound inequality contains *and*, the overlapping section that represents the compound inequality is an intersection.

- If a compound inequality contains *or*, its graph is a union; the solution is a solution of either inequality, not necessarily both.

Lesson 6-4

Solving Absolute Value Inequalities

- For a real number a, the inequality $|x| < a$ means the distance between x and 0 is less than a. The inequality $|x| > a$ means the distance between x and 0 is greater than a.

- When solving absolute value inequalities, there are two cases to consider. The first case is when the expression inside the absolute value symbols is nonnegative. The second case is when the expression inside the absolute value symbols is negative. The solution set is the intersection of the solutions of their union.

Lesson 6-5

Graphing Linear Inequalities in Two Variables

- To graph a linear Inequality, graph the boundary. Use a solid boundary when the inequality contains $\leq$ or $\geq$. Use a dashed boundary when the inequality contains $<$ or $>$. Then use a test point to determine which half-plane should be shaded. Finally, shade the half-plane that contains the solution.

Study Organizer

Foldables

Use your Foldable to review the module. Working with a partner can be helpful. Ask for clarification of concepts as needed.

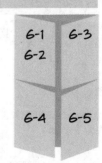

Test Practice

1. MULTIPLE CHOICE Select the graph that shows the solution set of $7 \leq n + 5$. (Lesson 6-1)

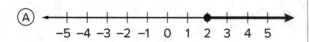
Ⓐ −5 −4 −3 −2 −1 0 1 2 3 4 5

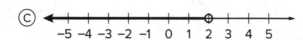
Ⓑ −5 −4 −3 −2 −1 0 1 2 3 4 5

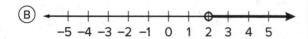
Ⓒ −5 −4 −3 −2 −1 0 1 2 3 4 5

Ⓓ −5 −4 −3 −2 −1 0 1 2 3 4 5

2. OPEN RESPONSE Eduardo is writing a historical novel. He wrote 16 pages today, bringing his total number of pages written to more than 50. How many pages p did Eduardo write before today? Complete the inequality that represents this situation. Then solve the inequality. (Lesson 6-1)

Inequality: (____) $+ p >$ (____)

Solution: $p >$ (____)

3. OPEN RESPONSE A farmer said that for every row of seeds he plants, he can harvest 6.5 bushels of tomatoes. The farmer needs to harvest at least 52 bushels of tomatoes. What is the least number of rows that the farmer will need to plant? (Lesson 6-1)

(_____)

4. OPEN RESPONSE Find the solution set of $3d - 8 < 4d + 2$. (Lesson 6-1)

(_____)

5. MULTIPLE CHOICE Which inequality has solutions represented by the graph? (Lesson 6-1)

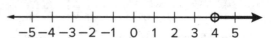
−5 −4 −3 −2 −1 0 1 2 3 4 5

Ⓐ $5 - x > 4$

Ⓑ $2x + 1 \geq 9$

Ⓒ $2x - 5 > 3$

Ⓓ $6x - 7 > 5$

6. TABLE ITEM Consider the inequality *Five plus two times a number n is less than or equal to eleven*. Indicate whether each representation is or is not a solution. (Lesson 6-2)

Representation	Solution?	
	Yes	No
$n \leq 3$		
$3 \leq n$		
$n \leq 8$		
$3 \geq n$		

7. OPEN RESPONSE Solve $8(t + 2) + 7(t + 2) - 3(t - 2) < 0$. Write the solution using set-builder notation. (Lesson 6-2)

(_____)

8. MULTIPLE CHOICE Solve $-\frac{4}{5}x - 3 \leq 17$. (Lesson 6-2)

Ⓐ $\{x | x \geq -25\}$

Ⓑ $\{x | x \geq -16\}$

Ⓒ $\{x | x \leq -25\}$

Ⓓ $\{x | x \leq -16\}$

9. MULTI-SELECT The science club is planning a car wash fundraiser. Halona writes and graphs an inequality to represent the number of cars c the science club needs to wash in order for the profits to cover their expenses. (Lesson 6-2)

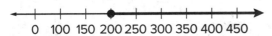

Which inequality could Halona have graphed for this situation? Select all that apply.

Ⓐ $2(c + 1) - 50 \geq 352$

Ⓑ $310 + c \leq 3(c - 30)$

Ⓒ $20(c - 1) \geq 4000$

Ⓓ $5(c + 1) - 200 \geq 250$

Ⓔ $50(c - 5) \geq 200$

10. OPEN RESPONSE Micaela wants to plant a square garden and enclose it with a fence. She has 120 feet of fencing available and she wants the sides of her garden to be at least 6 feet long. Complete the inequality to represent the possible side lengths. (Lesson 6-3)

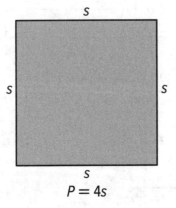

$P = 4s$

[ ] $\leq s \leq$ [ ]

11. OPEN RESPONSE Solve $4 + g \geq 3$ or $6g \geq -30$. Write the solution using set-builder notation. (Lesson 6-3)

[ ]

12. MULTIPLE CHOICE Which graph shows the solution of $3n - 1 < 5$ and $-2n + 3 < 11$? (Lesson 6-3)

Ⓐ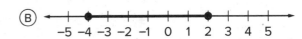

Ⓑ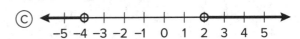

Ⓒ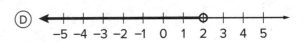

Ⓓ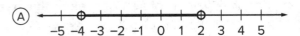

13. OPEN RESPONSE/GRAPH (Lesson 6-4)
Part A Solve $|h + 3| < 5$. Write the solution set using set-builder notation.

[ ]

Part B Graph the solution set on the number line provided.

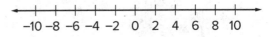

14. MULTIPLE CHOICE Which is NOT a true statement? (Lesson 6-4)

Ⓐ The empty set is the solution of $|k - 2| + 1 \leq -2$.

Ⓑ The solution of $|k - 2| \geq 2$ is $\{k | 4 \geq k \geq 0\}$.

Ⓒ The solution of $|3k + 3| \leq -9$ is $\{k | -4 \geq k \geq 2\}$.

Ⓓ $|k - 7| < -6$ has no solution.

15. MULTIPLE CHOICE The actual weight of a jar of peanuts tends to be within 0.5 ounce of its listed weight. Which graph shows the possible weights of a jar of peanuts that has a listed weight of 6.5 ounces? (Lesson 6-4)

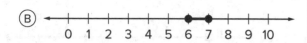

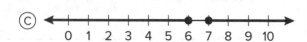

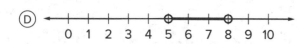

16. OPEN RESPONSE The equation for the boundary of the inequality graphed is $y = 3x - 1$. (Lesson 6-5)

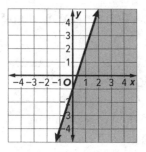

Write the inequality that represents the graph.

[   ]

17. TABLE ITEM Consider the graphs. Indicate which graph represents the solution to each of the inequalities listed in the table. (Lesson 6-5)

Graphs:

A. 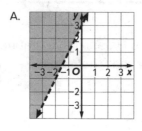 B.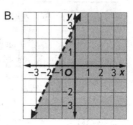

C. 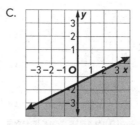 D.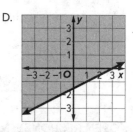

Inequality	Graphed Solution			
	A	B	C	D
$2y - x \leq -3$				
$2y - x \geq -3$				
$2x - y > -3$				
$2x - y < -3$				

18. GRAPH Graph the inequality $\frac{2}{3}x + 5 - y < 6$. (Lesson 6-5)

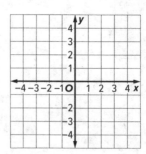

Module 1

Quick Check

1. $\frac{2}{3}$ **3.** $\frac{1}{13}$ **5.** 4   **7.** $\frac{3}{2}$ or $1\frac{1}{2}$

Lesson 1-1

1. $2 + 12 \div 4$ **3.** $3 \times 11 + 7$ **5.** $6 - 3 - 1$
7. $24 \div 6 + 7$ **9.** $116 \div 4 + 28 - 33$ **11.** $\frac{3+7}{2}$
13. $(4 + 9) \times 3$ **15.** $\frac{10}{4 \times 5}$ **17.** $\frac{1+2}{20}$ **19.** $\frac{12+16}{3+4}$
21. $\frac{36+14}{2 \times 5}$ **23.** $\frac{6+15}{13-9}$ **25.** 365×85
27. $2 \times 35 - 5$ **29.** 12×7 **31.** 2744   **33.** 11
35. 42   **37.** 7   **39.** 56   **41.** 22   **43.** 8
45. 324   **47.** 29   **49a.** $20 \times 5 + 9$
49b. 109 flies   **51a.** $2 \times \frac{1}{2}(30 + 50)24$
51b. 1920 in²   **53.** $8^4 + 6$ **55.** Tamara; when evaluating, first perform the multiplication and division from left to right, and then the addition and subtraction from left to right.
57. The cashier; Kelly should have entered the expression into her calculator as $3(18.95 - 2) + 2(11.50)$.   **59.** $10 \times 18 + 8 \times 18$; You can also find the length of each side of the apartment, 18 and $10 + 8$, and then multiply: $18(10 + 8)$.
61. The student should have added/subtracted from left to right. The correct value is 30.

Lesson 1-2

1. four times a number q **3.** 15 plus r
5. 3 times x squared   **7.** two times a plus six
9. twenty-five plus six times a number squared   **11.** three times a number raised to the fifth power divided by two   **13.** 5 times g to the sixth power   **15.** four minus five times h
17. 1 less than 7 times x cubed   **19.** 3 times n squared minus x **21.** 18 times the quantity p plus 5   **23.** $n - 35$ **25.** $\frac{1}{3}n$ **27.** $\frac{45}{r}$
29. $18 - 3d$ **31.** $\frac{20}{t^5}$ **33.** $(k + 2) - 15$

35. $2m + 6$ **37.** $\frac{6136}{y}$ **39.** $19.95 \times t - 10$ or $19.95t - 10$ **41.** $5f + 45h$ **43.** 1   **45.** 7
47. 149   **49.** $\frac{65}{4}$ **51.** 44   **53.** $\frac{1}{2}$ **55.** 18
57. 10   **59.** 16   **61.** 13   **63a.** $5t - 100$
63b. 1400 students   **65a.** $1.75 + 3.45m$
65b. $29.35   **67a.** Sample answer: the quotient of $5x$ and 2 plus y cubed; $5x$ divided by 2 plus y to the third power   **67b.** 18
69. 89   **71.** 52   **73.** Sample answer: the quotient of x minus 1 and 2; $\frac{x-1}{2}$
75a. $x - (36 \times 4)$ **75b.** $\frac{x - (36 \times 4)}{0.20}$
75c. 350 mi   **77.** Sample answer: An algebraic expression is a math phrase that contains one or more numbers or variables. To write an algebraic expression from a real-world situation, first assign variables. Then determine arithmetic operations done on the variables. Finally, put the terms in order.   **79.** Sample answer: Movie tickets cost $10 and a box of popcorn cost $5.25. You buy t movie tickets and a box of popcorn. What is the greatest number of movies tickets you can purchase with $50?

Lesson 1-3

1. Symmetric Property of Equality
3. Symmetric Property of Equality   **5.** 14
7. 34   **9a.** Exit 15 to Exit 8   **9b.** Symmetric Property of Equality

11.	$= (3 \div 2)\frac{2}{3}$	Multiplicative Identify
	$= \frac{3}{2} \cdot \frac{2}{3}$	Substitution
	$= 1$	Multiplicative Identify
13.	$= 2(5 - 5)$	Substitution
	$= 2(0)$	Substitution
	$= 0$	Multiplicative Property of Zero
15.	$= 2(2 - 1) \cdot \frac{1}{2}$	Substitution
	$= 2(1) \cdot \frac{1}{2}$	Substitution
	$= 2 \cdot \frac{1}{2}$	Multiplicative Identity
	$= 1$	Multiplicative Inverse

17.
$$= 4 + \frac{4}{9} + 7 + \frac{2}{9} \qquad \text{Substitution}$$
$$= 4 + 7 + \frac{4}{9} + \frac{2}{9} \qquad \text{Commutative (+)}$$
$$= 4 + 7 + \left(\frac{4}{9} + \frac{2}{9}\right) \quad \text{Associative (+)}$$
$$= 11 + \frac{6}{9} \qquad \text{Substitution}$$
$$= 11\frac{6}{9} \qquad \text{Substitution}$$
$$= 11\frac{2}{3} \qquad \text{Substitution}$$

19.
$$= (2 \cdot 8) \cdot (10 \cdot 2) \qquad \text{Associative (×)}$$
$$= 16 \cdot 20 \qquad \text{Substitution}$$
$$= 320 \qquad \text{Substitution}$$

21.
$$= \left(2\frac{3}{4} \cdot 1\frac{1}{8}\right) \cdot 32 \qquad \text{Associative (×)}$$
$$= \left(\frac{11}{4} \cdot \frac{9}{8}\right) \cdot 32 \qquad \text{Substitution}$$
$$= \frac{99}{32} \cdot 32 \qquad \text{Substitution}$$
$$= 99 \qquad \text{Substitution}$$

23.
$$= 2 \cdot 5 \cdot 4 \cdot 3 \qquad \text{Commutative (×)}$$
$$= (2 \cdot 5) \cdot (4 \cdot 3) \qquad \text{Associative (×)}$$
$$= 10 \cdot 12 \text{ or } 120 \qquad \text{Substitution}$$

25.
$$= \frac{4}{3} \cdot 3 \cdot 7 \cdot 10 \qquad \text{Commutative (×)}$$
$$= \left(\frac{4}{3} \cdot 3\right) \cdot (7 \cdot 10) \quad \text{Associative (×)}$$
$$= 4 \cdot 70 \text{ or } 280 \qquad \text{Substitution}$$

27. -64 **29.** -5 **31.** -9 **33.** Sample answer: Multiplicative Identity and Multiplicative Inverse **35.** 0; Additive Identity **37.** 1; Multiplicative Identity **39.** 5; Additive Identity **41.** 1; Multiplicative Inverse **43.** 3; Reflexive Property **45.** Yes; the Commutative and Associative Properties of Multiplication allow it to be rewritten. **47.** Sample answer: $126 + 28 + 52 = 126 + (28 + 52) = 126 + 80 = 206$ **49.** Sample answer: $5 = 3 + 2$ and $3 + 2 = 4 + 1$, so $5 = 4 + 1$; $5 + 7 = 8 + 4$ and $8 + 4 = 12$, so $5 + 7 = 12$. **51.** $4 \div 8 \neq 8 \div 4$ because $4 \div 8 = \frac{1}{2}$ and $8 \div 4 = 2$, so there is no Commutative Property for division. $16 \div (8 \div 4) \neq (16 \div 8) \div 4$ because $16 \div (8 \div 4) = 16 \div 2 = 8$ and $(16 \div 8) \div 4 = 2 \div 4 = \frac{1}{2}$, so there is no Associative Property for division. As long as neither number is 0, when the order of division of two numbers is switched, the results are multiplicative inverses of each other. **53a.** False; sample answer: $3 - 4 = -1$, which is not a whole number. **53b.** True **53c.** False; sample answer: $2 \div 3 = \frac{2}{3}$, which is not a whole number. **55.** $(2j)k = 2(jk)$; The other three equations illustrate the Commutative Property of Addition or Multiplication. This equation represents the Associative Property of Multiplication.

Lesson 1-4

1. $4(6) + 5(6)$; 54 **3.** $6(6) - 6(1)$; 30 **5.** $14(8) - 14(5)$; 42 **7a.** $39(23 + 2)$ **7b.** \$975
9a. $10\left(3\frac{3}{5}\right)$ **9b.** $10\left(3\frac{3}{5}\right) = 10\left(3 + \frac{3}{5}\right) = 10(3) + 10\left(\frac{3}{5}\right) = 30 + 6 = 36$ yards of fabric
11. $7(500 - 3)$; 3479 **13.** $36\left(3 + \frac{1}{4}\right)$; 117
15. $5(90 - 1)$; 445 **17.** $15(100 + 4)$; 1560
19. $12(100 - 2)$; 1176 **21.** $3(10 + 0.2)$; 30.6
23. $2(x) + 2(4)$; $2x + 8$ **25.** $4(8) + (-3m)(8)$; $32 - 24m$ **27.** $2(17) + (-4n)(17)$; $34 - 68n$
29. $\frac{1}{3}(27) + (-2b)(27)$; $9 - 54b$ **31.** $6(2c) + 6(-cd^2) + 6(d)$; $12c - 6cd^2 + 6d$ **33.** $3(m) + 3(n)$; $3m + 3n$ **35.** $\frac{1}{2}(14) + (6a)(14)$; $7 + 84a$
37. $0.3(9) + (-6x)(9)$; $2.7 - 54x$ **39.** $18r$
41. $2m + 7$ **43.** $13m + 5p$ **45.** $14m + 11g$
47. $12k^3 + 12k$ **49.** $18g$ **51.** $5a^2$ **53.** $2q^2 + q$
55a. $3a + 5(a - b)$

55b.
$$3a + 5(a - b) = 3a + 5a - 5b$$
$$\text{Distributive Property}$$
$$= (3a + 5a) - 5b \quad \text{Associative (+)}$$
$$= 8a - 5b \qquad \text{Substitution}$$

57. $24x + 28$ **59.** $18d + 20$ **61.** $7y^3 + y^4$
63. $4b$ **65.** $20x + 37y$ **67.** $2n + 2m$ and $2(n + m)$ **69.** No; sample answer: 10 pounds 5 ounces is $10(16) + 5 = 165$ ounces, but Ariana used the Distributive Property incorrectly. She should have written $8(20 + 2) = 8(20) + 8(2) = 160 + 16 = 176$ ounces. **71.** Sample answer: Algebraic expressions are helpful because they are easier to interpret and apply than verbal expressions. They can also be written in a more simplified form.

Lesson 1-5

1. $|p - t|$ and $|t - p|$ **3.** $|r - w|$ and $|w - r|$
5. 15 **7.** 22 **9.** 37 **11.** 32 **13.** -62 **15.** 11
17. 10 **19.** 5 **21.** 5 **23.** 6 **25.** -7.4 **27.** 8.4
29. -15 **31.** 22 **33.** 14.5 **35a.** $|g - d|$ and $|d - g|$ **35b.** 5 meters **37.** Sample answer: A meteorologist says that the high temperature is going to be 89 degrees. If the actual high temperature that day is x, then $|x - 89|$ represents the number of degrees the meteorologist is away from the actual high temperature.

39. False; sample answer: Suppose $a = 5$ and $b = -3$, then $|a + b| = |5 + (-3)| = |5 - 3| = |2| = 2$ and $|a| + |b| = |5| + |-3| = 5 + 3 = 8$. $2 \neq 8$, so Diaz's claim is not correct.

Lesson 1-6

1. 32 **3.** 11.4 seconds **5.** 0.512 **7.** Automatic Method: $6000; Exact Method: $5625 **9.** $2\frac{1}{3}$ snack bars **11.** 6 **13.** $333.33 **15.** $10,500
17. Because the number of students enrolled at Hartgrove High School can be counted, giving an exact enrollment is accurate. **19.** The map maker is probably accurate because the number of traffic lights in New York City is not very specific. **21.** Sample answer: Steve Nash would be selected as a free throw shooter. Michael Jordan would be selected as a free throw shooter. Shaquille O'Neal would not be selected as a free throw shooter.
23. light-years; Sample answer: The distance from Earth to the star is very great so using the largest distance unit is appropriate in this situation. **25.** Sample answer: The line represents the most accurate number of visitors at the zoo for a given temperature.
27. Sample answer: The number of visitors does not increase at the same rate for each average daily temperature. **29.** Sample answer: An employer might consider the number of sick days an employee takes or the amount of sales an employee generates.
31. Sample answer: $28.43 because $3.299 \times 8.618 = 28.430782$. The answer could be accurate to the thousands place, but it is only necessary to round to the nearest hundredths place because the penny is the smallest unit of money.

Module 1 Review

1. D **3.** $5(x + 7) - 4^3$ **5.** A **7.** B
9.

Statement	True	False
$4(6 - 2 \times 3) = 0$	X	
$11(3^2 - 9) + 2\left(\frac{1}{2}\right) = 0$		X
$4 \cdot 0 + 4^2 - 2^3 - (2 + 2 \cdot 3) = 0$	X	

11.

Expression	Yes	No
$-7(m^3 - 11)$		X
$-7(3m) - 7(-11)$	X	
$-21m - 77$		X
$-21m + 77$	X	
$-21m - 11$		X
$-7m^3 + 77$		X

13. C, E **15.** 67
17. No; sample answer: The manager rounded down, but actually spent much more than $4000. It would have been better to report a greater amount so that it was clear her budget was not overspent.

Module 2

Quick Check

1. $6n + 2$　**3.** $4b + 9$　**5.** 8　**7.** 32　**9.** 36

Lesson 2-1

1. $3m + 2 = 18$　**3.** $\frac{24}{x} = 14 - 2x$　**5.** $2 + 3h = 6$　**7.** $(48 + 33) + n = 107$　**9.** $2a + a^3 = b$
11. $x + x^2 = yz$　**13.** $A = \ell^2$　**15.** $P = 2\ell + 2w$　**17.** $I = prt$　**19.** The sum of j and sixteen is thirty-five.　**21.** Seven times the sum of p and twenty-three is the same as one hundred two.
23. Two-fifths of v plus three-fourths is identical to two-thirds of x squared.　**25.** g plus 10 is the same as 3 times g.　**27.** 4 times the sum of a and b is 9 times a.　**29.** Half of the sum of f and y is f minus 5.　**31.** Sample answer: The volume equals π times the radius squared times the height. The base is a circle so the expression πr^2 represents the area of the base.　**33.** Sample answer: The interest equals the product of the principal, the rate, and the time.　**35.** Sample answer: Force equals mass times acceleration. The expression ma represents the force on an object with mass m that is accelerating.
37. B　**39.** A　**41.** $y^2 - 12 = 5x$　**43.** $100 - 3b = 6b$　**45.** Four times n equals x times the difference of five and n.　**47.** The sum of y and the product of 3 and the square of x is 5 times x.　**49.** $V = \ell wh$　**51.** $m + 2m = 24$ or $3m = 24$　**53.** $c = 10w + 0.1(10w)$ or $c = 11w$
55a. It is correct. The product is squared, so parentheses are needed.　**55b.** It is not correct. One-half of a number means to multiply, not divide, by one-half. It should be $\frac{1}{2}n + 3 = n - 2$.　**57.** Sample answer: A teacher ordered 188 math books. The algebra books were packed in boxes of 12. The geometry books were packed in boxes of 10. He ordered one more box of algebra books than geometry books. How many books of each type book did he order? Let a = number of algebra books.
59. $S = 6\ell^2$　**61.** Sample answer: First, you should identify the unknown quantity or quantities for which you are trying to solve, and assign variables. Then, you should look for key words or phrases that can help you to determine operations that are being used. You can then write the equation using the numbers that you are given and the variables and operations that you assigned.

Lesson 2-2

1. 23　**3.** -43　**5.** -12　**7.** 73　**9.** -15
11. -54　**13.** $\frac{7}{20}$　**15.** $-\frac{7}{15}$　**17.** -937
19. -147　**21.** -25　**23.** $-\frac{9}{2}$　**25.** 15
27. 10　**29.** 64　**31.** 28　**33.** 18　**35.** 24
37. 27　**39.** 39　**41.** 64　**43.** 9　**45.** -12
47. 7　**49.** 64　**51.** -252　**53.** -52
55. $x + 33 = 2005$; $x = 1972$　**57.** $x - 21 = -9$; $x = 12°C$　**59a.** Let p = the number of players who signed up for the soccer league. If 13% of the players who signed up for the soccer league dropped out, then 100% -13%, or 87% of the players finished the season. So, $0.87p$ represents the number of players who finished the season.　**59b.** $0.87p = 174$　**59c.** $p = 200$; 200 players signed up for the soccer league
61. $\frac{2}{3} = -8n$; $-\frac{1}{12}$　**63.** $\frac{4}{5} = \frac{10}{16}n$; $\frac{32}{25}$
65. $4\frac{4}{5}n = 1\frac{1}{5}$; $\frac{1}{4}$　**67.** -77　**69.** $\frac{16}{3}$
71. -10　**73.** $-\frac{10}{7}$ or $-1\frac{3}{7}$　**75.** 18
77. 225　**79.** -14　**81.** 4　**83.** -49
85. 40　**87.** -15　**89.** $-\frac{8}{15}$　**91a.** $12x = 780$;
$x = 65$　**91b.** $20　**93.** $x = 216$; Multiplication Property of Equality　**95.** $y = -224$; Subtraction Property of Equality
97. $15 = b$; Division Property of Equality
99. $n - 16 = 29$ does not belong because for the other three, $n = 13$, and for this one $n = 45$.
101. Sample answer: $x - 4 = 10$　**103.** Sample answer: To solve $5x = 35$, I would divide each side by 5 to get $x = 7$. To solve $5 + x = 35$, I would subtract 5 from each side to get $x = 30$. In both equations I used properties of equality to isolate the variable. In the first equation I used the Division Property of Equality and I used the Subtraction Property of Equality in the second equation.

Lesson 2-3

1. -5 3. -5 5. 70 7. 27 9. 16
11. -61 13. $\frac{1}{2}a - 5.25 = 22.50$; \$55.50
15. $\frac{t-10}{15} = 4$; 70 treats
17. $71 = 2h - 1$; 36 inches
19. $\frac{18}{a}$ 21. $\frac{-35}{a}$ 23. $\frac{-24}{a}$ 25. $\frac{-14}{a}$
27. 7 29. 10 31. -16 33. -2
35. 18 37. $(n - 2) \div 3 = 30$; 92   39. Sample answer: Both are correct. Dividing by a number and multiplying by that number's reciprocal are equivalent operations.   41a. $x = \frac{-2}{a}$
41b. $x = 13a$ 41c. $x = \frac{10}{a}$ 43. Never; whenever three odd integers are added together, the sum is always odd.

Lesson 2-4

1. 6 3. 1 5. -2 7. -2 9. 14 11. 4
13. -5 15. 0 17. $7 + F = 4F + 1$; France won 2 gold medals and the U.S won 9 gold medals.   19. $38 + 4x = 45.5 + 2.5x$; 5 years   21. $180 - x = 10 + 2(90 - x)$; $10°$
23. $9(5 + x) = 15\frac{3}{7}x$; 7   25. no solution
27. identity   29. no solution   31. one solution
33. identity   35. no solution   37. no solution
39. all numbers   41. -25 43. 3
45. -2 47. 15 49a. Let $n =$ the first odd integer; $2(n + 2) = 3n - 13$ 49b. 17 and 19
51a. Let $k =$ the number; $4k - 3 = 2k + 5$
51b. $k = 4$ 51c. Substitute 4 for k in the expression for the perimeter of Figure 2, $2k + 5$. So the perimeter of Figure 2 is $2(4) + 5 = 8 + 5 = 13$.   51d. Substitute 4 for k in the expression for the perimeter of Figure 1, $4k - 3$. So the perimeter of Figure 1 is $4(4) - 3 = 16 - 3 = 13$. The perimeter for Figure 1 and Figure 2 is the same, so the value of k is correct.
53. Anthony is correct. When Patty added m to each side, she subtracted the terms instead of adding them.   55. Sample answer: $2(3x + 6) = 3(2x + 5)$ 57a. Incorrect; the 2 must be distributed over both g and 5, then 10 must be subtracted from each side; 6.

57b. Correct; the Subtraction Property was used to combine the variable terms on the left side of the equation. The Division Property was used to isolate the variable on one side.   57c. Incorrect; to eliminate $-6z$ on the left side of the equal sign, $6z$ must be added to each side of the equation; 1.   59. Sample answer: $2x + 1 = x + 9$

Lesson 2-5

1. $\{-2, 8\}$

3. $\varnothing$

5. $\{-3.25, 2\}$

7. $\varnothing$

9. $\{-8, 16\}$

11. $\{2, 5\}$ 13. $\{-6, 4\}$ 15. $\{0, 4\}$
17. $\{-5, -1\}$ 19. $|t - 400| = 15$; min $= 385°F$; max $= 415°F$ 21a. $|t - 20.9| = 5.3$ 21b. 15.6 to 26.2; 10.3 to 31.5

23. $\|x - 35\| = 0.5$	Absolute value equation
Case 1: $x - 35 = 0.5$	Definition of absolute value.
$x = 35.5$	Simplify
Case 2: $x - 35 = -0.5$	Definition of absolute value.
$x = 34.5$	Simplify

The bags of rock salt weigh no less than 34.5 pounds and no more than 35.5 pounds.
25. $|x| = 6$ 27. $|x + 2| = 4$ 29. $|x + 3| = 2$
31. $|x| = 4$
33. $\left\{-\frac{3}{2}, \frac{9}{2}\right\}$

35. $\{5.5, -5.5\}$

37. $\{2, -2\}$

39. $|x| = 1\frac{1}{2}$ 41. $\left|x - \frac{1}{4}\right| = \frac{1}{4}$ 43. $\left|x + \frac{1}{3}\right| = 1$
45a. $|x - 38| = 2$ 45b. $40°F$, $36°F$

47. $|x - 3| = 1$ **49.** $|a - b| + |b - c| = |a - c|$ **51.** Cami; The absolute value of a number cannot be negative. **53.** Sample answer: Let x = the temperature at night. Then the temperature is 4 ± 10 degrees.

Lesson 2-6

1. 40 **3.** 29.25 **5.** 9.8 **7.** 1.32 **9.** 0.84
11. 0.57 **13.** 6 **15.** 11 **17.** 18 **19.** 0.8
21. 11 **23.** −2 **25.** 1.44 **27.** −2.29
29. −2.2 **31.** 10 **33.** 3 **35.** −8.4
37. 12.5 gal **39.** $46.27 **41a.** 60 free throws **41b.** Sample answer: I assumed that Brent continues to make free throws at the same rate. **43.** 22.5 in. **45.** $3333.33
47. 204.55 mL **49.** 6 **51.** 10 **53.** 21 **55.** 8
57. 42 **59.** 27 **61.** 3 **63.** 15 **65.** −3
67. −0.4 **69.** −6 **71a.** Sample answer: 2.2 cm
71b. Sample answer: about 6.6 miles
71c. Sample answer: about 453.7 mi²
73a. $4.50; because 8 potatoes cost $1.50, multiply by 3 to get a cost of $4.50 for 24 potatoes. **73b.** $4.13; Sample answer: $4.13 is slightly less than $4.50, which aligns with my estimate. **73c.** 37 **73d.** $0.19 **75.** $\frac{2}{4}$ or $\frac{1}{2}$
77. Ratios and rates each compare two numbers by using division. However, rates compare two measurements that involve different units of measure. **79.** $\frac{x}{100} = \frac{z}{y}$

Lesson 2-7

1. $y = \frac{x - 1}{2}$ **3.** $f = \frac{5 - g}{7}$ **5.** $t = \frac{x}{7}$ **7.** $r = \frac{q}{2}$
9. $a = -\frac{b}{8}$ **11.** $v = \frac{u - z}{w}$ **13.** $g = \frac{10j + 9h}{f}$
15. $t = \frac{3}{2}(r - v)$ **17.** $a = \frac{-33 + x}{10c}$
19a. $\ell = \frac{P - 2w}{2}$ **19b.** 14 m
21a. $g = \frac{c - p}{13.50}$ **21b.** 6 games
25. ≈ 12.96 trillion pounds **27.** 0.44 ft
29. 24 miles **31.** 90.2 gallons
33. 82 students **35.** $c = \frac{2k - 3g}{b}$
37. $c = \frac{5p - 6j}{8}$ **39.** $c = x - 2d$
41. $t = \frac{w - 11v}{31}$ **43.** $c = \frac{-13 + f}{10 - d}$ **45.** $r = \frac{A}{P} - 1 = \frac{2182.25}{2150} - 1 = 1.015 - 1 = 0.015$; The

interest rate is 1.5%. **47a.** $y = \frac{mx + mt - z}{r}$
47b. Division by 0 is undefined, so in the original equation $m \neq 0$, and in the final equation $r \neq 0$. **49.** No; Sasha does not have a correct solution. When she multiplied F by $\frac{5}{9}$, she should have multiplied 32 by $\frac{5}{9}$.

Module 2 Review

1. A **3.** D **5.** D **7.** C **9.** B, D
11. C **13.** Let d = age of dogwood tree (in years); $3d - 2 = \frac{1}{2}(d + 8 + 8)$; The gingko tree is 10 years old, and the dogwood tree is 4 years old. **15.** A **17.** $b = 8$ **19.** B
21. D **23A.** $h = 3\frac{V}{B}$ **23B.** 12

Module 3

Quick Check

1, 3.

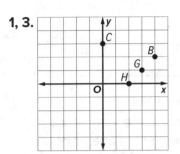

5. 20 **7.** −3

Lesson 3-1

1.

x	y
−1	−1
1	1
2	1
3	2

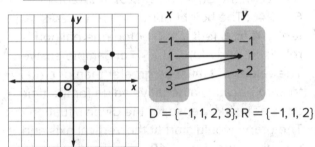

D = {−1, 1, 2, 3}; R = {−1, 1, 2}

3.

x	y
3	−2
1	0
−2	4
3	1

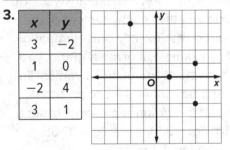

D = {−2, 1, 3}; R = {−2, 0, 1, 4}

5a. independent: price of item, dependent: number of items purchased **5b.** As the price of an item increases, the number of items purchased decreases.

7.

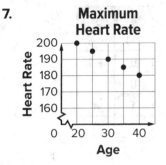

9. The x-axis represents the time in seconds. The y-axis represents the height of the elevator in feet. The x-axis has a scale of 1 mark = 1 second. The y-axis has a scale of 1 mark = 10 feet. The origin (0, 0) represents a height of 0 feet in 0 seconds. **11.** {(1, 7), (3, 45), (5, 11), (13, 15)}
13. {(2, 5), (5, 0), (7, 8), (7, 10), (10, 2)} **15.** {(2, 80), (3, 120), (6, 240), (8, 320)}; The x-axis represents the number of gallons of syrup. The y-axis represents the number of gallons of sap. The x-axis has a scale of 1 mark = 1 gallon of syrup. The y-axis has a scale of 1 mark = 40 gallons of sap. The origin (0, 0) represents 0 gallons of sap makes 0 gallons of syrup.
17a. {(0, 12), (1, 8), (2, 23), (3, 28), (4, 11), (5, 11)}
17b. {0, 1, 2, 3, 4, 5} **17c.** {8, 11, 12, 23, 28}
19. sample answer:

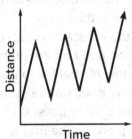

21. Tim drives away from the pizzeria, stops to make a delivery, continues to drive away from the pizzeria, stops to make another delivery, and then returns to the pizzeria. **23.** Disagree; The intersection point represents a time when Tim and Lauren were both at the same distance from the pizzeria. **25.** Disagree; sample counterexample: In the relation {(1, 2), (1, 3)}, the domain is {1}, so it has one element, while the range is {2, 3}, which has two elements. **27a.** Sample answer: {(−1, −3), (0, −3), (0, −1), (1, 4), (2, 5)}

27b. sample answer:

x	y
−1	−3
0	−1
0	−3
1	4
2	5

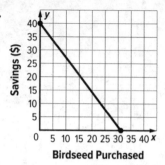

29. Sample answer: A dependent variable is determined by the independent variable for a given relation.

Lesson 3-2

1. Yes; for each element of the domain, there is only one element of the range.   **3.** No; the element 4 in the domain is paired with both 2 and 5 in the range.   **5.** No; the element 5 in the domain is paired with both −3 and 2 in the range.   **7.** Yes; for each element of the domain, there is only one element of the range.
9. Yes; for any value x, the vertical line passes through no more than one point on the graph.
11. Yes; for any value x, the vertical line passes through no more than one point on the graph.
13. No; for $x > 0$, the vertical line passes through more than one point on the graph.
15a.

Year	2014	2015	2016	2017
Value ($)	254,000	293,000	338,000	372,000

15b. Domain: {2014, 2105, 2016, 2017}; Range: {254,000; 293,000; 338,000; 372,000}
15c. For each element of the domain, there is only one element of the range. So, this relation is a function.   **17.** 26   **19.** 2   **21.** 42   **23.** 4
25. 6   **27.** $9b^2 − 3b$ **29.** $f(3.5) = 12.25$, which means the area of a square with a side of length 3.5 units is 12.25 square units.
31. $f(12) = \$435$, which is the cost of a gym membership for 12 months, or 1 year.   **33.** −1
35. 14   **37.** −4   **39.** $−8y − 3$ **41.** $−2c −8$
43. $−10d − 15$

45a.

45b. Yes; for any value x, the vertical line passes through no more than one point on the graph.   **45c.** $f(3) = 36.25$, which means if Aisha buys 3 pounds of birdseed, she saves $36.25; $f(18) = 17.50$, which means if Aisha buys 18 pounds of birdseed, she saves \$17.50; $f(36) = −5$, which means if Aisha wants to buy 36 pounds of birdseed, she needs \$5 extra.   **45d.** 8 pounds   **47a.** $h(20) = 46$; The height of the balloon 20 seconds after it is released is 46 feet.   **47b.** 2 minutes is $2(60) = 120$ seconds, so calculate $h(120)$ by substituting $t = 120$ in the equation; $h(120) = 2(120) + 6 = 246$; The height of the balloon is 246 feet.   **47c.** 6 feet; $t = 0$ before the balloon is released, and $h(0) = 6$.   **47d.** Sample answer: The values of t must be greater than or equal to zero because a negative value for the time does not make sense for the given situation. The graph would start at the vertical axis and go only to the right.   **49.** Sample answer: You can determine whether each element of the domain is paired with exactly one element of the range. For example, if given a graph, you could use the vertical line test; if a vertical line intersects the graph more than once, then the relation that the graph represents is not a function.   **51.** $f(g + 3.5) = −4.3g − 17.05$
53. Sample answer: $f(x) = 3x + 2$

Lesson 3-3

1. Neither; because the function has continuous sections but is not a single line or curve, it is neither continuous or discrete   **3.** Discrete; because the function is made up entirely of individual points, it is discrete.   **5.** Continuous; because the function is graphed with a single line, it is continuous.   **7.** Discrete; because the function is made up entirely of individual points, it is discrete.   **9.** discrete   **11.** discrete

13. linear   **15.** nonlinear   **17.** nonlinear
19. nonlinear   **21.** nonlinear   **23.** linear
25a. linear

25b.

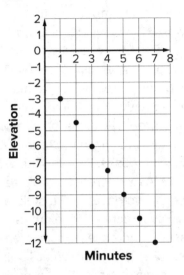

27a. nonlinear

27b.

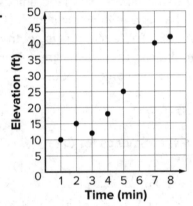

29. continuous; nonlinear   **31.** neither;
nonlinear   **33.** discrete; nonlinear
35. Sample answer: A studio charges musicians
to use the space and recording equipment by
the hour, rounding a fraction of an hour up. So,
for up to 1 hour, the studio charges $100, but
for up to 2 hours, the studio charges $200, and
so on. The function that models this situation is
neither discrete nor continuous.

Lesson 3-4

1. x-intercept: $(-0.75, 0)$ y-intercept: $(0, 2)$
positive: when $x > -0.75$ negative: when
$x < -0.75$ **3.** x-intercepts: $(0, 0)$ and $(2, 0)$
y-intercept: $(0, 0)$ positive: when $x < 0$ and
when $x > 2$ negative: $0 < x < 2$
5. x-intercepts: $(-2, 0)$ y-intercept: $(0, 4)$
positive: when $x > -2$ negative: when $x < -2$

7. x-intercepts: $(-5, 0)$ and $(3, 0)$ y-intercept:
$(0, 3)$ positive: $-5 < x < 3$ negative: $x < -5$
and when $x > 3$
9. x-intercepts: none y-intercept: $(0, -3)$
positive: never negative: always
11. The x-intercept is 0. The y-intercept is 0.
This means that Ryan earns $0 for working
0 hours. The function is positive when x is
greater than 0, which means that Ryan
earns money for working. No portion of the
graph shows that the function is negative.
13a. The x-intercept is 6. The y-intercept is
1950.   **13b.** The x-intercept means that after
6 months, Javier's remaining balance will be
$0, or it will take Javier 6 months to repay his
parents. The y-intercept means that Javier
owes his parents $1950 after 0 months, or
Javier initially borrowed $1950 from his parents.

15. Sample graph; no solution

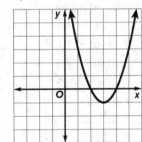

17. 2, 4

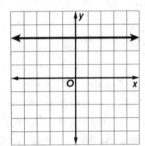

19. −3, 2

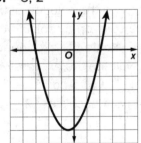

21. The zero of the function is at 32. This
represents that after tying ribbon on 32 gift
bags, Juanita will have no ribbon left.

23. *x*-intercepts: (−2, 0) and (2, 0) *y*-intercept: (0, 4) positive: −2 < *x* < 2 negative: when *x* < −2 and when *x* > 2   **25.** The *x*-intercepts are 3 and 7. That means that the bird will be at sea level at 3 seconds and at 7 seconds. The *y*-intercept is 4.5. This means that at time 0, the bird was at a height of 4.5 feet. The function is positive when *x* is less than 3 and when *x* is greater than 7, which means that the bird is above sea level from 0 to 3 seconds and after 7 seconds. The function is negative when *x* is between 3 and 7, which means that the bird is below sea level, or under water, for 4 seconds.

27. 1; *x*-int: 1; *y*-int: −15

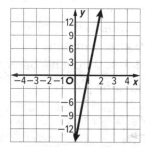

29. To find the *x*-intercept in a graph, find the place where the function crosses the *x*-axis. To find the *y*-intercept in a graph, find the place where the function crosses the *y*-axis. To find the *x*-intercept in a table, find the *x*-value when the *y*-value is 0. To find the *y*-intercept in a table, find the *y*-value when the *x*-value is 0.

31. Find the related function. Subtract 16 from each side: $0 = x + 4 + (2^4 − 6) − 16$. Evaluate the exponent: $0 = x + 4 + (16 − 6) − 16$. Evaluate the expression in parentheses: $0 = x + 4 + 10 − 16$. Add and subtract: $0 = x − 2$. Replace 0 for *f*(*x*). The related function is *f*(*x*) = *x* − 2. The graph of the related function intersects the *x*-axis at 2. This is the *x*-intercept, or zero. So the solution of the equation is 2. Check the solution by solving the equation algebraically. Evaluate the exponent: $16 = x + 4 + (16 − 6)$. Evaluate the expression in parentheses: $16 = x + 4 + 10$. Add: $16 = x + 14$. Subtract 14 from each side: $2 = x$.

Lesson 3-5

1. This function is symmetric in the line *x* = −1.
3. This function is symmetric in the line *x* = 2.5.

5. The graph is symmetric in the line *x* = 5. In the context of the situation, the symmetry of the graph tells you that the area is the same when the width is a number less than or greater than 5.   **7.** always decreasing
9. decreasing: *x* < 1.5; increasing: *x* > 1.5
11. extrema: *B* and *D*; rel min: *D*; rel max: *B*
13. extrema: *B* and *D*; rel min: *B*; rel max: *D*
15. Point *A* is a relative maximum. Point *A* represents the greatest height of the golf ball given the distance from the tee.   **17.** As *x* decreases, *y* increases. As *x* increases, *y* increases.   **19.** As *x* decreases, *y* decreases. As *x* increases, *y* increases.   **21.** no line symmetry; always decreasing; extrema: none; As *x* decreases, *y* increases. As *x* increases, *y* decreases.   **23.** The approximate point (2.5, 114) is a relative maximum. This represents the greatest height of the rock given the time.
25. The graph has one relative minimum at about (−2.25, −16); This statement is not true because there are two relative minimums: one at about (−2.25, −16) and one at about (2.25, −16).

Lesson 3-6

1.

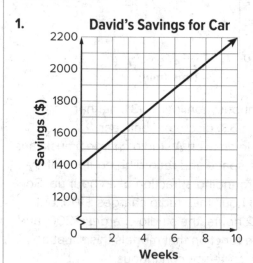

David's Savings for Car

3.

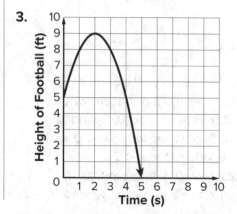

5.

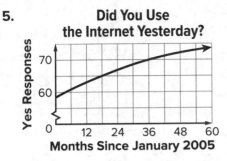

**Did You Use
the Internet Yesterday?**

7. Sample answer: Internet use at home initially has a higher number of users than Internet use away from home. Both Internet use at home and Internet use away from home increase after 36 months since March 2004. Neither Internet use at home nor Internet use away from home reaches 0 users.

9.

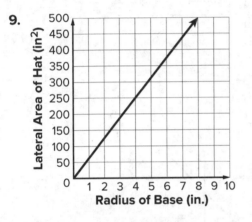

11. Sample answer: The graph on the calculator and the graph I sketched are both linear, increasing, and have an x- and y-intercept at 0.
13. $P(x) = 28x - 840$; $P(x)$ is Aidan's profit from fixing and selling x bicycles. **15.** The x-intercept; To find the x-intercept, locate the point on the graph when $P(x) = 0$, which is 30. So, when 30 bicycles are bought and sold, Aidan makes a profit of $0.

Module 3 Review

1. B, C, F
3. C
5. Sample answer: The element −4 in the domain is paired with both 8 and 13 in the range. This relation is not a function. **7.** B
9. (−1, 0) and (0, −1) **11.** 40; 60 **13.** Sample answer: In 1900 the population of Ohio was nearly 4 million more than the population of Florida. Both populations grew between 1900 and 1950. At this point, the population of Ohio exceeded that of Florida by approximately 5 million, indicating a greater growth rate for Ohio than Florida during those decades. Then from 1950 to 2000, the population of Ohio grew by about 3.4 million, whereas the population of Florida grew by about 13 million, indicating a significantly greater growth rate for Florida during those decades. In fact, by 2000, the population of Florida surpassed Ohio by more than 4 million people.

Module 4

Quick Check

1, 3, 5.

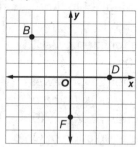

7. $y = -3x + 1$ **9.** $y = \frac{5}{2}x - 6$ **11.** $y = -10x + 6$

Lesson 4-1

1.

x	y
−2	0
−2	1
−2	2

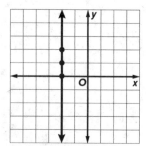

3.

x	y
−1	8
0	0
1	−8

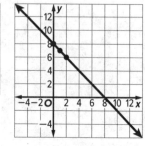

5.

x	y
0	8
1	7
2	6

7.

x	y
0	1
2	2
4	3

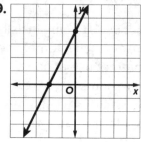

9.

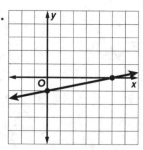

11.

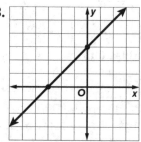

13.

15a. The *x*-intercept is 6. This means that after 6 weeks, Amanda will have $0 in her school lunch account. The *y*-intercept is 210. This means that there was initially $210 in Amanda's school lunch account.

15b.

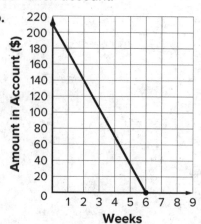

17.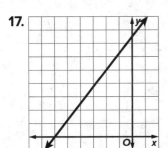

19.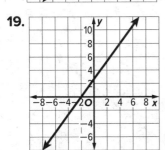

21. x-int: 7; y-int: -2

23. x-int: $1\frac{1}{3}$; y-int: 4

25. x-int: $-1\frac{1}{2}$; y-int: 1

27. $y = 1.7x + 40$; The y-intercept is 40. This means that it would cost $40 to hook up the car.

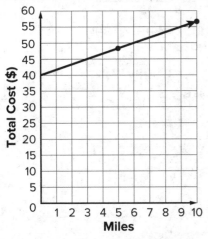

29. Sample answer: The x-intercept is -4. The x-intercept is not reasonable because the football team cannot lose -4 games. The y-intercept is 4. The y-intercept is reasonable because the y-intercept means that if the football team won 4 games, they lost 0 games. **31.** No; sample answer: A horizontal line only has a y-intercept and a vertical line only has an x-intercept. **33.** In the equation, let $y = 0$ to find the x-intercept: $2x + (0) = 4$. So the x-intercept is 2. In the equation, let $x = 0$ to find the y-intercept: $2(0) + y = 4$. So the y-intercept is 4. Robert graphed points at $(2, 0)$ and $(0, 4)$ and connected the points with a line. **35.** Sample answer: $y = 8$; horizontal line

37. Sample answer: $x - y = 0$; line through $(0, 0)$

Lesson 4-2

1. $\frac{1}{5}$ **3.** increased about 1.9 people per square mile **5a.** -5; This means the temperature decreased 5°F per hour from 6 A.M. to 7 A.M. **5b.** -5; This means the temperature decreased 5°F per hour from 1 P.M. to 2 P.M. **7.** linear; $-\frac{1}{1}$ or -1 **9.** not linear **11.** $-\frac{3}{5}$ **13.** 1 **15.** 0 **17.** $\frac{1}{6}$ **19.** $\frac{4}{3}$ **21.** undefined **23.** 1 **25.** undefined **27.** 2 **29.** undefined **31.** -1 **33.** undefined **35.** $-\frac{7}{2}$ **37.** $\frac{5}{2}$ **39.** $\frac{3}{4}$ **41.** 6 **43.** 8 **45.** 11 **47.** $\frac{1}{20}$ **49.** $-\frac{1}{2}$ **51.** $\frac{1}{3}$ **53.** $\frac{1}{2}$ **55.** -1 **57.** $\frac{7}{4}$ **59.** 3 **61.** After drawing a graph, use the two points on the graph to determine the slope. This can be done by counting squares for the rise and run of the line or by using the coordinates of the points in the slope formula. **63.** The rate of change is $2\frac{1}{4}$ inches of growth per week. **65.** Step 1; she reversed the order of the x-coordinates in the formula. **67.** The difference in the x- values is always 0, and division by 0 is undefined.

Lesson 4-3

1. $y = 5x - 3$ **3.** $y = -6x - 2$ **5.** $y = 3x + 2$ **7.** $y = x - 12$ **9.** $y = 5x + 6$ **11.** $y = \frac{1}{3}x - 2$ **13.** $y = -0.25x - 3$ **15.** $y = 25x + 100$ **17.** $y = 0.12x + 9$

19.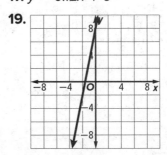

21.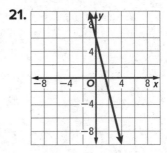

23.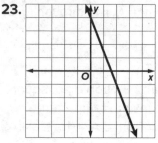

25.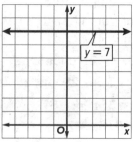

$y = 7$

27.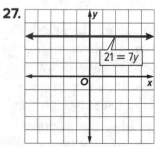

$21 = 7y$

29a. $c = 13 + 8p$

29b.

Streaming Television Plan

Total Cost ($)

Premium Channels

29c. $37   **31.** $y = \frac{1}{2}x - 3$

33.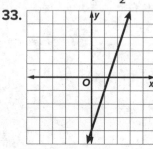

35.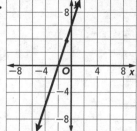

37. $y = 2x - 3$ **39.** $y = -x - 1$
41a. $T = 10x + 80$

41b.

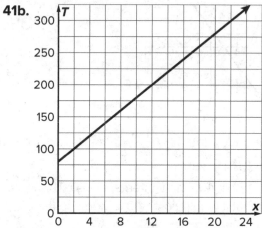

41c. 300°F   **41d.** $T = -5x + 300$
43a. $y = -4.25x + 25.5$

43b.

Charity Walk

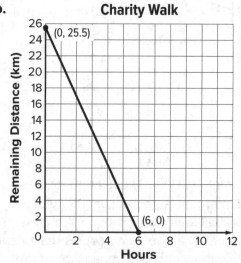

Remaining Distance (km)

Hours

(0, 25.5)

(6, 0)

43c. The x-intercept (6) represents the number of hours it will take Jazmin to complete the walk. The y-intercept (25.5) represents the length of the walk.   **43d.** Using the graph, I can determine the value of x when y equals $-17 + 25.5$ or 8.5 km, and use the value of the x-intercept. The value of x is 4 when $y = 8.5$ and the x-intercept is 6. Therefore, Jazmin has $6 - 4$ or 2 hours more to walk.   **45.** Yes; you can find the value of x on the graph when $y = 0$; $x = \frac{1}{2}$.

47. Sample answer: $y = 25x + 200$; I have $200 in savings and will save $25 per week until I have enough money to buy a new phone. I can predict how much money I'll have after x number of weeks.

Lesson 4-4

1. $g(x)$ is a translation of the parent function 11 units up   **3.** $g(x)$ is a translation of the parent function 7 units right   **5.** $g(x)$ is a translation of the parent function 10 units left and 1 unit down   **7.** $g(x) = 4x$; $g(x)$ is the translation of $f(x)$ 3.5 units down.   **9.** $g(h) = 8h + 15$; $g(h)$ is the translation $f(h)$ of 5 units up.   **11.** $g(x)$ is a vertical compression of the parent function by a factor of $\frac{1}{3}$ **13.** $g(x)$ is a horizontal compression of the parent function by a factor of $\frac{1}{3}$ **15.** $g(x)$ is a horizontal stretch of the parent function by a factor of 2.5   **17.** $g(x)$ is a vertical stretch of the parent function by a factor of 8 and a reflection across the x-axis   **19.** $g(x)$ is a horizontal stretch of the parent function by a factor of $\frac{5}{4}$ and a reflection across the y-axis   **21.** $g(x)$ is a horizontal compression of the parent function by a factor of $\frac{2}{3}$ and a reflection across the y-axis   **23.** $g(x)$ is a translation of the parent function 2 units right and 8 units down   **25.** $g(x)$ is a vertical compression of the parent function by a factor of $\frac{1}{5}$ **27.** $g(x)$ is a horizontal compression of the parent function by a factor of 0.4   **29.** $g(x) = x - 7$ **31.** $g(x) = 1.5x$; The graph of $g(x) = 1.5x$ is the graph of $f(x) = 0.50x$ stretched vertically by a factor of 3.   **33a.** $g(x) = 1.29x$ **33b.** The graph of $g(x) = 1.29x$ is the graph of $f(x) = x$ stretched vertically by a factor of 1.29.   **35.** $y = \frac{1}{a}x$; The function is horizontally stretched by a factor of a.

Lesson 4-5

1. This sequence has a common difference of 4 between its terms. This is an arithmetic sequence.   **3.** This sequence does not have a common difference between its terms. This is not an arithmetic sequence.   **5.** This sequence does not have a common difference between its terms. This is not an arithmetic sequence.

7. This sequence has a common difference of 3 between its terms. This is an arithmetic sequence.   **9.** 1.06; 4.26, 5.32, 6.38   **11.** −2; 13, 11, 9   **13.** $\frac{1}{3}$; $3\frac{2}{3}$, 4, $4\frac{1}{3}$ **15.** 4; 19, 23, 27   **17.** 2; −5, −3, −1   **19.** $a_n = -5n + 2$; −33   **21.** $a_n = -4n - 7$; −35   **23a.** $f(n) = -4n + 128$

23b.

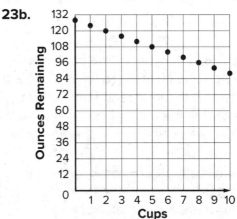

23c. 72 ounces   **25a.** $f(n) = 0.17n + 0.71$

25b.

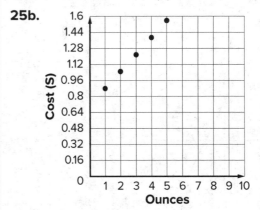

25c. 8 ounces   **27a.** $f(n) = 2n + 28$

27b.

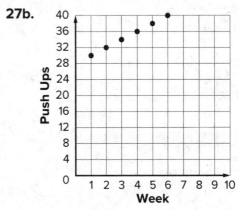

27c. 11th week   **29.** This sequence does not have a common difference between its terms. This is not an arithmetic sequence.   **31.** This sequence has a common difference of 2 between its terms. This is an arithmetic sequence.

33. $a_n = -4n + 34$

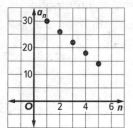

35a. $a_n = 3000 + 500n$ **35b.** $15,000
37a. $a_n = 3n - 1$ **37b.** 59 **39.** Sample answer: 5, 3, 8, 6, 11, 9, 14, ...; The pattern is to subtract 2 from the first term to find the second term, then add 5 to the second term to find the third term.**41.** Sample answer: 2, −8, −18, −28, ... **43a.** Sample answer: $a_n = -2 - 3n$ **43b.** $a_n = -19 + 7n$ **43c.** $a_n = 12 - 2n$ **45.** On day 9, Andre has read 270 pages while Sam has 270 pages left to read. The table shows that both functions have a value of 270 when $x = 9$.

Lesson 4-6

1.

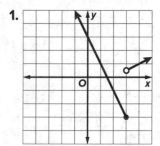

D = all real numbers,
R = $f(x) \geq -3$

3.

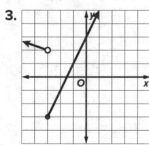

D = all real numbers,
R = $f(x) \geq -3$

5.

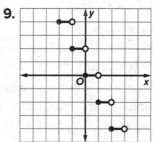

D = all real numbers,
R = $f(x) \geq -2.5$

7.

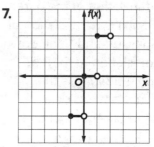

D = all real numbers,
R = all integer multiples of 3

9.

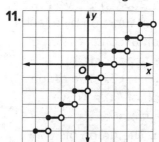

D = all real numbers,
R = all even integers

11.

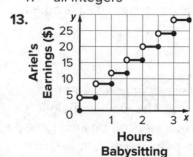

D = all real number,
R = all integers

13.

Ariel's Earnings ($)

Hours Babysitting

15.
$$f(x) = \begin{cases} 16.20 & \text{if } 0 < x \le 1 \\ 19.30 & \text{if } 1 < x \le 2 \\ 22.40 & \text{if } 2 < x \le 3 \\ 25.50 & \text{if } 3 < x \le 4 \\ 28.60 & \text{if } 4 < x \le 5 \end{cases}$$

$D = \{x \mid 0 < x \le 5\};$
$R = \{16.20, 19.30, 22.40, 25.50, 28.60\}$

17. $g(x) = \begin{cases} 2x + 1 & \text{if } x \le 2 \\ x - 2 & \text{if } x > 2 \end{cases}$

19. \$133.00

21a.

x	0	2	4	6	8
f(x)	0	75	175	275	375

21b. $f(x) = 25 + 50[[x]]$

21c.

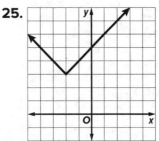

21d. $2 < x \le 3$

23. Sample answer:
$$y = \begin{cases} -x & \text{if } x < -4 \\ 2x & \text{if } -4 \le x \le 2 \\ x - 2 & \text{if } x > 2 \end{cases}$$

25. A step function has different constants over different intervals of its domain. A piecewise-defined function can have different algebraic rules over different intervals of its domain.

27. $f(x) = \begin{cases} \frac{1}{2}x - 3 & \text{if } x > 6 \\ -\frac{1}{2}x + 3 & \text{if } x \le 6 \end{cases}$

29. $R = f(x) \ge 0$

31. 2.4

Lesson 4-7

1. The graph of $g(x)$ is the parent function translated 5 units down.   **3.** The graph of $g(x)$ is the parent function translated 2 units right and 7 units up.   **5.** The graph of $g(x)$ is the parent function translated 1 unit up.

7. $f(x) = |x + 2|$ **9.** $f(x) = |x| - 3$
11. $f(x) = |x| + 1$ **13.** The graph of $g(x)$ is a horizontal compression of the parent function.
15. The graph of $g(x)$ is a vertical stretch of the parent function.   **17.** The graph of $g(x)$ is a horizontal stretch of a parent function.
19. The graph of $g(x)$ is a reflection of the parent function across the x-axis and a vertical stretch.   **21.** The graph of $g(x)$ is a reflection of the parent function across the y-axis and a horizontal stretch.   **23.** The graph of $g(x)$ is a reflection of the parent function across the y-axis and a horizontal compression.

25.

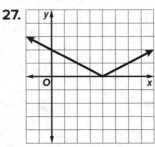

D = all real numbers,
$R = g(x) \ge 3$

27.

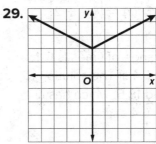

D = all real numbers,
$R = f(x) \ge 0$

29.

D = all real numbers,
$R = f(x) \ge 2$

31.

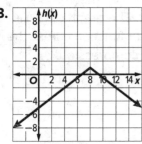

D = all real numbers,
R = $f(x) \leq -3$

33.

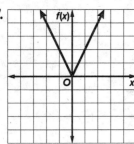

D = all real numbers,
R = $h(x) \leq 1$

35. $y = 65|10 - x|$

37.

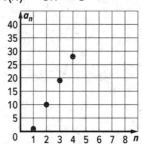

D = all real numbers,
R = $f(x) \geq 0$
The graph of $f(x)$ is the parent function horizontally compressed by a factor of $\frac{1}{2}$.

39. $f(x) = |-3x - 5|$ **41.** $f(x) = \left|\frac{1}{3}x + 2\right|$

43. $x = |s - 16|$

45. $x = |t - 21.7|$; The range of times is twice the value of x, $3.2(2) = 6.4$ s; The solution to the equation is 24.9 and 18.5, which has a range of $24.9 - 18.5 = 6.4$ s.

47. $x = |b - 12|$

49. To get the graph of $h(x)$, the parent absolute value function is reflected in the x-axis, then translated 2 units left and 3 units down.

51. $f(x) = \begin{cases} -x + 5 & \text{if } x < 3 \\ x - 1 & \text{if } x \geq 3 \end{cases}$

Module 4 Review

1.

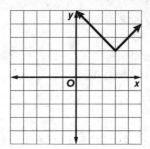

3. −500 gallons/hr   **5.** B   **7.** A

9. dilation   **11.** C

13. $f(n) = 9n - 8$

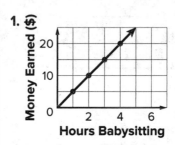

15.

Hours Worked, x	Money Earned, $f(x)$
30	270
35	315
40	360
45	427.5
50	495

17. B

19. Sample answer: It is translated 5 units up.

21. $f(x) = -|x - 4| + 3$

Module 5

Quick Check

1. $y = 5 - x$ **3.** $y = x + 5$ **5.** (4, 2)   **7.** (2, −4)
9. (−3, −3)

Lesson 5-1

1. $y = \frac{1}{2}x$ **3.** $-\frac{3}{4}x + \frac{17}{2}$ **5.** $y = \frac{1}{2}x + 1$
7. $d = 3t + 12$ **9.** $C = 2.54y + 62.38$
11. $y = -4$ **13.** $y = \frac{4}{3}x - \frac{1}{3}$ **15.** $y = -\frac{3}{2}x - \frac{9}{2}$
17. $y = -\frac{4}{11}x + \frac{58}{11}$ **19.** $y = -\frac{1}{2}x - \frac{9}{2}$
21. $y = \frac{1}{6}x + \frac{19}{24}$ **23.** $C = 10d + 12$
25. $T = -4.5x + 103$ **27.** $y = 3x - 1$
29. $y = -x - 4$ **31.** $y = -x + 3$ **33.** No;
substituting 3 and −1 for x and y results in an
equation that is not true.   **35.** Yes; substituting
15 and −13 for x and y results in an equation
that is true.   **37.** Sample answer: (3, −3)
39. Sample answer: (0, −5)   **41.** Sample
answer: (0, 4)   **43.** C; x represents the number
of plane tickets per order and y represents the
total cost of an order.   **45.** A; x represents the
number of hours and y represents the oil level
in the tank, in inches.   **47a.** $y = x + 2.5$
47b. 10   **47c.** $y = x + 1.5$ **49a.** $y = 7.5x + 1$
49b. 1; Koby's puppy weighed 1 pound at birth
(0 months)   **49c.** 7.5; Koby's puppy gained 7.5
pounds a month for the first 6 months.
51. Jacinta; Tess switched the x- and
y-coordinates on the point that she entered in
Step 3.   **53.** Sample answer: Let y represent
the number of quarts of water in a pitcher, and
let x represent the time in seconds that water
is pouring from the pitcher. As time increase
by 1 second, the amount of water in the pitcher
decrease by $\frac{1}{2}$ qt. An equation is $y = -\frac{1}{2}x + 4$.
The slope is the rate at which the water is
leaving the pitcher, $\frac{1}{2}$ quart per second. The
y-intercept represents the amount of water in
the pitcher when it is full, 4 qt.

Lesson 5-2

1. $y + 3 = -1(x + 6)$

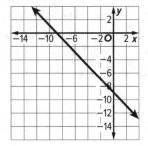

3. $y - 11 = \frac{4}{3}(x + 2)$

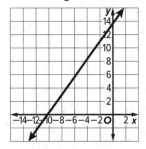

5. Sample answer: $y + 3 = -4(x - 1)$
7. Sample answer: $y - 3 = \frac{4}{3}(x - 3)$
9. $y = -6x - 47$ **11.** $y = \frac{1}{6}x - \frac{8}{3}$
13. $y - 18 = 3.5(x - 5)$ **15.** $2x - y = 6$
17. $x + 6y = -7$ **19.** $x - y = -1$
21. $2x + 3y = -13$
23. $3x + y = -3$
25. Sample answer: $y = x - 5; y = -x + 1$
27. Sample answer: $y = -5x + 2; y = \frac{1}{5}x + 2$
29. Sample answer:
$$y = -\frac{3}{4}x + \frac{3}{2}; y = \frac{4}{3}x + \frac{17}{3}$$
31. neither
33. perpendicular   **35.** neither
37. $5x + 4y = 20$
39. $y = 9x + 5; 9x - y = -5$
41. $y = -6x - 45; 6x + y = -45$
43. $y = \frac{9}{10}x - 4\frac{3}{10}; 9x - 10y = 43$
45. Yes; sample answer: The line that
represents one of the ceiling walls has a slope
of $-\frac{1}{4}$ and the line that represents the other
ceiling wall has a slope of 4.

47a. Sample answer: $y - 0 = 0.5(x - 0)$
47b. $y = 0.5x$ **47c.** $x - 2y = 0$
49. Sample answer: You need to know the
slope of the line and the y-intercept of the line,
the slope and the coordinates of another point
on the line, or the coordinates of two points on
the line.

51. No; the line through (7, −10) and (3, −2) has a slope of −2 and $2x − y = −5$ has a slope of 2.

53. Sample answer: $y − g = \dfrac{j − g}{h − f}(x − f)$

55. Sample answer: Jocari spent $18 to go to a carnival and play games. The price she paid included admission. The games cost $2 each; $y − 18 = 2(x − 5)$, $y = 2x + 8$.

Lesson 5-3

1. Positive; as time spent exercising increases, the more Calories are burned.   **3.** Negative; as weight increases, the number of repetitions decreases.   **5a.** $y = −328.275x + 3142.15$

5b. about 187.675 million

7a. There is a positive correlation between the child's age and annual cost.

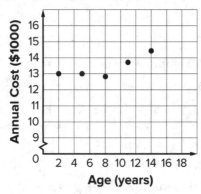

7b. $y = 270x + 10{,}640$ **7c.** about $15,230
9. no correlation
11a.

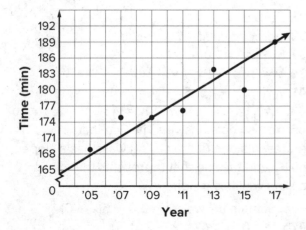

11b. Sample answer: x represents the number of years since 2005, so year 2005 is represented by $x = 0$ and year 2020 is

represented by $x = 15$. Two points on the line of fit are (4, 175) and (12, 189). Use these two points to find the slope to be 1.75 and the equation of the line of fit to be
$y = 1.75x + 168$.   **11c.** Sample answer: about 196 minutes   **11d.** Sample answer: Not all of the data points are close to the line of fit, so there is not a consistent trend regarding the length of games. Therefore, the predicted game length may or may not be accurate.
13a. positive correlation; As the number of years since 2007 increases, the price of a ticket increases.   **13b.** $y = 2.87x + 64.24$
13c. about $133.12

15a.

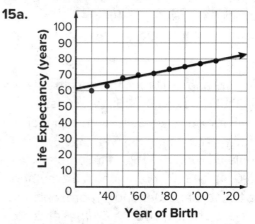

15b. Sample answer: About 81.8; The data show a positive correlation, so as the years increase, the life expectancy also increases. Therefore, the life expectancy should be higher than that of a baby born in 2010.   **15c.** Sample answer: I assumed that the trend continues, so as the year increases, the life expectancy also increases.
17. Sample answer: The salary of an individual and the years of experience that he or she has could be modeled using a scatter plot. This would be a positive correlation because the more experience an individual has, the higher the salary would likely be.
19. Neither; line g has the same number of points above the line and below the line. Line f is close to 2 of the points; but for the rest of the data, there are 3 points above and 3 points below the line.
21. Sample answer: You can visualize a line to determine whether the data has a positive or negative correlation. The graph shows

the ages and heights of people. To predict a person's age given his or her height, write a linear equation for the line of fit. Then substitute the person's height and solve for the corresponding age. You can use the pattern in the scatter plot to make decisions.

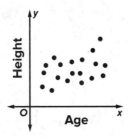

Lesson 5-4

1a.

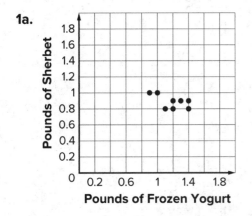

1b. Negative; as the number of pounds of frozen yogurt consumed increases, the number of pounds of sherbet consumed decreases.
1c. The relationship may be a causation. Since both are frozen desserts, eating more frozen yogurt may cause people to decrease the amount of sherbet they eat. Other things that might influence the data are an increase in frozen yogurt stores and a decrease in popularity or availability of sherbet.
3. Correlation, sample answer: Having a wider palm does not cause someone to watch less television. **5.** Causation; sample answer: An increase in the price of cereal likely causes customers to buy less cereal.

7a.

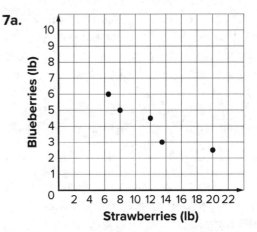

7b. Negative; as the number of pounds of strawberries produced increases, the number of pounds of blueberries produced decreases.
7c. The relationship is a correlation, but not a causation. A better yield of strawberries does not cause the blueberries to grow poorly. Other factors, such as temperature and rain, could be affecting the plants that week.

9. positive correlation and causation; Sample answer: Because pizzas are topped with cheese, an increase in the number of pizzas made cause more cheese to be used.
11. Sample answer: Two elements can have a strong correlation, but it does not mean that one causes the other. There could be an unknown factor affecting the elements.
13. Sample answer: Correlation does not mean causation. Even though there is a strong correlation that does not mean buying swimsuits causes the use of air conditioners. Another factor, like the temperature, could be affecting both swimsuit sales and use of air conditioners.

Lesson 5-5

1a. $y = -1.31x + 50.95$ **1b.** $r \approx -0.714$; The equation models the data fairly well. Its negative value means that as the years since 2010 increase, the total number of goals the soccer team scores each season decreases.
3a. $y = 8.52x + 3.18$
3b. $r \approx 0.999$; The equation models the data very well. Its positive value means that as the years since 2010 increase, sales, in millions of dollars, increase.

5a. $y = 103.77x + 108.06$ **5b.** about $3221.16
7a. $y = 0.59x + 1.51$ **7b.** The residuals are randomly scattered and are centered about the line $y = 0$. So, the best-fit line models that data well.   **9a.** $y = 0.26x + 21.21$
9b. $r \approx 0.359$; The equation does not model the data well. Its value means that as the years since the 2011−2012 school year increase, the percentage of students in public school who met all six of California's physical fitness standards each year varies.   **9c.** Because the data on the students who meet all six standards is reported as a percentage, it cannot exceed 100.   **11a.** $y = 140.4x + 13.8$
11b. $r \approx 0.999$; The equation models the data very well. Its positive value means that as the number of games increases, the cumulative number of yards increases.
11c. Sample answer: Because the data have a positive correlation, the total number of yards will increase as the number of games increases. So, the running back will have run for 950 yards between games 6 and 9.
11d. during game 7
13a. $y = 9619x + 443,918.8$ **13b.** $r \approx 0.999$; The equation models the data very well. Its positive value means that as the number of years since the 2010-2011 school year increases, the number of student athletes participating in college athletics each year increases.
13c. The residuals are randomly scattered and are centered about the line $y = 0$. So, the best-fit line models that data well.
13d. about 684,394
15. Apply a linear regression model to the data. Use the number of each test as the independent variable. If there is no correlation, the r-value will not be close enough to 1 or −1. If this is the case, the line of fit could not be used to predict the scores of the other students.
17a. $y = 84,345.0x + 5,003,868.3$
17b. about 7,365,528

Lesson 5-6

1. $\{(-1, -9), (-4, -7), (-7, -5), (-10, -3), (-13, -1)\}$

3. $\{(-2, -4), (-1, -2), (1, 0), (0, 2), (2, 4)\}$
5. $\{(-3, 5), (-9, 2), (-15, -1), (-21, -4)\}$
7. $\{(16, -1), (12, -2), (8, -3), (4, -4)\}$
9. $\{(-49, -4), (35, 8), (-28, -1), (7, 4)\}$

11.
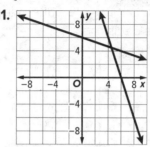

13.
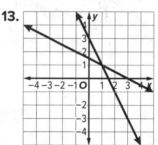

15.
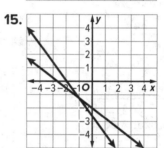

17. $f^{-1}(x) = \frac{x}{6} - 7$
19. $f^{-1}(x) = \frac{5}{2}(x + 16)$ **21.** $f^{-1}(x) = \frac{1 - 5x}{4}$
23a. $P^{-1}(x) = \frac{x + 36}{7.6}$ **23b.** x represents Alisha's profit and represents the number of dozens of brownies sold.   **23c.** 5
25a. $C^{-1}(x) = \frac{x - 125}{16}$ **25b.** 108 feet
27. $f^{-1}(x) = \frac{1}{4}x + 6$ **29.** $f^{-1}(x) = 6x - 42$
31. $f^{-1}(x) = \frac{7}{2}x - 14$ **33.** $f^{-1}(x) = \frac{1}{7}x - \frac{6}{7}$
35. $f^{-1}(x) = 2x - 22$ **37.** B   **39.** A
41. $\{(-k, b), (p, -g), (-m, -w), (q, r)\}$ **43.** The slopes are reciprocals. For example, if the slope of one line is $\frac{2}{3}$, then the slope of the inverse function is $\frac{3}{2}$.   **45.** Sample answer: This claim is incorrect. The −1 in the inverse function notation is not an exponent. As an example, the inverse function for $y = x + 1$ is found by switching x and y and solving for y, which gives $y = x - 1$. $y = x - 1$ is not the same as $y = \frac{1}{(x + 1)}$, which is not a line. This method does not work.

47. $a = 2$; $b = 14$

49. sometimes; Sample answer: $f(x)$ and $g(x)$ do not need to be inverse functions for $f(a) = b$ and $g(b) = a$. For example, if $f(x) = 2x + 10$, then $f(2) = 14$ and if $g(x) = x - 12$, then $g(14) = 2$, but $f(x)$ and $g(x)$ are not inverse functions. However, if $f(x)$ and $g(x)$ are inverse functions, then $f(a) = b$ and $g(b) = a$.

51. Sample answer: A situation may require substituting values for the dependent variable into a function. By finding the inverse of the function, the dependent variable becomes the independent variable. This makes the substitution an easier process.

Module 5 Review

1. A **3.** $y = 1.5x + 11$ **5.** A
7. $y - 4 = 2.5(x - 2)$ **9.** A
11. a positive correlation **13.** C
15. $f^{-1}(x) = -2x + 1$;

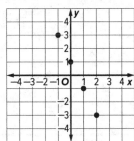

17. A

Module 6

Quick Check

1. 4 **3.** −2 **5.** $\{-29, 7\}$ **7.** $\{-1, 15\}$

Lesson 6-1

1.

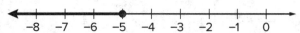

3.

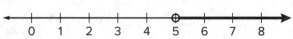

5.

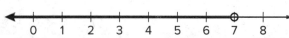

7. $t < -1$ **9.** $w < 5$ **11.** $b \geq -5$
13. $\{m \mid m < 7\}$ **15.** $\{r \mid r \leq 15\}$ **17.** $\{b \mid b \geq 2\}$
19. $\{c \mid c \leq -4\}$ **21.** $\{m \mid m \geq 4\}$
23. $\{r \mid r \geq 22\}$ **25.** $\{a \mid a \leq -4\}$
27. $\{w \mid w \geq -5\}$ **29.** $\{x \mid x \leq 5\}$
31. $\frac{3}{10}x \leq 4.50,\ x \leq \$15$ **33.** no more than 2.1 pounds per day **35.** at least 500 pieces

37. $\{m \mid m \leq -68\}$

39. $\{c \mid c > 121\}$

41. $\{x \mid x \leq 20\}$

43. $\{h \mid h > 21\}$

45. $\{n \mid n \geq 108\}$

47. $\{r \mid r < 16\}$

49. $\{t \mid t > -1\}$

51. $\{z \mid z \geq 11\}$

53. $\{d \mid d > -2\frac{1}{2}\}$

55. d **57.** a **59.** b
61. Sample answer: Let n = the number.
$n + 7 \leq -18;\ \{n \mid n \leq -25\}$
63. Sample answer: Let n = the number.
$n + 2 \leq 1;\ \{n \mid n \leq -1\}$
65. Sample answer: Let n = the number.
$-12n \leq 84;\ \{n \mid n \geq -7\}$

67. $\{g \mid g > 4\}$

69. $\{x \mid x < 36\}$

71. $\{m \mid m < 5.4\}$

73. $\{c \mid c \geq 3.7\}$

75. $\$22.23$.

77. Sample answer: Let x represent the decibel level of the calls of a blue whale; $x - 83 \leq 105;\ x \leq 188$. The calls of a blue whale are less than or equal to 188 decibels.
79. $-\frac{x}{2} < 1$ **81a.** $x < \frac{7}{a}$ **81b.** $x \geq \frac{12}{a}$
81c. $x > 3$ **81d.** $x \geq \frac{1}{4}$

Lesson 6-2

1a. $15 + 2h \leq 35$ **1b.** $h \leq 10$; 10 hours
3a. $1.50 + 0.25(5x - 1) \leq 3.75$ **3b.** $x \leq 2$; 2 mi
3c. Because the service charges per $\frac{1}{a}$ mile, multiply a by the number of miles x to find the number of $\frac{1}{a}$ miles.

Copyright © McGraw-Hill Education

Subtract 1 from the total number of $\frac{1}{a}$ miles, ax, to find the number of additional $\frac{1}{a}$ miles. Multiply the difference by the cost per additional $\frac{1}{a}$ mile, 0.25, and add the cost for the first $\frac{1}{a}$ mile, 1.50. This sum is less than or equal to 3.75, so $1.50 + 0.25(ax - 1) \leq 3.75$.

5a. $100 + 40x \leq 250$

5b. $x \leq 3.75$; 3 people

7. $21 > 15 + 2x$; $x < 3$

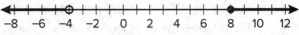

9. $\frac{x}{8} - 13 > -6$; $x > 56$

11. $37 < 7 - 10x$; $x < -3$

13. $-\frac{5}{4}x + 6 < 12$; $x > -\frac{24}{5}$

15. $15x + 30 < 10x - 45$; $x < -15$

17. $\{a \mid a \leq 11\}$

19. $\{b \mid b$ is a real number.$\}$

21. $\{a \mid a \geq -9\}$

23. $\{x \mid x \geq \frac{1}{2}\}$ **25.** $\{m \mid m \geq 18\}$

27. $\{w \mid w > -2\}$ **29.** $\{x \mid x \leq 8\}$

31. $\{x \mid x > -6\}$ **33.** $\{x \mid x \geq 1.5\}$

35. $\{p \mid p \leq 1\frac{1}{9}\}$

37a. $2x + 4 \leq 13$; $x \leq 4.5$

37b. 4.5 ft **37c.** 5 ft

39. Eric does not have any pencils. Based on his statement, the inequality is $6p + 15 < 20$, where p is the number of pencils. The solution of the inequality is $p < \frac{5}{6}$. However, the number of pencils must be a whole number, so $p = 0$.

41. $10n - 7(n - 2) > 5n - 12$
(Original inequality)

$10n - 7n - 14 > 5n - 12$
(Distributive Property)

$3n - 14 > 5n - 12$
(Combine like terms.)

$3n - 14 - 5n > 5n - 12 - 5n$
(Subtract $5n$ from each side.)
$-2n - 14 > -12$
(Simplify.)

$-2n - 14 + 14 > -12 + 14$
(Add 14 to each side.)

$-2n > 2$
(Simplify.)

$\frac{-2n}{-2} < \frac{2}{-2}$
Divide each side by -2. Change $>$ to $<$.

$n < -1$
(Simplify.)

The solution set is $\{n \mid n < -1\}$.

43. $\frac{76 + 80 + 78 + x}{4} \geq 82$; $x \geq 94$; Mei needs a score of at least 94 on the next exam.

45. Sample answer: $2(2x - 1) < 10$

47. Let $c =$ the number of baseball cards Ted has; $4c > 5c - 15$; $15 > c$; Ted has fewer than 15 cards. **49.** $\emptyset$; If the inequality is always true, the opposite inequality will always be false.

51. Sample answer: The solution set for the inequality that results in a false statement is the empty set, as in $12 \geq 15$. The solution set for an inequality in which any value of x results in a true statement is all real numbers, as in $12 \leq 12$.

Lesson 6-3

1. $\{f \mid 6 < f < 11\}$

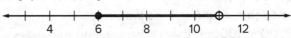

3. $\{y \mid y \geq 8$ or $y < -4\}$

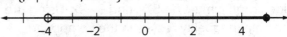

5. $\{p \mid -4 < p \leq 5\}$

7. $\{h \mid 2 \le h < 3\}$

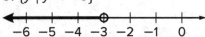

9. $\{y \mid y < -3\}$

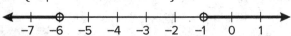

11. $\{b \mid 4 < b \le 5\}$

13. $\{m \mid m < -6 \text{ or } m > -1\}$

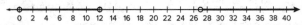

15. $\{m \mid 2 \le m < 4\}$

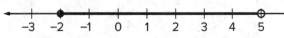

17a. $x + 8 < 20$ or $x + 8 > 35$ **17b.** $0 < x < 12$ or $x > 27$; Because the combined height of the sign and pole cannot be negative, the value of x must be greater than 0.

17c.

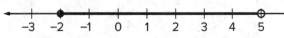

19. $-3 < x \le 3$ **21.** $x < -2$ or $x \ge 1$

23. $b > 3$ or $b \le 0$ **25.** $y < -1$ or $y \ge 1$

27. $f \mid -2 < f < -1\}$

29. $\{b \mid -2 < b < 6\}$

31. $\{a \mid -2 \le a < 5\}$

33. Sample answer: Let n = the number. $n - 2 \le 4$ or $n - 2 \ge 9$; $\{n \mid n \le 6 \text{ or } n \ge 11\}$
35. $54° \le x \le 68°$ **37.** The minimum is 67, since the solution of the inequality $2000 \le 1000 + 15x$ is $66\frac{2}{3} \le x$, and the number of students must be a whole number. The maximum is 100, since the solution of the inequality $1000 + 15x \le 3000$ is $x \le 133\frac{1}{3}$, but the bus can only hold 100 students. **39a.** The side lengths must be 5, x, and $9 - x$. Using the Triangle Inequality results in the compound inequality $x + 5 > 9 - x$ and $14 - x > x$.

39b. The solution of the compound inequality is $2 < x < 7$, so each of the lengths must be greater than 2 m but less than 7 m. The sum of the two lengths must be 9 m.

41. $\{x \mid -2 < x < 5\}$

43a. $\$400 \le x \le \$800$ **43b.** $\$428 \le x \le \$856$
45. B
47. $x > -1$ or $x \ge 4$; This can be written as $x > -1$ because this is the union of two graphs.
49. The union of the two graphs is the graph on the left, so the graph on the left is the graph of the solution set for **Exercise 47**. The intersection of the two graphs is the graph on the right, so the graph on the right is the graph of the solution set for **Exercise 48**.
51a. $x > -\frac{4}{a}$ and $x \le \frac{4}{a}$
51b. $x < -\frac{6}{a}$ or $x > 5a$
53. Sometimes; The graph of $x > 2$ or $x < 5$ includes the entire number line.

Lesson 6-4

1. $\{x \mid -24 < x < 8\}$

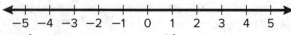

3. $\{c \mid -3 \le c \le 4\}$

5. $\{\varnothing\}$

7. $\{r \mid r < -8 \text{ or } r > 4\}$

9. $\{h \mid h \le -3 \text{ or } h \ge 6\}$

11. $\{v \mid v \text{ is a real number.}\}$

13. $\left\{n \mid n \le -5\frac{1}{4} \text{ or } n \ge 3\frac{3}{4}\right\}$

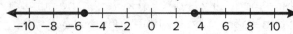

15. $\left\{ h \mid -5\frac{2}{3} < h < 5 \right\}$

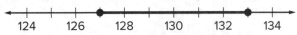

17. $\{\emptyset\}$

19. $\left\{ r \mid -2 < r < \frac{2}{3} \right\}$

21. $\{x \mid 58.5 \le x \le 61.5\}$ **23a.** $|p - 130| \le 3.05$

23b.

25a. $|x - 515| \le 114$ **25b.** 287 to 743
27. $|n + 2| \ge 1$ **29.** $|w - 2| < 2$ **31.** $|x| > 1$
33. d   **35.** c
37. $|x - 92| \le 8$; $\{x \mid 84 \le x \le 100\}$

39. $\{x \mid -1 \le x \le 3\}$

41. $\{x \mid -7 \le x \le 3\}$

43. $\{x \mid x > 18 \text{ or } x < -17\}$

45. By definition, the absolute value is always greater than a negative number. Therefore, no matter what number is chosen, it will always be greater than −1 when evaluated in the absolute value inequality given.   **47a.** Set the absolute value of an unknown variable, x, minus the recommended weight, 516, to be less than or equal to the variance of 4. So, the inequality $|x - 516| \le 4$ represents the situation.
47b. Write two inequalities, one for each case: $x - 516 \le 4$ and $-(x - 516) \le 4$. For the first case, add 516 to both sides: $x \le 520$. For the second case, distribute the negative on the left side, subtract 516 from both sides and divide by a negative 1 remembering to switch the inequality sign: $x \ge 512$. This means a box of cereal should have a minimum weight of 512 g and a maximum weight of 520 g.

49. The solution set for $|x - 2| > 4$ is $\{x \mid x < -2 \text{ or } x > 6\}$. The solution set for $-2x < 4$ or $x > 6$ is $\{x \mid x > -2\}$. One includes numbers greater than −2, and the other includes numbers less than −2 or greater than 6. These solution sets are not the same.
51. Jordan is correct. Chloe did not distribute the negative to both x and 3.
53. $(-8 \le n < -3)$ or $(1 < n \le 6)$. To solve this compound inequality, split it into two inequalities. The first one to solve is $|n + 1| > 2$ and the second one is $|n + 1| \le 7$. The solution set of the entire problem is the overlap of the individual solutions.

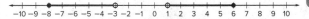

55. No; Sample answer: Lucita forgot to change the direction of the inequality sign for the negative case of the absolute value.
57. Sample answer: If $t = 0$, then the absolute value is equal to 0, not greater than 0.
59. Sample answer: When an absolute value is on the left and the inequality symbol < or ≤, the compound sentence uses *and*, and if the inequality symbol is > or ≥, the compound sentence uses *or*. To solve, if $|x| < n$, then set up and solve the inequalities $x < n$ and $x > -n$, and if $|x| > n$, then set up and solve the inequalities $x > n$ or $x < -n$.

Lesson 6-5

1.

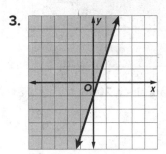

3.

5.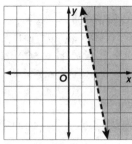

7.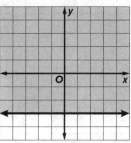

9.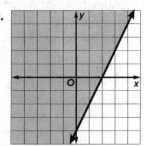

11.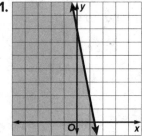

13a. $y < 1240x + 48{,}200$

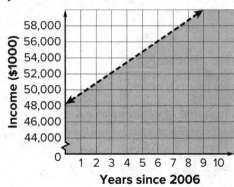

13b. no, no, yes, no

15.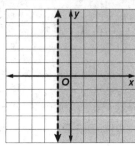

17.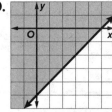

19.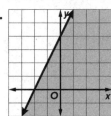

21.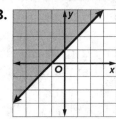

23.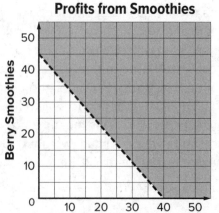

25a. $2.25p + 2b > 90$; $p \geq 0$ and $b \geq 0$

25b.

25c. The café sold more than 22 berry smoothies. **25d.** 41; Sample explanation: 40 peach smoothies results in a profit of exactly $90, so to make a profit of more than $90, the café must have sold 41 smoothies. **27.** The value of c must be positive. Since (0, 0) is a solution of the inequality, $a(0) + b(0) < c$ must be a true statement, so $0 < c$. **29.** Sample answer: $y < -x + 1$ **31.** Sample answer: The inequality $y > 10x + 45$ represents the cost of a monthly smartphone data plan with a one-time fee of $45, plus $10 per GB of data used. Both the domain and range are nonnegative real numbers because the GB used, and the total cost cannot be negative.

Module 6 Review

1. A

3. 8 rows

5. C

7. $\{t \mid t < -3\}$

9. A, B

11. $\{g \mid -5 \leq g\}$

13A. $\{h \mid -8 < h < 2\}$

13B.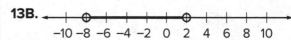

15. B

17.

Inequality	Graphed Solution			
	A	B	C	D
$2y - x \leq -3$			X	
$2y - x \geq -3$				X
$2x - y > -3$		X		
$2x - y < -3$	X			

English	Español

A

absolute value **(Lesson 1-5)** The distance a number is from zero on the number line.

valor absoluto La distancia que un número es de cero en la línea numérica.

absolute value function **(Lesson 4-7)** A function written as $f(x) = |x|$, in which $f(x) \geq 0$ for all values of x.

función del valor absoluto Una función que se escribe $f(x) = |x|$, donde $f(x) \geq 0$, para todos los valores de x.

accuracy **(Lesson 1-6)** The nearness of a measurement to the true value of the measure.

exactitud La proximidad de una medida al valor verdadero de la medida.

additive identity **(Lesson 1-3)** Because the sum of any number a and 0 is equal to a, 0 is the additive identity.

identidad aditiva Debido a que la suma de cualquier número a y 0 es igual a, 0 es la identidad aditiva.

additive inverses **(Lesson 1-3)** Two numbers with a sum of 0.

inverso aditivos Dos números con una suma de 0.

algebraic expression **(Lesson 1-2)** A mathematical expression that contains at least one variable.

expresión algebraica Una expresión matemática que contiene al menos una variable.

arithmetic sequence **(Lesson 4-5)** A pattern in which each term after the first is found by adding a constant, the common difference d, to the previous term.

secuencia aritmética Un patrón en el cual cada término después del primero se encuentra añadiendo una constante, la diferencia común d, al término anterior.

asymptote **(Lesson 9-1)** A line that a graph approaches.

asíntota Una línea que se aproxima a un gráfico.

axis of symmetry **(Lesson 11-1)** A line about which a graph is symmetric.

eje de simetría Una línea sobre la cual un gráfica es simétrico.

B

bar graph **(Lesson 12-2)** A graphical display that compares categories of data using bars of different heights.

gráfico de barra Una pantalla gráfica que compara las categorías de datos usando barras de diferentes alturas.

base **(Lesson 1-1)** In a power, the number being multiplied by itself.

base En un poder, el número se multiplica por sí mismo.

best-fit line **(Lesson 5-5)** The line that most closely approximates the data in a scatter plot.

línea de ajuste óptimo La línea que más se aproxima a los datos en un diagrama de dispersión.

bias **(Lesson 12-3)** An error that results in a misrepresentation of a population.

sesgo Un error que resulta en una tergiversación de una población.

binomial (Lesson 10-1)   The sum of two monomials.

bivariate data (Lesson 5-3)   Data that consists of pairs of values.

boundary (Lesson 6-5)   The edge of the graph of an inequality that separates the coordinate plane into regions.

box plot (Lesson 12-4)   A graphical representation of the five-number summary of a data set.

binomio   La suma de dos monomios.

datos bivariate   Datos que constan de pares de valores.

frontera   El borde de la gráfica de una desigualdad que separa el plano de coordenadas en regiones.

diagram de caja   Una representación gráfica del resumen de cinco números de un conjunto de datos.

C

categorical data (Lesson 12-1)   Data that can be organized into different categories.

causation (Lesson 5-4)   When a change in one variable produces a change in another variable.

closed (Expand 1-3)   If for any members in a set, the result of an operation is also in the set.

closed half-plane (Lesson 6-5)   The solution of a linear inequality that includes the boundary line.

coefficient (Lesson 1-4)   The numerical factor of a term.

coefficient of determination (Lesson 11-8)   An indicator of how well a function fits a set of data.

common difference (Lesson 4-5)   The difference between consecutive terms in an arithmetic sequence.

common ratio (Lesson 9-5)   The ratio of consecutive terms of a geometric sequence.

completing the square (Lesson 11-5)   A process used to make a quadratic expression into a perfect square trinomial.

compound inequality (Lesson 6-3)   Two or more inequalities that are connected by the words *and* or *or*.

compound interest (Lesson 9-3)   Interest calculated on the principal and on the accumulated interest from previous periods.

conditional relative frequency (Lesson 12-7)   The ratio of the joint frequency to the marginal frequency.

datos categóricos   Datos que pueden organizarse en diferentes categorías.

causalidad   Cuando un cambio en una variable produce un cambio en otra variable.

cerrado   Si para cualquier número en el conjunto, el resultado de la operación es también en el conjunto.

semi-plano cerrado   La solución de una desigualdad linear que incluye la línea de limite.

coeficiente   El factor numérico de un término.

coeficiente de determinación   Un indicador de lo bien que una función se ajusta a un conjunto de datos.

diferencia común   La diferencia entre términos consecutivos de una secuencia aritmética.

razón común   El razón de términos consecutivos de una secuencia geométrica.

completar el cuadrado   Un proceso usado para hacer una expresión cuadrática en un trinomio cuadrado perfecto.

desigualdad compuesta   Dos o más desigualdades que están unidas por las palabras *y* u *o*.

interés compuesto   Intereses calculados sobre el principal y sobre el interés acumulado de períodos anteriores.

frecuencia relativa condicional   La relación entre la frecuencia de la articulación y la frecuencia marginal.

consistent (Lesson 7-1) A system of equations with at least one ordered pair that satisfies both equations.

consistente Una sistema de ecuaciones para el cual existe al menos un par ordenado que satisfice ambas ecuaciones.

constant function (Lesson 4-3) A linear function of the form $y = b$.

función constante Una función lineal de la forma $y = b$.

constant term (Lesson 1-2) A term that does not contain a variable.

término constante Un término que no contiene una variable.

constraint (Lesson 2-1) A condition that a solution must satisfy.

restricción Una condición que una solución debe satisfacer.

continuous function (Lesson 3-3) A function that can be graphed with a line or an unbroken curve.

función continua Una función que se puede representar gráficamente con una línea o una curva ininterrumpida.

correlation coefficient (Lesson 5-5) A measure that shows how well data are modeled by a regression function.

coeficiente de correlación Una medida que muestra cómo los datos son modelados por una función de regresión.

cube root (Lesson 8-5) One of three equal factors of a number.

raíz cúbica Uno de los tres factores iguales de un número.

curve fitting (Lesson 11-8) Finding a regression equation for a set of data that is approximated by a function.

ajuste de curvas Encontrar una ecuación de regresión para un conjunto de datos que es aproximado por una función.

D

decreasing (Lesson 3-5) Where the graph of a function goes down when viewed from left to right.

decreciente Donde la gráfica de una función disminuye cuando se ve de izquierda a derecha.

define a variable (Lesson 1-2) To choose a variable to represent an unknown value.

definir una variable Para elegir una variable que represente un valor desconocido.

degree of a monomial (Lesson 10-1) The sum of the exponents of all its variables.

grado de un monomio La suma de los exponents de todas sus variables.

degree of a polynomial (Lesson 10-1) The greatest degree of any term in the polynomial.

grado de un polinomio El grado mayor de cualquier término del polinomio.

dependent (Lesson 7-1) A consistent system of equations with an infinite number of solutions.

dependiente Una sistema consistente de ecuaciones con un número infinito de soluciones.

dependent variable (Lesson 3-1) The variable in a relation, usually y, with values that depend on x.

variable dependiente La variable de una relación, generalmente y, con los valores que depende de x.

descriptive modeling (Lesson 1-6) A way to mathematically describe real-world situations and the factors that cause them.

modelado descriptivo Una forma de describir matemáticamente las situaciones del mundo real y los factores que las causan.

difference of two squares (Lesson 10-7)   The square of one quantity minus the square of another quantity.

diferencia de dos cuadrados   El cuadrado de una cantidad menos el cuadrado de otra cantidad.

dilation (Lesson 4-4)   A transformation that stretches or compresses the graph of a function.

homotecia   Una transformación que estira o comprime el gráfico de una función.

dimensional analysis (Lesson 2-7)   The process of performing operations with units.

análisis dimensional   El proceso de realizar operaciones con unidades.

discrete function (Lesson 3-3)   A function in which the points on the graph are not connected.

función discreta   Una función en la que los puntos del gráfico no están conectados.

discriminant (Lesson 11-6)   In the Quadratic Formula, the expression under the radical sign that provides information about the roots of the quadratic equation.

discriminante   En la Fórmula cuadrática, la expresión bajo el signo radical que proporciona información sobre las raíces de la ecuación cuadrática.

distribution (Lesson 12-5)   A graph or table that shows the theoretical frequency of each possible data value.

distribución   Un gráfico o una table que muestra la frecuencia teórica de cada valor de datos posible.

domain (Lesson 3-1)   The set of the first numbers of the ordered pairs in a relation.

dominio   El conjunto de los primeros números de los pares ordenados en una relación.

dot plot (Lesson 12-2)   A diagram that shows the frequency of data on a number line.

gráfica de puntos   Una diagrama que muestra la frecuencia de los datos en una línea numérica.

double root (Lesson 11-3)   Two roots of a quadratic equation that are the same number.

raíces dobles   Dos raíces de una función cuadrática que son el mismo número.

E

elimination (Lesson 7-3)   A method that involves eliminating a variable by combining the individual equations within a system of equations.

eliminación   Un método que consiste en eliminar una variable combinando las ecuaciones individuales dentro de un sistema de ecuaciones.

end behavior (Lesson 3-5)   The behavior of a graph at the positive and negative extremes in its domain.

comportamiento extremo   El comportamiento de un gráfico en los extremos positivo y negativo en su dominio.

equation (Lesson 2-1)   A mathematical statement that contains two expressions and an equal sign, $=$.

ecuación   Un enunciado matemático que contiene dos expresiones y un signo igual, $=$.

equivalent equations (Lesson 2-2)   Two equations with the same solution.

ecuaciones equivalentes   Dos ecuaciones con la misma solución.

equivalent expressions (Lesson 1-4)   Expressions that represent the same value.

expresiones equivalentes   Expresiones que representan el mismo valor.

evaluate (Lesson 1-1)   To find the value of an expression.

evaluar   Calcular el valor de una expresión.

explicit formula (Lesson 9-6)   A formula that allows you to find any term a_n of a sequence by using a formula written in terms of n.

fórmula explícita   Una fórmula que le permite encontrar cualquier término a_n de una secuencia usando una fórmula escrita en términos de n.

exponent (Lesson 1-1) When n is a positive integer in the expression x^n, n indicates the number of times x is multiplied by itself.

exponente Cuando n es un entero positivo en la expresión x^n, n indica el número de veces que x se multiplica por sí mismo.

exponential decay function (Lesson 9-1) A function in which the independent variable is an exponent, where $a > 0$ and $0 < b < 1$.

función exponenciales de decaimiento Una ecuación en la que la variable independiente es un exponente, donde $a > 0$ y $0 < b < 1$.

exponential equation (Lesson 8-7) An equation in which the variable occur as exponents.

ecuación exponencial Una ecuación en la cual las variables ocurren como exponentes.

exponential function (Lesson 9-1) A function in which the independent variable is an exponent.

función exponencial Una función en la que la variable independiente es el exponente.

exponential growth function (Lesson 9-1) A function in which the independent variable is an exponent, where $a > 0$ and $b > 1$.

función de crecimiento exponencial Una función en la que la variable independiente es el exponente, donde $a > 0$ y $b > 1$.

extrema (Lesson 3-5) Points that are the locations of relatively high or low function values.

extrema Puntos que son las ubicaciones de valores de función relativamente alta o baja.

extreme values (Lesson 12-5) The least and greatest values in a set of data.

valores extremos Los valores mínimo y máximo en un conjunto de datos.

factoring (Lesson 10-5) The process of expressing a polynomial as the product of monomials and polynomials.

factorización El proceso de expresar un polinomio como el producto de monomios y polinomios.

factoring by grouping (Lesson 10-5) Using the Distributive Property to factor some polynomials having four or more terms.

factorización por agrupamiento Utilizando la Propiedad distributiva para factorizar polinomios que possen cuatro o más términos.

family of graphs (Lesson 4-4) Graphs and equations of graphs that have at least one characteristic in common.

familia de gráficas Gráficas y ecuaciones de gráficas que tienen al menos una característica común.

five-number summary (Lesson 12-4) The minimum, quartiles, and maximum of a data set.

resumen de cinco números El mínimo, cuartiles y máximo de un conjunto de datos.

formula (Lesson 2-7) An equation that expresses a relationship between certain quantities.

fórmula Una ecuación que expresa una relación entre ciertas cantidades.

function (Lesson 3-2) A relation in which each element of the domain is paired with exactly one element of the range.

función Una relación en que a cada elemento del dominio de corresponde un único elemento del rango.

function notation (Lesson 3-2) A way of writing an equation so that $y = f(x)$.

notación functional Una forma de escribir una ecuación para que $y = f(x)$.

G

geometric sequence (Lesson 9-5) A pattern of numbers that begins with a nonzero term and each term after is found by multiplying the previous term by a nonzero constant r.

secuencia geométrica Un patrón de números que comienza con un término distinto de cero y cada término después se encuentra multiplicando el término anterior por una constante no nula r.

greatest integer function (Lesson 4-6) A step function in which $f(x)$ is the greatest integer less than or equal to x.

función entera más grande Una función del paso en que $f(x)$ es el número más grande menos que o igual a x.

H

half-plane (Lesson 6-5) A region of the graph of an inequality on one side of a boundary.

semi-plano Una región de la gráfica de una desigualdad en un lado de un límite.

histogram (Lesson 12-2) A graphical display that uses bars to display numerical data that have been organized in equal intervals.

histograma Una exhibición gráfica que utiliza barras para exhibir los datos numéricos que se han organizado en intervalos iguales.

I

identity (Lesson 2-4) An equation that is true for every value of the variable.

identidad Una ecuación que es verdad para cada valor de la variable.

identity function (Lesson 4-4) The function $f(x) = x$.

función de identidad La función $f(x) = x$.

inconsistent (Lesson 7-1) A system of equations with no ordered pair that satisfies both equations.

inconsistente Una sistema de ecuaciones para el cual no existe par ordenado alguno que satisfaga ambas ecuaciones.

increasing (Lesson 3-5) Where the graph of a function goes up when viewed from left to right.

crecciente Donde la gráfica de una función sube cuando se ve de izquierda a derecha.

independent (Lesson 7-1) A consistent system of equations with exactly one solution.

independiente Un sistema consistente de ecuaciones con exactamente una solución.

independent variable (Lesson 3-1) The variable in a relation, usually x, with a value that is subject to choice.

variable independiente La variable de una relación, generalmente x, con el valor que sujeta a elección.

index (Lesson 8-4) In nth roots, the value that indicates to what root the value under the radicand is being taken.

índice En enésimas raíces, el valor que indica a qué raíz está el valor bajo la radicand.

inequality (Lesson 6-1) A mathematical sentence that contains $<$, $>$, $\leq$, $\geq$, or $\neq$.

desigualdad Una oración matematica que contiene uno o más de $<$, $>$, $\leq$, $\geq$, o $\neq$.

interquartile range (Lesson 12-4) The difference between the upper and lower quartiles of a data set.

rango intercuartil La diferencia entre el cuartil superior y el cuartil inferior de un conjunto de datos.

intersection (Lesson 6-3) The graph of a compound inequality containing *and*.

intersección La gráfica de una desigualdad compuesta que contiene la palabra y.

interval **(Expand 4-3)** The distance between two numbers on the scale of a graph.

intervalo La distancia entre dos números en la escala de un gráfico.

inverse functions **(Lesson 5-6)** Two functions, one of which contains points of the form (a, b) while the other contains points of the form (b, a).

funciones inversas Dos funciones, una de las cuales contiene puntos de la forma (a, b) mientras que la otra contiene puntos de la forma (b, a).

inverse relations **(Lesson 5-6)** Two relations, one of which contains points of the form (a, b) while the other contains points of the form (b, a).

relaciones inversas Dos relaciones, una de las cuales contiene puntos de la forma (a, b) mientras que la otra contiene puntos de la forma (b, a).

J

joint frequencies **(Lesson 12-7)** Entries in the body of a two-way frequency table.

frecuencias articulares Entradas en el cuerpo de una tabla de frecuencias de dos vías.

L

leading coefficient **(Lesson 10-1)** The coefficient of the first term when a polynomial is in standard form.

coeficiente inicial El coeficiente del primer término cuando un polinomio está en forma estándar.

like terms **(Lesson 1-4)** Terms with the same variables, with corresponding variables having the same exponent.

términos semejantes Términos con las mismas variables, con las variables correspondientes que tienen el mismo exponente.

line of fit **(Lesson 5-3)** A line used to describe the trend of the data in a scatter plot.

línea de ajuste Una línea usada para describir la tendencia de los datos en un diagrama de dispersión.

line symmetry **(Lesson 3-5)** A figure has line symmetry if each half of the figure matches the other side exactly.

simetría de línea Una figura tiene simetría de línea si cada mitad de la figura coincide exactamente con el otro lado.

linear equation **(Lesson 3-3)** Equations that can be written in the form $Ax + By = C$ with a graph that is a straight line.

ecuación lineal Ecuaciones que puede escribirse de la forma $Ax + By = C$ con un gráfico que es una línea recta.

linear extrapolation **(Lesson 5-3)** The use of a linear equation to predict values that are outside the range of data.

extrapolación lineal El uso de una ecuación lineal para predecir valores que están fuera del rango de datos.

linear function **(Lesson 3-3)** A function with a graph that is a line.

función lineal Una función con un gráfico que es una línea.

linear interpolation **(Lesson 5-3)** The use of a linear equation to predict values that are inside the range of data.

interpolación lineal El uso de una ecuación lineal para predecir valores que están dentro del rango de datos.

linear regression **(Lesson 5-5)** An algorithm used to find a precise line of fit for a set of data.

regresión lineal Un algoritmo utilizado para encontrar una línea precisa de ajuste para un conjunto de datos.

linear transformation (Lesson 12-6)   One or more operations performed on a set of data that can be written as a linear function.

literal equation (Lesson 2-7)   A formula or equation with several variables.

lower quartile (Lesson 12-4)   The median of the lower half of a set of data.

transformación lineal   Una o más operaciones realizadas en un conjunto de datos que se pueden escribir como una función lineal.

ecuación literal   Un formula o ecuación con varias variables.

cuartil inferior   La mediana de la mitad inferior de un conjunto de datos.

M

mapping (Lesson 3-1)   An illustration that shows how each element of the domain is paired with an element in the range.

marginal frequencies (Lesson 12-7)   The totals of each subcategory in a two-way frequency table.

maximum (Lesson 11-1)   The highest point on the graph of a curve.

measurement data (Lesson 12-1)   Data that have units and can be measured.

measures of center (Lesson 12-1)   Measures of what is average.

measures of spread (Lesson 12-4)   Measures of how spread out the data are.

median (Lesson 12-4)   The beginning of the second quartile that separates the data into upper and lower halves.

metric (Lesson 1-6)   A rule for assigning a number to some characteristic or attribute.

minimum (Lesson 11-1)   The lowest point on the graph of a curve.

monomial (Lesson 8-1)   A number, a variable, or a product of a number and one or more variables.

multiplicative identity (Lesson 1-3)   Because the product of any number a and 1 is equal to a, 1 is the multiplicative identity.

multiplicative inverses (Lesson 1-3)   Two numbers with a product of 1.

cartografía   Una ilustración que muestra cómo cada elemento del dominio está emparejado con un elemento del rango.

frecuencias marginales   Los totales de cada subcategoría en una tabla de frecuencia bidireccional.

máximo   El punto más alto en la gráfica de una curva.

medicion de datos   Datos que tienen unidades y que pueden medirse.

medidas del centro   Medidas de lo que es promedio.

medidas de propagación   Medidas de cómo se extienden los datos son.

mediana   El comienzo del segundo cuartil que separa los datos en mitades superior e inferior.

métrico   Una regla para asignar un número a alguna caracteristica o atribuye.

mínimo   El punto más bajo en la gráfica de una curva.

monomio   Un número, una variable, o un producto de un número y una o más variables.

identidad multiplicativa   Dado que el producto de cualquier número a y 1 es igual a, 1 es la identidad multiplicativa.

inversos multiplicativos   Dos números con un producto es igual a 1.

multi-step equation (Lesson 2-3)   An equation that uses more than one operation to solve it.

ecuaciones de varios pasos   Una ecuación que utiliza más de una operación para resolverla.

N

negative (Lesson 3-4)   Where the graph of a function lies below the x-axis.

negativo   Donde la gráfica de una función se encuentra debajo del eje x.

negative correlation (Lesson 5-3)   Bivariate data in which y decreases as x increases.

correlación negativa   Datos bivariate en el cual y disminuye a x aumenta.

negative exponent (Lesson 8-3)   An exponent that is a negative number.

exponente negativo   Un exponente que es un número negativo.

negatively skewed distribution (Lesson 12-5)   A distribution that typically has a median greater than the mean and less data on the left side of the graph.

distribución negativamente sesgada   Una distribución que típicamente tiene una mediana mayor que la media y menos datos en el lado izquierdo del gráfico.

no correlation (Lesson 5-3)   Bivariate data in which x and y are not related.

sin correlación   Datos bivariados en los que x e y no están relacionados.

nonlinear function (Lesson 3-3)   A function in which a set of points cannot all lie on the same line.

función no lineal   Una función en la que un conjunto de puntos no puede estar en la misma línea.

nth root (Lesson 8-4)   If $a^n = b$ for a positive integer n, then a is the nth root of b.

raíz enésima   Si $a^n = b$ para cualquier entero positive n, entonces a se llama una raíz enésima de b.

nth term of an arithmetic sequence (Lesson 4-5)   The nth term of an arithmetic sequence with first term a_1 and common difference d is given by $a_n = a_1 + (n - 1)d$, where n is a positive integer.

enésimo término de una secuencia aritmética   El enésimo término de una secuencia aritmética con el primer término a_1 y la diferencia común d viene dado por $a_n = a_1 + (n - 1)d$, donde n es un número entero positivo.

numerical expression (Lesson 1-1)   A mathematical phrase involving only numbers and mathematical operations.

expresión numérica   Una frase matemática que implica sólo números y operaciones matemáticas.

O

open half-plane (Lesson 6-5)   The solution of a linear inequality that does not include the boundary line.

medio plano abierto   La solución de una desigualdad linear que no incluye la línea de limite.

outlier (Lesson 12-5)   A value that is more than 1.5 times the interquartile range above the third quartile or below the first quartile.

parte aislada   Un valor que es más de 1,5 veces el rango intercuartílico por encima del tercer cuartil o por debajo del primer cuartil.

P

parabola (Lesson 11-1)   The graph of a quadratic function.

parábola   La gráfica de una función cuadrática.

parallel lines (Lesson 5-2)   Nonvertical lines in the same plane that have the same slope.

parameter (Lesson 4-3)   A value in the equation of a function that can be varied to yield a family of functions.

parent function (Lesson 4-4)   The simplest of functions in a family.

percentile (Lesson 12-1)   A measure that tells what percent of the total scores were below a given score.

perfect cube (Lesson 8-5)   A rational number with a cube root that is a rational number.

perfect square (Lesson 8-5)   A rational number with a square root that is a rational number.

perfect square trinomials (Lesson 10-7)   Squares of binomials.

perpendicular lines (Lesson 5-2)   Nonvertical lines in the same plane for which the product of the slopes is −1.

piecewise-defined function (Lesson 4-6)   A function defined by at least two subfunctions, each of which is defined differently depending on the interval of the domain.

piecewise-linear function (Lesson 4-6)   A function defined by at least two linear subfunctions, each of which is defined differently depending on the interval of the domain.

polynomial (Lesson 10-1)   A monomial or the sum of two or more monomials.

population (Lesson 12-3)   All of the members of a group of interest about which data will be collected.

positive (Lesson 3-4)   Where the graph of a function lies above the x-axis.

positive correlation (Lesson 5-3)   Bivariate data in which y increases as x increases.

positively skewed distribution (Lesson 12-5)   A distribution that typically has a mean greater than the median.

prime polynomial (Lesson 10-6)   A polynomial that cannot be written as a product of two polynomials with integer coefficients.

líneas paralelas   Líneas no verticales en el mismo plano que tienen pendientes iguales.

parámetro   Un valor en la ecuación de una función que se puede variar para producir una familia de funciones.

función basica   La función más fundamental de un familia de funciones.

percentil   Una medida que indica qué porcentaje de las puntuaciones totales estaban por debajo de una puntuación determinada.

cubo perfecto   Un número racional con un raíz cúbica que es un número racional.

cuadrado perfecto   Un número racional con un raíz cuadrada que es un número racional.

trinomio cuadrado perfecto   Cuadrados de los binomios.

líneas perpendiculares   Líneas no verticales en el mismo plano para las que el producto de las pendientes es −1.

función definida por piezas   Una función definida por al menos dos subfunciones, cada una de las cuales se define de manera diferente dependiendo del intervalo del dominio.

función lineal por piezas   Una función definida por al menos dos subfunciones lineal, cada una de las cuales se define de manera diferente dependiendo del intervalo del dominio.

polinomio   Un monomio o la suma de dos o más monomios.

población   Todos los miembros de un grupo de interés sobre cuáles datos serán recopilados.

positiva   Donde la gráfica de una función se encuentra por encima del eje x.

correlación positiva   Datos bivariate en el cual y aumenta a x disminuye.

distribución positivamente sesgada   Una distribución que típicamente tiene una media mayor que la mediana.

polinomio primo   Un polinomio que no puede escribirse como producto de dos polinomios con coeficientes enteros.

principal square root (Lesson 8-5) The nonnegative square root of a number.

raíz cuadrada principal La raíz cuadrada no negativa de un número.

proportion (Lesson 2-6) A statement that two ratios are equivalent.

proporción Una declaración de que dos proporciones son equivalentes.

Q

quadratic equation (Lesson 11-3) An equation that includes a quadratic expression.

ecuación cuadrática Una ecuación que incluye una expresión cuadrática.

quadratic expression (Lesson 10-3) An expression in one variable with a degree of 2.

expresión cuadrática Una expresión en una variable con un grado de 2.

quadratic function (Lesson 11-1) A function with an equation of the form $y = ax^2 + bx + c$, where $a \neq 0$.

función cuadrática Una función con una ecuación de la forma $y = ax^2 + bx + c$, donde $a \neq 0$.

quartiles (Lesson 12-4) Measures of position that divide a data set arranged in ascending order into four groups, each containing about one fourth or 25% of the data.

cuartiles Medidas de posición que dividen un conjunto de datos dispuestos en orden ascendente en cuatro grupos, cada uno de los cuales contiene aproximadamente un cuarto o el 25% de los datos.

R

radical expression (Lesson 8-5) An expression that contains a radical symbol, such as a square root.

expresión radicales Una expresión que contiene un símbolo radical, tal como una raíz cuadrada.

radicand (Lesson 8-4) The expression under a radical sign.

radicando La expresión debajo del signo radical.

range (Lesson 3-1) The set of second numbers of the ordered pairs in a relation.

rango El conjunto de los segundos números de los pares ordenados de una relación.

range (Lesson 12-4) The difference between the greatest and least values in a set of data.

rango La diferencia entre los valores de datos más grande o menos en un sistema de datos.

rate of change (Lesson 4-2) How a quantity is changing with respect to a change in another quantity.

tasa de cambio Cómo cambia una cantidad con respecto a un cambio en otra cantidad.

rational exponent (Lesson 8-4) An exponent that is expressed as a fraction.

exponente racional Un exponente que se expresa como una fracción.

reciprocals (Lesson 1-3) Two numbers with a product of 1.

recíprocos Dos números con un producto de 1.

recursive formula (Lesson 9-6) A formula that gives the value of the first term in the sequence and then defines the next term by using the preceding term.

formula recursiva Una fórmula que da el valor del primer término en la secuencia y luego define el siguiente término usando el término anterior.

reflection (Lesson 4-4) A transformation in which a figure, line, or curve, is flipped across a line.

reflexión Una transformación en la que una figura, línea o curva, se voltea a través de una línea.

relation (Lesson 3-1)   A set of ordered pairs.

relative frequency (Lesson 12-7)   The ratio of the number of observations in a category to the total number of observations.

relative maximum (Lesson 3-5)   A point on the graph of a function where no other nearby points have a greater y-coordinate.

relative minimum (Lesson 3-5)   A point on the graph of a function where no other nearby points have a lesser y-coordinate.

residual (Lesson 5-5)   The difference between an observed y-value and its predicted y-value on a regression line.

root (Lesson 3-4)   A solution of an equation.

relación   Un conjunto de pares ordenados.

frecuencia relativa   La relación entre el número de observaciones en una categoría y el número total de observaciones.

máximo relativo   Un punto en la gráfica de una función donde ningún otro punto cercano tiene una coordenada y mayor.

mínimo relativo   Un punto en la gráfica de una función donde ningún otro punto cercano tiene una coordenada y menor.

residual   La diferencia entre un valor de y observado y su valor de y predicho en una línea de regresión.

raíces   Una solución de una ecuación.

S

sample (Lesson 12-3)   A subset of a population.

scale (Lesson 3-1)   The distance between tick marks on the x- and y-axes.

scatter plot (Lesson 5-3)   A graph of bivariate data that consists of ordered pairs on a coordinate plane.

sequence (Lesson 4-5)   A list of numbers in a specific order.

set-builder notation (Lesson 6-1)   Mathematical notation that describes a set by stating the properties that its members must satisfy.

simplest form (Lesson 1-4)   An expression is in simplest form when it is replaced by an equivalent expression having no like terms or parentheses.

slope (Lesson 4-2)   The rate of change in the y-coordinates (rise) to the corresponding change in the x-coordinates (run) for points on a line.

solution (Lesson 2-2)   A value that makes an equation true.

muestra   Un subconjunto de una población.

escala   La distancia entre las marcas en los ejes x e y.

gráfica de dispersión   Una gráfica de datos bivariados que consiste en pares ordenados en un plano de coordenadas.

secuencia   Una lista de números en un orden específico.

notación de construción de conjuntos   Notación matemática que describe un conjunto al declarar las propiedades que sus miembros deben satisfacer.

forma reducida   Una expresión está reducida cuando se puede sustituir por una expresión equivalente que no tiene ni términos semejantes ni paréntesis.

pendiente   La tasa de cambio en las coordenadas y (subida) al cambio correspondiente en las coordenadas x (carrera) para puntos en una línea.

solución   Un valor que hace que una ecuación sea verdadera.

square root (Lesson 8-5) One of two equal factors of a number.

raíz cuadrada Uno de dos factores iguales de un número.

standard deviation (Lesson 12-4) A measure that shows how data deviate from the mean.

desviación tipica Una medida que muestra cómo los datos se desvían de la media.

standard form of a polynomial (Lesson 10-1) A polynomial that is written with the terms in order from greatest degree to least degree.

forma estándar de un polinomio Un polinomio que se escribe con los términos en orden del grado más grande a menos grado.

statistic (Lesson 12-3) A measure that describes a characteristic of a sample.

estadística Una medida que describe una característica de una muestra.

step function (Lesson 4-6) A type of piecewise-linear function with a graph that is a series of horizontal line segments.

función de paso Un tipo de función lineal por piezas con un gráfico que es una serie de segmentos de línea horizontal.

substitution (Lesson 7-2) A process of solving a system of equations in which one equation is solved for one variable in terms of the other.

sustitución Un proceso de resolución de un sistema de ecuaciones en el que una ecuación se resuelve para una variable en términos de la otra.

symmetric distribution (Lesson 12-5) A distribution in which the mean and median are approximately equal.

distribución simétrica Un distribución en la que la media y la mediana son aproximadamente iguales.

system of equations (Lesson 7-1) A set of two or more equations with the same variables.

sistema de ecuaciones Un conjunto de dos o más ecuaciones con las mismas variables.

system of inequalities (Lesson 7-5) A set of two or more inequalities with the same variables.

sistema de desigualdades Un conjunto de dos o más desigualdades con las mismas variables.

T

term (Lesson 1-2) A number, a variable, or a product or quotient of numbers and variables.

término Un número, una variable, o un producto o cociente de números y variables.

term of a sequence (Lesson 4-5) A number in a sequence.

término de una secuencia Un número en una secuencia.

transformation (Lesson 4-4) The movement of a graph on the coordinate plane.

transformación El movimiento de un gráfico en el plano de coordenadas.

translation (Lesson 4-4) A transformation in which a figure is slid from one position to another without being turned.

translación Una transformación en la que una figura se desliza de una posición a otra sin ser girada.

trend (Lesson 5-3) A general pattern in the data.

tendencia Un patrón general en los datos.

trinomial (Lesson 10-1) The sum of three monomials.

trinomio La suma de tres monomios.

two-way frequency table (Lesson 12-7) A table used to show frequencies of data classified according to two categories, with the rows indicating one category and the columns indicating the other.

tabla de frecuencia bidireccional Una tabla utilizada para mostrar las frecuencias de los datos clasificados de acuerdo con dos categorías, con las filas que indican una categoría y las columnas que indican la otra.

two-way frequency table (Lesson 12-7)   A table used to show frequencies of data classified according to two categories, with the rows indicating one category and the columns indicating the other.

two-way relative frequency table (Lesson 12-7)   A table used to show frequencies of data based on a percentage of the total number of observations.

tabla de frecuencia bidireccional   Una tabla utilizada para mostrar las frecuencias de los datos clasificados de acuerdo con dos categorías, con las filas que indican una categoría y las columnas que indican la otra.

tabla de frecuencia relativa bidireccional   Una tabla usada para mostrar las frecuencias de datos basadas en un porcentaje del número total de observaciones.

U

union (Lesson 6-3)   The graph of a compound inequality containing *or*.

univariate data (Lesson 12-1)   Measurement data in one variable.

upper quartile (Lesson 12-4)   The median of the upper half of a set of data.

unión   La gráfica de una desigualdad compuesta que contiene la palabra *o*.

datos univariate   Datos de medición en una variable.

cuartil superior   La mediana de la mitad superior de un conjunto de datos.

V

variable (Lesson 1-2)   A letter used to represent an unspecified number or value.

variable (Lesson 12-1)   Any characteristic, number, or quantity that can be counted or measured.

variable term (Lesson 1-2)   A term that contains a variable.

variance (Lesson 12-4)   The square of the standard deviation.

vertex (Lesson 4-7)   Either the lowest point or the highest point of a function.

vertex form (Lesson 11-2)   A quadratic function written in the form $f(x) = a(x - h)^2 + k$.

variable   Una letra utilizada para representar un número o valor no especificado.

variable   Cualquier característica, número, o cantidad que pueda ser contada o medida.

término variable   Un término que contiene una variable.

varianza   El cuadrado de la desviación estándar.

vértice   El punto más bajo o el punto más alto en una función.

forma de vértice   Una función cuadrática escribirse de la forma $f(x) = a(x - h)^2 + k$.

X

x-intercept (Lesson 3-4)   The *x*-coordinate of a point where a graph crosses the *x*-axis.

intercepción x   La coordenada *x* de un punto donde la gráfica corte al eje de *x*.

Z

zero (Lesson 3-4)   An *x*-intercept of the graph of a function; a value of *x* for which $f(x) = 0$.

cero   Una intersección *x* de la gráfica de una función; un punto *x* para los que $f(x) = 0$.

Index

Copyright © McGraw-Hill Education